PALAEONTOLOGICA
FIELD GUIDES TO F(
Second edition, revis

Fossils
of the Chalk

Edited by

ANDREW B. SMITH
DAVID J. BATTEN

THE PALAEONTOLOGICAL ASSOCIATION
LONDON
2002

ISBN 0 901702 78 1
ISSN 0962 5321

Series Editor David J. Batten
Institute of Geography and Earth Sciences,
University of Wales, Aberystwyth
Ceredigion SY23 3DB, UK

Front cover
Chalk cliffs at Beachy Head, Sussex

Printed in Great Britain by Henry Ling Limited, at the Dorset Press, Dorchester, DT1 1HD

EDITORS

Smith, Andrew B., Department of Palaeontology, The National History Museum, London SW7 5BD, UK.

Batten, David J., Institute of Geography and Earth Sciences, University of Wales, Aberystwyth SY23 3DB, UK.

LIST OF CONTRIBUTORS

Cleevely, R. J. Department of Palaeontology, The Natural History Museum, London SW7 5BD

Collins, J. S. H. 8, Shaws Cottages, Perry Rise, London SE23 2QN.

Doyle, P. School of Earth and Environmental Sciences, University of Greenwich, Medway University Campus, Chatham Maritime, Kent ME4 4TB

Gale, A. S. School of Earth and Environmental Sciences, University of Greenwich, Medway University Campus, Chatham Maritime, Kent ME4 4TB, and Department of Palaeontology, The Natural History Museum, London SW7 5BD

Kennedy, W. J. Department of Earth Sciences, University of Oxford, Oxford OX1 3PW

Longbottom, A. Department of Palaeontology, The Natural History Museum, London SW7 5BD

Milner, A. C. Department of Palaeontology, The Natural History Museum, London SW7 5BD

Morris, N. J. Department of Palaeontology, The Natural History Museum, London SW7 5BD

Owen, E. Department of Palaeontology, The Natural History Museum, London SW7 5BD

Patterson, C. (deceased) Department of Palaeontology, The Natural History Museum, London SW7 5BD

Smith, A. B. Department of Palaeontology, The Natural History Museum, London SW7 5BD

Taylor, P. D. Department of Palaeontology, The Natural History Museum, London SW7 5BD

Wood, R. Schlumberger Cambridge Research, High Cross, Madingley Road, Cambridge CB3 0EL

Wright, C. W. Department of Earth Sciences, University of Oxford, Oxford OX1 3PW

CONTENTS

vi   *Contents*

PREFACE TO FIRST EDITION

The characteristic white limestones that form the Chalk were deposited across much of northern Europe during the Upper Cretaceous, 100 to 65 million years ago. Although the Chalk is not normally rich in fossils, the preservation is often exquisite and specimens are relatively easy to prepare out from the matrix using needles and brushes or an air abrasive. This explains why the Chalk has been, and continues to be, a favourite formation of many fossil collectors. The extensive collections housed in museums are the result of persistent collecting over more than 150 years, though few were collected with the necessary detailed stratigraphical horizoning that is so essential to today's study of the Chalk fauna.

This book is aimed at the amateur fossil collector and sets out to provide a pocket field guide to the range of macrofossils found in the Chalk of Britain. It covers most of the important groups that are likely to be encountered in field collecting, with the exception of serpulid worms and corals which unfortunately could not be covered (neither group is either diverse or important in the Chalk). It must be emphasized that the fossils illustrated here are only a small selection of the total fauna likely to be found in the Chalk, though authors have tried to include all of the common species and genera. In some cases more detailed and comprehensive monographs on particular groups have been published and the more important works are generally listed at the start of each chapter. In other cases the primary systematic work has yet to be done and no such publications exist.

Almost 400 Chalk fossils are illustrated, most of which are housed in the British Museum (Natural History). For each entry there is a brief description to aid identification, together with the stratigraphical range of the species. In addition, where similar species exist that might be confused, but which have not been illustrated, these are sometimes compared and contrasted in the text. It is beyond the scope of this book to deal comprehensively with the terminology applied to the fossilized hard parts of so many different groups and a reasonable background knowledge has had to be assumed of the reader. Nor is it possible to discuss the taxonomic framework that each author has adopted. Any standard palaeontological text-book, or the *Treatise on invertebrate paleontology* ought to be consulted on these matters.

Probably more than for any other stratigraphical formation, amateurs have played a major role in furthering our understanding of the Chalk faunas and their distribution. Notable amateurs include the Reverend Thomas Wiltshire, J. Starkie Gardner, Henry Willett, A. W. Rowe, R. M. Brydone, C. T. A. Gaster and E. V. and C. W. Wright. Without their diligence our knowledge of the Chalk fauna would be very much the poorer. I sincerely hope that future generations of amateurs will continue this tradition and that this book will provide encouragement to their pursuit.

Andrew B. Smith
17th February, 1987

PREFACE TO SECOND EDITION

Since the first edition of Fossils of the Chalk appeared 14 years ago a great deal of new information on the taxonomy of the fauna and the nature of the Chalk environment has been published, and refinements to the biostratigraphical and lithostratigraphical framework have been made. This new edition is consequently revised and enlarged to bring it up to date with current thinking. There is an expanded Introduction that outlines the depositional environment of Chalk sediment and a more detailed stratigraphical framework. Three new chapters have been added, covering corals, serpulid worms and nautiloid molluscs, groups that were omitted from the first edition. Other chapters from the original edition have been updated, some with new text-figures or plates, to take into account work published since 1987. There are now descriptions and illustrations of 434 species in 67 plates, providing even better coverage of the Chalk fauna.

Andrew B. Smith
David J. Batten

1. INTRODUCTION

by A. S. GALE *and* W. J. KENNEDY

The Chalk is one of the most widespread and distinctive geological formations of the British succession (Text-fig. 1.1). Virtually identical Cretaceous deposits, even containing the same fossils, extend across northern Europe and into central Asia, as far east as the Aral Sea, and are found as far afield as Texas and Western Australia. In modern oceans, chalk forms mostly in the deep-sea from accumulation of calcareous plankton rain, the calcite skeletons of myriad single-celled organisms. But as sea levels rose during the Late Cretaceous, pelagic chalks spread far onto the continental shelf and epicontinental seas of northern Europe. Widespread chalk deposition continued for about 40 million years, until the end of the Early Palaeocene, and hundreds of metres of chalk formed a blanket-like cover over vast areas. Exactly how deep the Chalk Sea was in the British area is hard to ascertain, but maximum values of 300 m above present-day sea level are likely. Salinities were normal, and the sea floor was well-oxygenated for virtually all Chalk deposition in the British area (the possible exception is a thin laminated black shaly unit, the Black Band, at the base of the Turonian Welton Formation in Yorkshire and Lincolnshire). The sea floor was also below the limit of light penetration, and there is a striking absence of those organisms that dominate in shallow-water limestones that formed in the photic zone. At its maximum extent, the Chalk Sea probably covered all of the British Isles except the Scottish highlands. The Cretaceous world also differed in that there is no evidence of polar ice-caps, and warm temperate vegetation extended to 85° North. Oxygen isotope ratios from the English Chalk give sea-surface temperatures in the range of 20–30°C, substantially warmer than at present.

THE SEDIMENTARY ENVIRONMENT

Composition and deposition of Chalk

Chalk is composed largely of the remains of minute planktonic photo-synthesizing organisms called coccolithophores. The building blocks of the skeleton of these organisms are calcite tablets or platelets 0·5–1 μm across. These are arranged in rings and rosettes, generally 3–5 μm across

TEXT-FIG. 1.1. Map of the British Isles, showing the principal surface outcrops of Upper Cretaceous Chalks and correlative deposits.

and known as coccoliths, which are in turn arranged into a hollow coccosphere that encloses the soft parts of the organism. Platelets and coccoliths are abundant in Chalks, but complete coccospheres are rare.

Each individual produces coccoliths throughout its life, and these overlap to form the coccosphere, and may even be shed periodically. On death, the coccospheres usually decay, and coccoliths are liberated to the sediment.

A typical coccolith is composed of a lower and upper disc, each disc being made up of radially oriented platelets. A central cylinder usually connects the discs. Platelets in the upper and lower discs are often, but by no means always, joined by a platelet that is part of the central cylinder. In such cases, electron diffraction has shown that the whole three-piece unit is, in fact, a single crystal of low magnesian calcite. The c-axis of the crystal runs along the length of the platelets, the a (or b) axis at right angles to it. Surfaces of platelets are therefore presumed to be (010) planes.

There is great variety in coccolith form, although most are composed of discs of radial platelets. The most obvious difference between the main coccolith groups is the shape of the platelet and the way they are bound together. Species with strongly bound platelets resist mechanical break-up and chemical attack much better than those that are weakly bound. Thus, in conditions of high mechanical abrasion or solution only certain species will survive, giving a false impression of the original floral diversity. On the other hand, a single species of coccolithophore can produce coccoliths of different morphology as it progresses through its life-cycle.

Other locally abundant constituents of chalk are tests of foraminiferans, globular calcispheres, tiny prismatic fragments of the outer calcitic shell layer of the bivalve *Inoceramus* and calcitic debris of echinoids and crinoids. Although white chalk is commonly 98 per cent pure calcium carbonate, the grey-coloured lower part of the succession contains up to 40 per cent of clay minerals, together with a little sand and silt grade quartz, and the green authigenic mineral glauconite. The Chalk was originally deposited as plankton rain, and probably sedimented in the form of copepod faecal pellets. Primary production was uniform over a very wide area, and mean rates of sedimentation (derived from division of thickness by time) are low everywhere in the Anglo-Paris Basin; a mean deposition rate of 25 mm per thousand years (kyr) is typical for the area.

Large areas of the Chalk sea floor had an uneven topography, comprising a series of domes and ridges separated by basins and troughs; extensive flat platforms also developed. Palaeotopography was controlled by relative subsidence over downfaulted blocks which produced considerable lateral thickness variations. As primary productivity was uniform over a very wide area, resedimentation of the plankton rain blanket was evidently a widespread depositional process.

Large-scale mass movement (e.g. slumps, slides) on the chalk sea floor was not a common phenomenon in the Anglo-Paris Basin although it has been described from the Chalk of southern England. This type of mass movement is rendered conspicuous by slump folding of beds, picked out by hardgrounds and layers of flint nodules, and more rarely by debris flows. Mass movement probably only occurred when a relatively steep slope created great instability, as over an active fault line.

Resuspension of chalk over topographic highs by currents was probably the commonest process by which sediment transfer took place. Much of White Chalk deposition would have taken place beneath storm wave base, although wave-scouring was probably responsible for flat, regionally extensive hardgrounds formed during Mid Turonian time in southern England. Bottom currents, augmented by the presence of highs, were thus responsible for much of the resuspension and subsequent redistribution of sediments on the Chalk sea floor. The continual nature of this process, perhaps seasonally moderated, meant that there was little difference in age between sediment eroded and sediment deposited, so that climatic cycles (see below) were faithfully preserved even in redeposited basinal chalks.

Rhythmicity in the Chalk: a record of climatic change

One of the most conspicuous features of any chalk cliff is the regular repetition of beds on the scale of 1 m or less. In the Lower Chalk, the alternations are of more and less marly chalk, the harder, more calcareous beds weathering proud of the outcrop. In the lower part of the White Chalk above, the alternations are of nodular chalk and recessing thin, wispy marls. In the overlying flinty chalks, beds of flint nodules or sheets of black flint occur with striking regularity about every metre. The average periodicity of these rhythms can be calculated simply by taking an interval of known duration and dividing by the number of beds present; results are in the order of 20–40 kyr. These values fall in the periodicity of Milankovitch climatic cycles, caused by variations in the shape of the Earth's orbit, which alter the amount of solar radiation reaching a given point on the surface of the Earth. The major Milankovitch frequencies are at 21, 40 and 100 kyr, and have been identified and used as a time-scale in Pleistocene sediments, where they acted as a pacemaker for the development of ice-ages. The chalk rhythms are thus a record of Cretaceous Milankovitch cycles, but the exact means by which climate change led to the formation of more and less marly beds or determined exactly where flints formed, remains unresolved.

Trace fossils and bioturbation

Virtually all of the Chalk succession shows evidence of intense burrowing, often to the extent that all trace of other primary depositional features has been destroyed. The commoner traces present are summarised in Text-figure 1.2; *Chondrites, Planolites, Taenidium* and *Thalassinoides* occur throughout the succession, with rare *Zoophycos* at a few levels only.

Burrows are an important aid to the interpretation of the depositional and diagenetic history of chalks. They provide a key to the differentiation of depositional or synsedimentary primary features (e.g. hardgrounds,

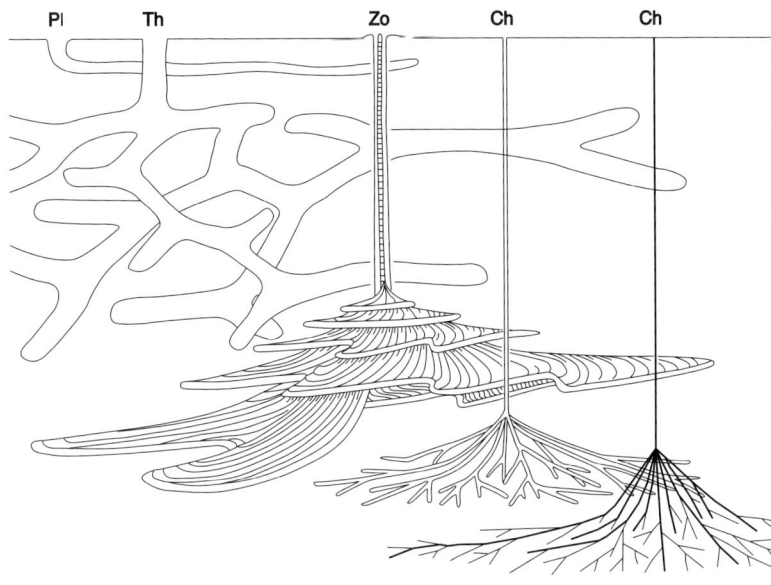

TEXT-FIG. 1.2. The Chalk trace fossils association, modified after Bromley and Ekdale (1986). Trace fossils represented are:

Ch, *Chondrites*: plant-like, ramifying burrow systems, with a few vertical, but many horizontal or inclined tubes, rarely more than a few millimetres in diameter. Burrows are circular in cross-section, and of a constant diameter, with smooth walls. Branching is common, lateral, never equal, with a faint radial symmetry visible in some bedding-parallel sections. The producer of this system is unknown, but was presumably vermiform. The structures are a feeding trace, and occur throughout the Chalk, commonly appearing as groups of small black or white dots in section.

Pl, *Planolites*: larger (up to 10 mm diameter) irregular, predominantly horizontal or gently inclined burrows that may branch. They appear to have been produced by some vermiform animal, the burrow filling having passed through the gut before being packed into the tube behind.

Zo, *Zoophycos*: spreiten burrows produced by reworking of sediment to give a sheet of material with characteristic lunate structure in section, the sheet forming a tabulate or complex helicoid form, rather like a spiral staircase, and sometimes with a marginal tube. The overall structure appears most commonly as a series of parallel ribbons, the ribbons being sections through helical systems that were originally some tens of centimetres high and perhaps half a metre across. Even when short lengths only are visible, the alternating light and dark lunae are quite distinctive.

Th, *Thalassinoides*: burrow systems consisting of a three-dimensional network of vertical shafts and horizontal galleries that may extend downwards to a depth of more than a metre. When well-preserved, *Thalassinoides* show a very charac-teristic Y-shaped branching pattern, and by analogy with Recent examples, structures of this type were produced by decapod or other crustaceans.

bedding, lamination) and later, diagenetic ones. Broadly speaking if burrows have piped one sediment type into another, then this difference must have been a primary one. Equally, if burrows cut lamination, then this too is primary. If burrowers avoided areas of hardened chalk, then they are either reworked intraclasts or the result of early diagenesis nodule and/or hardground formation. Equally, if structures cut or offset burrows (e.g. solution seams, diagenetic lamination), these are of a relatively late-stage origin.

Burrows also provide a means of gauging relative compaction. This is graphically demonstrated in the Lower Chalk where there is differential compaction of the more and less argillaceous parts of small-scale cycles.

Where present, burrowers indicate oxygenated bottom waters at the level of the interface from which they arise. Their depth of penetration is usually only a few centimetres or decimetres, but the arthropods responsible for *Thalassinoides* excavated down to more than 1 m below discontinuity surfaces. Where burrow disturbance of bedding is absent and lamination survives, this may indicate low oxygen levels in bottom water, as has been suggested in other ancient pelagic sediments.

Intense bioturbation had an important effect on substrate conditions. Most burrowers of the type represented in chalks fed on the sediment, turning it into fecal pellets. These had quite different hydrodynamic properties when compared to their parent ooze, and were sometimes the focus of early cementation and diagenesis (relevant as a possible origin of glauconite, phosphate and micrite pellets in winnowed beds).

Intense burrowing also fluidised sediment, which may account for the rarity of larger benthonic organisms at some levels, and has been claimed as the reason for the specialised adaptations of some elements of the Chalk benthos. The sea bed during much of the deposition of the chalk may be envisaged as consisting of: (1) a granular, thixotropic, readily suspended, fecal pellet-rich layer 5–10 mm thick, with up to 60 per cent water; (2) below, a zone of high water content (50 per cent +) and also thixotropic; and (3) progressively firmer sediment, de-watered by burrowing organisms, and behaving plastically.

The consequence for chalk sedimentation was considerable. The granular surface layer would have been easily resuspended as clouds of sediment-charged water, and could easily have flowed down-slope, exposing wide areas of firm sea floor. Where dewatered, well below the surface, it could have supported open burrow systems, and being plastic, would have recorded deformation and slumping.

The differences between surface, thixotropic muds and deeper, plastic muds can be detected in ancient sediments. When thixotropic muds were burrowed, the movements of the animals produced a zone of deformation adjacent to the burrow margin as grains slipped past each other to accommodate the animals' body, and the burrow outline became diffuse. In contrast, in firmer, compacted plastic muds, sharp burrow outlines were the result.

Some outcrops show that deeper burrows, such as *Zoophycos, Chondrites* and *Thalassinoides*, have sharp outlines and were excavated in plastic sediment some tens of centimetres or more below the sea bed. These are superimposed upon a blurred background bioturbation produced by burrowers excavating the top few centimetres of thixotropic sediment: burrow systems are tiered (Bromley and Ekdale, 1986). Deep burrows, especially those of arthropods (*Thalassinoides*) provided conduits through which sea-water flushed (in Recent shallow-water environments they may remain open for many hundreds of years) down several metres into the sediment. It is, therefore, scarcely surprising that they were a focus of early diagenesis, nodule and hardground formation, as discussed below.

Early carbonate diagenesis

Early lithification of chalks is now a widely accepted and documented phenomenon. Hardgrounds and nodular chalks are successive products of a sequence of events summarised in Text-figure 1.3. The succession began very simply with a slowing or pause in sedimentation. The sea floor was stabilised, and a burrowing infauna developed, superimposing relatively deep, permanent burrows on more diffuse, background traces. Renewed sedimentation buried this surface, which may be preserved as an 'omission surface'. As an alternative, a pause in sedimentation could be marked by erosion of the soft sea bed, and concentration of coarse grains and fossils on top of an 'erosion surface'.

If the phase associated with an omission surface was prolonged, pure chalks generally underwent cementation in a zone up to 1 m thick extending from just below the sediment-water interface. Cementation started at scattered foci throughout the sediment, and a nodular chalk developed. The sediment between nodules remained soft, and was reworked by burrowing animals which avoided the nodules, their burrows becoming increasingly distorted and restricted as nodule growth progressed. If such a development was followed by a phase of intraformational erosion which winnowed away soft, uncemented matrix, then a conglomerate resulted, itself evidence of penecontemporaneous cementation, and an indicator of hard, pebbly, sea-floor conditions.

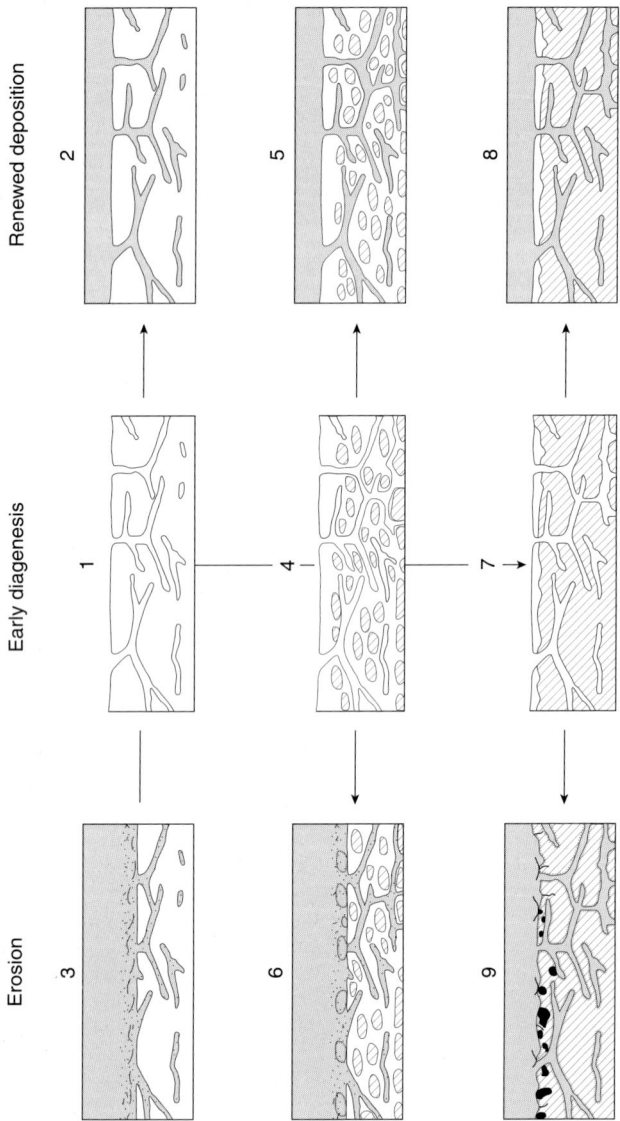

Early diagenesis   Renewed deposition

Erosion

If the process of nodule growth continued, nodules eventually joined up to produce a three-dimensional framework of lithified chalk, in which the only soft material survived in the infills of burrows that remained open as the chalk cemented around them. These 'incipient hardgrounds' generally possess diffuse upper surfaces, and grade down into nodular chalk.

If such an incipient hardground was subjected to scour, soft sediment was stripped from the top of the lithified layer, which having been exposed as a rocky sea floor, qualifies as a true hardground. Indications of lithific-ation and exposure include the presence of borings and encrusting organ-isms on the hardground surface, phosphate and glauconite mineralisation, and the development of reworked pebble beds. Lithification appears to have proceeded rapidly in many hardgrounds, and burrow systems (generally *Thalassinoides*) survived as open tunnels, the chalk lithifying around them while still inhabited by crustaceans. These burrows then functioned as conduits of soft sediment, commonly re-excavated by later generations of burrowers, and used as pathways to explore the soft sediment below the lithified layer.

Hardground surfaces vary from highly convoluted to planar, and may represent palaeontologically undetectable pauses in sedimentation, or breaks of tens to hundreds of thousands (rarely millions) of years. They provided a lithified sea floor that represented the other extreme substrate from the semi-fluid, thixotropic ooze produced by the activities of burrowing infauna.

The early cementation of chalk ooze into nodular chalks and hard-grounds sometimes led to the preservation of originally aragonitic fossils,

---

TEXT-FIG. 1.3. Flow diagram showing relationship between diagenesis, erosion, and burial of nodular chalks and hardgrounds. 1, a pause in sedimentation leads to the development of an omission suite of trace fossils (*Thalassinoides*). 2, if buried, this is preserved as an omission surface. 3, with scour, an erosion surface is formed; burrows are infilled with calcarenitic chalk and the surface is overlain by a shelly lag. 4, early diagenesis associated with a longer pause in deposition leads to the growth of calcareous nodules in soft sediment; burrow systems are extended, the animals avoiding the nodules. 5, burial at this stage leads to a nodular chalk. 6, if eroded, nodules are reworked and burrows truncated; the nodular chalk is overlain by a terminal intraformational conglomerate, often with mineralized and bored pebbles; burrows are infilled by, and pebbles embedded in, winnowed calcarenitic chalk. 7, prolonged diagenesis leads to a link up of nodules to form a continuous lithified subsurface layer; later burrows are entirely restricted to sites of pre-lithification burrows. 8, if buried, this rock band, with no signs of superficial mineralization, becomes an incipient hardground. 9, with erosion, the rock band is exposed on the sea floor, and a true hardground develops; it may become bored (borings shown in black), encrusted by epizoans, or mineralized; all of these processes also affect the walls of burrows.

the aragonite having long since vanished. They can then be preserved as undeformed moulds showing fine surface detail, as in the Upper Turonian Chalk Rock hardground-complex, or the Campanian hardgrounds in the Norfolk Chalk and Irish White Limestone. With repetition of the sedimentation/cementation/erosion cycle, composite hardgrounds, with many generations of early lithification, developed. The classic example of this is the English Chalk Rock hardground complex, documented by Bromley (1965), Bromley and Gale (1982) and others.

Reworking into pebble beds and exposure as rocky bottoms, lack of compactional fabrics in thin section, boring, encrustation, superficial mineralisation, and avoidance by burrowers incapable of penetrating hardened limestone, all show the above phenomena to have been syn- or immediately post-depositional. The abundance of hardgrounds and nodular chalks at some levels suggests that these events may represent a short period of time only (i.e. hundreds to thousands of years). This is supported by carbon- and oxygen-isotope data, which indicate formation of cement from pore fluids with the composition of normal seawater.

Flints and their origin

Nodules and tabular sheets of black chert, commonly surrounded by a white patina, are abundant and conspicuous through much of the White Chalk. Nodular flints are concentrated in discrete, laterally continuous beds, which occur rhythmically every metre or so through the succession; tabular flints, usually laterally discontinuous, may be parallel to bedding or occupy fractures at low or high angles to bedding. Flints have posed numerous questions to even the casual observer; for example, why do they assume such odd and complex shapes? How do they form? Why do they occur in evenly spaced beds?

Flints are of complex shape because they commonly replace, and overgrow, the burrows that are ubiquitously present in the Chalk. One of the commonest burrows is the arthropod trace *Thalassinoides*, which is made up of horizontal, polygonal networks joined by vertical tubes, all 1–10 cm in diameter (Text-fig. 1.2). Flints have commonly replaced part of this network and consequently have elongated or branching forms. The surface of flints may show clearly the burrowed fabric of the chalk they have replaced, including the root-like millimetre-scale *Chondrites* and spiral sheets of *Zoophycos.*

Flints are concretions; that is, they have grown within the sediment after its deposition by the precipitation of silica, replacing chalk, at specific sites. Silica for flint formation was certainly provided by siliceous sponges, and probably by siliceous planktonic micro-organisms (diatoms, radiolarians). Fossil sponges, now replaced by iron sulphides or oxides, are common in the chalk. Silica secreted by organisms (Opal-CT), was readily soluble in the pore fluids of sediments on and below the chalk sea floor and re-precipitated some centimetres beneath the sea floor at the lower limit of occurrence of aerobic bacteria, in burrows which would have contained small concentrations of decaying organic matter. When first formed, flints were loose aggregations of tiny spheres of the silica mineral cristobalite, which recrystallised at depths of tens of metres to form finely crystalline quartz nodules and sheets with a characteristic conchoidal fracture. Flints are commonly absent in the clay-rich lower levels of the Chalk, where silica is present as abundant lepispheres, having never undergone secondary migration to form concretionary masses.

Fossil preservation

The majority of Chalk invertebrate fossils were originally siliceous (some sponges), aragonitic (ammonites, nautiloids, part or all of some gastropods and bivalve shells) or calcitic (e.g. parts of some gastropods and bivalves, brachiopods, echinoderms, belemnites, some sponges, bryozoans). The original opaline silica of sponges is lost, and may have been replaced by a stable quartz polymorph or iron sulphide (pyrites; marcasite is uncommon), the latter generally oxidised to iron oxide/hydroxide complexes (limonite, goethite). Aragonite, well-preserved in the underlying Albian Gault Clay, is never preserved in the Chalk. It may be replaced by calcite, silica or phosphate (fluorcarbonate-apatite, when original nacreous sheen may survive), but has generally disappeared. Then internal and external moulds are usually combined into a composite mould, so that in, for example, ammonites, ornament of the shell exterior is superimposed on features of the internal mould such as the sutures, leading to the apparent presence of organisms that actually encrusted the now-vanished shell attached to the 'internal' (actually composite) mould. Such composite moulds commonly bear a film of iron sulphide or green glauconite (often oxidised to brown limonite), and may show signs of plastic deformation without indications of cracking, suggesting that dissolution of the shell took place early in the diagenetic history of the Chalk.

Originally aragonitic fossils are common only in the lower, clay-rich Lower Chalk and its correlatives in southern England. They occur at several levels in the Plenus Marls and Middle Chalk, and are well-preserved in the hardgrounds of the Chalk Rock and nodular correlatives of the Upper Turonian. In the White Chalk above, the only evidence for originally aragonitic organisms is as impressions on the attachment scars of originally calcitic organisms such as oysters, or as the calcitic parts of the organism only, as with the calcitic jaws (aptychi) of ammonites. This raises the probability that originally aragonitic shells dissolved on the sea floor or just below it, and that typical White Chalk faunas, dominated by originally calcitic organisms are 'preservation faunas' in which the aragonitic component has been lost.

Condensed horizons at the base of, and within, the chalk sequence (Glauconitic Marl, Chalk Basement Beds, Totternhoe Stone, Chalk Rock) yield moulds of originally aragonitic and calcitic fossils as variably worn phosphatic and glauconitised remanié fossils that may be substantially older than their enclosing matrix. In the Chalk Basement Beds of Dorset, Devon and Somerset, originally aragonite and calcite shell material may be phosphatized. In the White Limestone of Northern Ireland, originally aragonitic fossils occur at restricted, condensed levels as well-preserved uncrushed moulds with glauconitic coatings, sometimes laminated and of microbial origin. Originally chitinous fossils, such as arthropods, are generally rare in the Chalk, although their presence is indicated by the ubiquitous trace-fossil *Thalassinoides*. Their exoskeletons have been destroyed by chitin-consuming (chitinoclastic) bacteria.

Preservation of the chalk benthic fauna provides some information about the nature of depositional processes. The fact that complete, unbored bivalves, brachiopods, epifaunal echinoids and even fish with scales in place are found is itself evidence that periodic rapid deposition took place. If the real rate of deposition was a few cm/kyr disarticulation would have occurred and fossils would have been destroyed by the boring organisms (notably the sponge *Cliona*), which are widespread in calcite fossils (Bromley 1972). Rarely, entire specimens of epifaunal asteroids and echinoids are found in white chalks. Such specimens retain all the spines and small plates that would normally fall away after a few days or weeks of decomposition, and must, therefore, have been both killed and preserved by burial in a substantial layer of sediment that was deposited very rapidly. It is highly likely that turbidity currents were responsible for this type of preservation, but turbidites have not been described from the English Chalk, although they are common in the Central Graben of the North Sea. The absence of identifiable turbidites could have several possible causes. Firstly, the Chalk is thoroughly and pervasively bioturbated by deeper tier burrowers, such as *Thalassinoides*, which have obliterated shallow traces (Bromley 1992) and could have completely destroyed primary sedimentary structures. Secondly, the fine sediments contain few larger particles and, therefore, show very little evidence of grading. In the Madeira abyssal plain, mud turbidites formed by high-density, non-turbulent currents show neither grading nor lamination.

The nature of the floor of the Chalk sea

Several early palaeoecological studies on Chalk fossils were based on the premise that the Chalk sea floor was a uniform ooze. Consequently, morphological features in various groups have been interpreted as adaptations to such substrates. In particular, mention must be made of the work of Carter (1968, 1972) on Chalk bivalves. He interpreted the spines developed on the margins of the lower valves in *Arctostrea colubrina* (Lamarck), *Spondylus spinosus* (J. Sowerby) and *Plicatula inflata* J. de C. Sowerby as structures that prevented these species from sinking into the soft ooze-like sea floor. However, other groups of chalk benthos do not display any structures that are unequivocally adaptations to very soft substrates. Chalk *Terebratulina* may have possessed root-like pedicles, but today these are present in forms inhabiting both mud and sand bottoms (G. Curry, pers. comm. 1999).

A number of lines of evidence suggest that relatively firm substrates existed, with a sharp sediment-water interface. The presence of cutting and compacting spines on certain burrowing irregular echinoids implies that the sediment was sufficiently cohesive to require this adaptation. These infaunal echinoids lived with apices near the sediment-water interface, so even the highest few centimetres of sediment must have been able to support mucus-lined burrows (Gale and Smith 1982). The abundance of minute suspension-feeding brachiopods and bivalves throughout the White Chalk, and the frequency of encrusting epifauna on even very small substrates, suggests that the bottom was stable, albeit covered by a millimetre or so of easily resuspendable material.

Large, vagile benthonic animals were able to move about on the sea bed, for example, cidarid echinoids, asteroids and lobsters. From this it can be reasonably inferred that the substrate was capable both of holding their weight and providing sufficient resistance to spines and limbs to allow movement.

Rocky sea floors (the exposed surfaces of exhumed nodules in the form of conglomerates), or continuous lithified surfaces (hardgrounds) were the end member of substrate conditions. That they really were exposed is shown by the ubiquitous occurrence of borings of endolithic organisms such as clionid sponges and bivalves, and the presence more rarely of encrusting bivalves, corals, bryozoans and serpulid polychaetes. Surprisingly, such surfaces are not generally associated with concentrations of vagile benthos adapted for hard substrates such as epifaunal echinoids and byssate bivalves.

CHALK STRATIGRAPHY

Text-figure 1.4 shows the standard divisions of the Chalk in terms of stage, and classic zonal and lithostratigraphic division as used in the following chapters. This greatly simplified nomenclature conceals a mass of often conflicting detail of both litho- and biostratigraphy, and numerous local terms.

Lithostratigraphy

Early lithological subdivisions of the Chalk were provided by John Phillips (Kent), Gideon Mantell (Sussex) and Samuel Woodward (Norfolk). All were based primarily on clay content and the presence, absence and relative abundance of flints. These early classifications had considerable general agreement, although the actual boundaries chosen by various authors differed considerably. All authors had a marly lowest unit called variously 'Grey Chalk', 'Grey Chalk Marl' or 'Chalk Marl' overlain by a 'Lower Chalk', distinguished by its hardness and absence of flints. The 'Upper Chalk' was characterised by its numerous flints. Phillips and Woodward separated off a unit with few flints at the base of the 'Upper Chalk' as 'Medial Chalk' or 'Chalk with few flints'. It is interesting to note that the Lower White/Middle Chalk boundary, nowadays deemed to be so distinctive, did not figure importantly in any of these schemes. The use of a 'Lower Chalk' (generally incorporating the Chalk Marl/Grey Chalk) without flints and an 'Upper Chalk' with flints continued through much of the nineteenth century.

In 1853 Sharpe revised this *status quo* by placing the Chloritic Marl and the Chalk Marl within the Lower Chalk, and establishing a Middle Chalk, characterised by a sparsity of fossils. In 1880 Jukes-Browne described the Melbourn Rock in Cambridgeshire, and subsequently used this unit to support the introduction of a 'Middle Chalk', conveniently separated from the Lower Chalk by the thin but feature-forming Melbourn Rock. A higher hard band, the Chalk Rock formed an equivalent feature with which to separate the Middle and Upper Chalk. This tripartite division was accepted by the British Geological Survey (BGS) and mapped widely through the twentieth century.

TEXT-FIG. 1.4. Stage, zonal, and lithostratigraphic divisions of the Chalk used in this book. Time scale from Gradstein *et al.* (1994).

AGE (Ma)	STAGE	BIOSTRATIGRAPHY					Southern England	Northern England
		Inoceramid Zone	Cephalopod Zone	Assemblage Zone				
	MAASTRICHTIAN		*B. sumensis*	*Ostrea lunata*			Nomenclature uncertain	
70			*B. obtusa*					
			B. pseudobtusa					
			B. lanceoata					
	CAMPANIAN		*B. minor* II	*Belemnitella mucronata*				
			B. minor I					
75			*B. woodi*					
			B. mucronata				Portsdown Formation	
80				*Gonioteuthis quadrata*	UPPER CHALK	WHITE CHALK GROUP	Culver Formation	
		?		*Offaster pilula*			Newhaven Formation	Flamborough Formation
				M. testudinarius / U. socialis				
85	SANTONIAN	*Sphenoceramus* spp.		*Micraster coranguinum*			Seaford Formation	
		C. undulatoplicatus						
		V. involutus						
	CONIACIAN	*V. koeneni*	?	*Micraster cortestudinarium*			Lewes Formation	Burnham Formation
		C. schloenbachi						
		C. erectus	*Forresteria petrocoriense*					
		C. waltersdorfensis						
90		*M. fiegei*	*Subprionocyclus neptuni*	*Sternotaxis plana*			New Pit Formation	
		I. securiformis						
	TURONIAN	*I. lamarcki*	*Collignoniceras woollgari*	*Terebratulina lata*	MID. CHALK			Welton Formation
		I. cuvieri						
		M. subhercynicus		*Mytiloides labiatus*			Holywell Formation	
		Mytiloides spp.	*M. nodosoides*					
			F. catinus					
		Inoceramus pictus	*N. juddii / M. geslinianum*	*A. plenus*	4		Zig Zag Chalk Formation	
95			*C. guerangeri*	*H. subglobosus*	3	LOWER CHALK		Ferriby Formation
	CENOMANIAN		*A. jukesbrownei*			GREY CHALK GROUP		
		I. atlanticus	*A. rhotomagense*				West Melbury Formation	
			C. inerme					
		I. schoendorfi	*M. dixoni*	*S. varians*	2			
		I. virgatus						
		I. crippsi crippsi	*M. mantelli*		1			
		I. anglicus						

Arthur Rowe did not recognise the separate Middle and Upper Chalk divisions because they were based on what he regarded as an ephemeral and diachronous lithological feature (the Chalk Rock) and instead used the term 'White Chalk' to include both. In 1986 Mortimore formally named the 'Sussex Chalk Formation', which overlies a Lower Chalk Formation. In the same year Robinson, working in the North Downs, divided the Chalk into five formations, based on diverse criteria, including the presence/absence of rhythmicity, flint, and clay content.

In 1997 Bristow *et al.* reintroduced the threefold Lower, Middle and Upper Chalk formations, which they justified because the base of the Lewes Chalk Member of Mortimore (1983) can be widely mapped in southern England and corresponds roughly to the old Middle/Upper Chalk boundary. Gale and Hancock (1999) argued that this represented an arbitrary division and, thus defined, the Middle and Upper Chalk lacked any lithological integrity.

The first attempt this century to separate the litho- and biostratigraphy of the Chalk was made in 1978 by Wood and Smith, when they erected an entirely separate lithostratigraphical scheme for Yorkshire and Lincolnshire. They established the Ferriby Formation for the dominantly white chalk with marly partings that overlies the Red Chalk and underlies the thin but distinctive Black Band at the base of the overlying Welton Chalk Formation. This paper is particularly important in that it established the distinctive and separate lithological succession of the Chalk in north-east England for the first time.

A compromise lithostratigraphical scheme was recently agreed (at Keyworth in November, 1999) by the BGS, the Stratigraphy Commission of the Geological Society of London and other interested parties. This is summarised in Text-figure 1.4. A separate lithostratigraphy for the Chalk Group was recognised in the Northern Province (Yorkshire, Lincolnshire, north Norfolk), comprising three formations (Welton, Burnham, Flamborough). In the Southern Province (southern England) the Chalk was split into a Grey Chalk Group (=Lower Chalk; divided into West Melbury and Zigzag formations) and a White Chalk Group (=Middle and Upper Chalk; divided into Holywell, New Pit, Lewes, Seaford, Newhaven, Culver, and Portsdown formations). The 'transitional zone' between the Southern and Northern Provinces in East Anglia was deemed to require further study. This nomenclature is now used by BGS in all new maps and memoirs, but substantial problems still exist. In particular, the new subdivision of the former 'Lower Chalk' appears to have little, if any, real lithological significance.

UPPER CRETACEOUS STAGE BOUNDARIES

The boundaries of the Upper Cretaceous stages used here follow, wherever possible, those proposed at the Second International Symposium on Cretaceous Stage boundaries held in Brussels during 8–16 September 1995 (Rawson *et al.*, 1996) or ratified subsequently. These are as follows:

Cenomanian: the first occurrence of the planktonic foraminiferan *Rotalipora globotruncanoides*, for which the first appearance of the ammonites *Mantelliceras* and *Schloenbachia* are useful proxies in the Chalk sequence.

Turonian: the first occurrence of the ammonite *Watinoceras devonense*, for which the bivalve *Mytiloides puebloensis* is a good proxy in the Chalk sequence.

Coniacian: the first occurrence of the bivalve *Cremnoceramus deformis erectus* [this is the prior name of *Cremnoceras rotundatus* (*sensu* Tröger *non* Fiege), the species referred to at the Brussels meeting]. This is very rare in the British Chalk sequence.

Santonian: first appearance of the bivalve *Cladoceramus undulatoplicatus*.

Campanian: last occurrence of the crinoid *Marsupites testudinarius*.

Maastrichtian: first occurrence of the ammonite *Pachydiscus neubergicus*, for which the first occurrence of the belemnite *Belemnella lanceolata* is a good proxy in the Chalk succession.

Biostratigraphy

For most of this century, the Chalk has been subdivided into assemblage zones based mostly upon calcite macrofossils, particularly belemnites, echinoids, brachiopods and bivalves. This scheme was developed in the Paris Basin in northern France by Edmond Hébert, and was almost entirely empirical in its approach. It was extended and augmented by Charles Barrois, who in 1876 published a detailed zonal account of the entire Chalk of England and Ireland, based on six months of field work on horseback. Barrois' scheme was reviewed by Jukes-Browne in 1880, and later applied to the cliffs at Dover by William Hill. However, zonation of the Chalk was really championed by Arthur Rowe, a Margate doctor who was a keen amateur palaeontologist, well known for his work on the evolution of the echinoid *Micraster*. In a series of papers (1900–08) on the 'Zones of the White Chalk of the English coast', he elaborated his strongly opinionated views about Chalk zonation, which were to stand as the *status quo* with very little change for 60 years. The main subsequent changes were the addition of an *Offaster pilula* Zone (for the lower part of the *G. quadrata* Zone), and separation of the upper part of his *B. mucronata* Zone into the Maastrichtian belemnite zones of *B. lanceolata*, *B. pseudobtusa*, *B. obtusa* and *B. sumensis*. Text-figure 1.5 shows the progressive development towards the zonal scheme that is currently used. The Cenomanian and Turonian have been zoned with ammonites.

At the present, the traditional macrofossil assemblage zones are neither highly regarded nor widely used by research workers on the English Chalk. This is because they are rather arbitrary, poorly defined units which are coarse by modern biostratigraphical standards and many do not have very widespread applicability. For example, the *B. mucronata* Zone of Rowe and others is over 150 m thick and represents

Hébert 1863	Barrois 1876	Rowe 1900	Brydone 1912	Rawson *et al.* 1978	
B. mucronata	*B. mucronata*	*B. mucronata*	*O. lunata*	*B. occidentalis*	Maast
				B. lanceolata	
			B. mucronata	*B. mucronata*	Campanian
B. quadrata	*B. quadrata*	*A. quadratus*	*A. quadratus*	*G. quadrata*	
			O. pilula	*O. pilula*	
M. coranguinum	*Marsupites*	*Marsupites* band	*Marsupites* s-zone	*M. testudinarius*	Santonian
		Uintacrinus band	*Uintacrinus* s-zone	*U. socialis*	
	M. coranguinum	*M. coranguinum*	*M. coranguinum*	*M. coranguinum*	Coniac
M. cortestudinarium	*M. cortestudinarium*	*M. cortestudinarium*	*M. cortestudinarium*	*M. cortestudinarium*	
H. planus	*H. planus*	*H. planus*	*H. planus*	*H. planus*	
I. labiatus	*T. gracilis*	*T. gracilis*	*T. gracilis*	*T. lata*	Turonian
	I. labiatus	*R. cuvieri*	*R. cuvieri*	*I. labiatus*	
	B. plenus	*A. plenus*	*A. plenus*	*S. gracile*	
	H. subglobosus	*H. subglobosus*	*H. subglobosus*	*C. naviculare*	Cenomanian
				A. rhotomagense	
		A. varians	*S. varians*	*M. mantelli*	

TEXT-FIG. 1.5. Evolution of Chalk zonation.

approximately 5–6 myr of Late Cretaceous time, although Christensen (1994) has subdivided this into a number of zones on the basis of belemnites. Research on Chalk faunas is nowadays based on highly accurate collection of specimens against detailed lithological logs. Assemblage zones continue to be used in this book because they provide a widely used and convenient reference for the broad distribution of Chalk fossils. However, it is very important to note that the zonal usage has evolved over time, and thus the meaning of individual zones has changed significantly (see Text-fig. 1.5).

Long-distance, high-resolution biostratigraphical correlation for the Late Cretaceous (in the absence of ammonites, and the scarcity or absence of critical planktic foraminiferans for much of the post-Cenomanian succession) relies particularly on inoceramid bivalves. Their application in the UK is currently undergoing refinement; a provisional scheme is provided in Text-figure 1.4.

REGIONAL VARIATIONS IN UPPER CRETACEOUS CHALK AND CORRELATIVE FACIES AND FAUNAS

Cenomanian (Text-fig. 1.6A).

In south-east England (Kent, Surrey, Sussex, Hampshire, Isle of Wight) the Cenomanian Stage is represented by more or less rhythmically bedded marly chalks, overlying a basal condensed calcareous glauconitic sand, the Glauconitic Marl (1–5 m). These basal sands locally contain phosphatised fossils and rest with strong disconformity upon the Albian Upper Greensand (west) or Gault Clay (east). The Glauconitic Marl is overlain by the clay-rich Chalk Marl (20–40 m), above which lies the more carbonate-rich Grey Chalk (20–35 m). A thin (2–5 m) more clay-rich unit, the Plenus Marl, surmounts the Lower Chalk. The basal part of the overlying White Chalk Formation is of Cenomanian age (the so-called 'Melbourn Rock'), and comprises nodular calcisphere chalks with wispy marls. In Berkshire, Oxfordshire and through the Chilterns into East Anglia, a marked erosional disconformity develops between the Lower and Middle Cenomanian, on which rests the Totternhoe Stone, a metre-thick calcarenite containing abundant phosphatised and unphosphatised fossils. In North Norfolk, Lincolnshire and Yorkshire, the Cenomanian is represented by a succession (<30 m) of clay-poor, thinly-bedded white chalks called the Ferriby Chalk Formation, which is overlain by a thin (decimetre), laminated, organic-rich level, the Black Band, of latest Cenomanian age. In Dorset, the Lower Chalk thins, and the succession

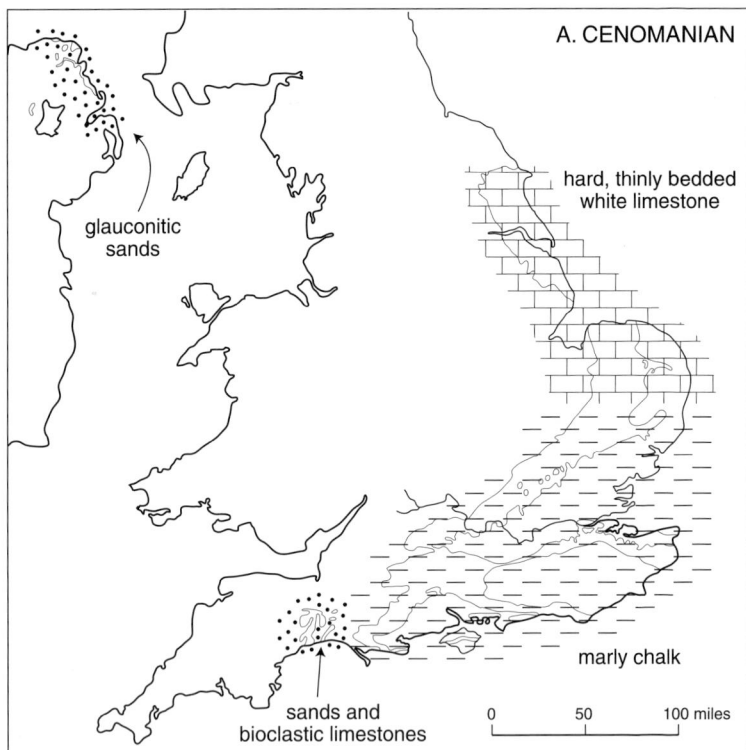

TEXT-FIG. 1.6A. Facies distributions during Chalk deposition in the UK: Cenomanian.

progressively onlaps the underlying Albian, such that the age of the base of the Chalk becomes younger westwards. The thin 'Basement Beds' are quartzose sandy chalks that contain abundant phosphatised fossils at some localities. In South Devon, the Cenomanian is represented by the richly fossiliferous, quartzose, bioclastic, and locally conglomeratic Beer Head Limestone.

Turonian (Text-fig. 1.6B).

The Turonian Chalk succession is consistent across the Southern Province. At the base comes the Holywell Formation, comprising nodular, intraclastic-calcisphere and inoceramid-prism chalks yielding a rather low-diversity fauna of small brachiopods and echinoids, and abundant

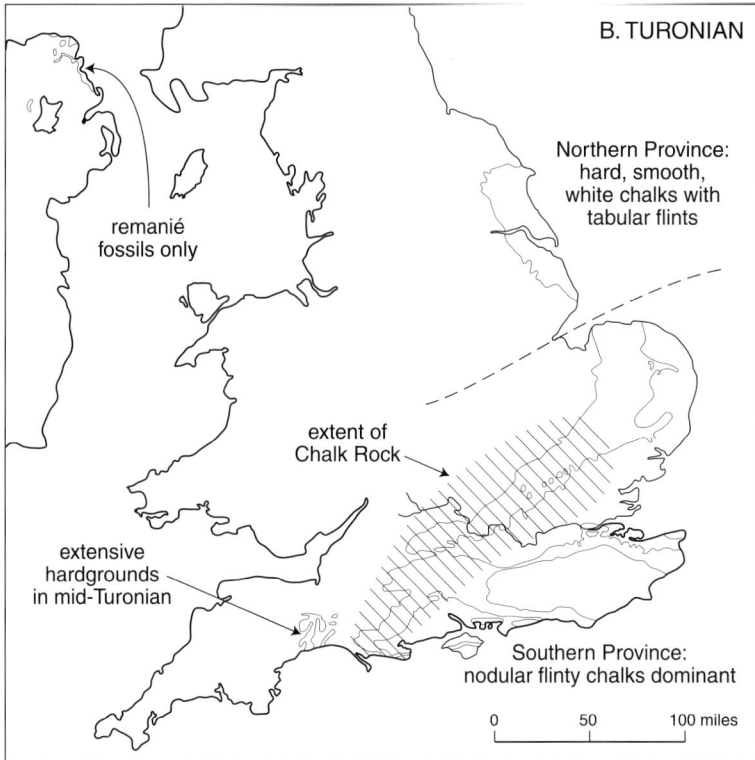

TEXT-FIG. 1.6B. Facies distributions during Chalk deposition in the UK: Turonian.

Mytiloides. The overlying New Pit Formation includes finer coccolith chalks, with thin marl beds and a few flints. The Lewes Formation is made up of calcarenite, nodular, and flinty fossiliferous chalks that yield abundant echinoids, brachiopods and bivalves. The highly condensed Chalk Rock is developed in the New Pit and Lewes chalks and is a group of mature, mineralised hardgrounds that extend across England from Hertfordshire to Dorset. Locally the Chalk Rock yields beautifully preserved moulds of originally aragonitic fossils, including diverse ammonites. In the Northern Province, the Turonian is represented by white chalks of the Burnham Formation, which contain thin marls and flint layers; they are sparsely fossiliferous. The Turonian is absent in Northern Ireland except for phosphatised Upper Turonian fossils in the Upper Glauconitic Sandstone.

Coniacian-Santonian (Text-fig. 1.6c)

In the Southern Province, the Lower Coniacian is represented by fossiliferous nodular and flinty chalks of the upper Lewes Formation, which locally contain numerous hardgrounds. In some areas (e.g. Chilterns, East Anglia), these amalgamate into a single, highly condensed group of closely spaced mineralised hardgrounds called the Top Rock. The Upper Coniacian and Santonian in southern England comprise soft, coccolith chalks of the Seaford Formation containing numerous beds of nodular flints (approximately equivalent to the *M. coranguinum* Zone), which are generally rather sparsely fossiliferous, with the exception of fragments of large inoceramid bivalves. Rarely, beautifully preserved echinoids and asteroids are found, and air-weathered surfaces yield numerous small calcite fossils. The Upper Santonian is also made up of soft, fine chalks of the lower Newhaven

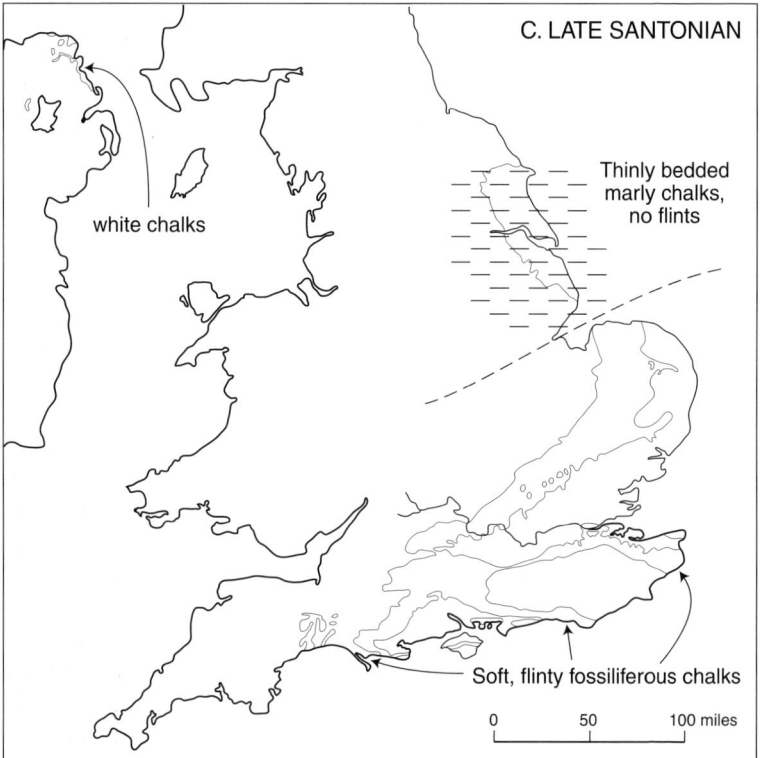

TEXT-FIG. 1.6c. Facies distributions during Chalk deposition in the UK: Late Santonian.

Formation that locally contain marl seams. Abundant plates of the stemless benthic crinoids *Uintacrinus socialis* and *Marsupites testudinarius* characterise the Upper Santonian everywhere in southern England.

In the Northern Province (Yorkshire), the Coniacian and Lower Santonian are represented by the hard flinty chalks of the upper Burnham Formation, and the Upper Santonian by the lower Flanborough Formation of flintless chalks containing thin marls. On the Causeway Coast of Antrim, Northern Ireland, the Upper Santonian is represented by flinty chalks of the lowest part of the White Limestone Formation.

Campanian (Text-fig. 1.6D).

In the Southern Province (Sussex, Hampshire, Isle of Wight, Dorset, Wiltshire), the Campanian is represented by white nannofossil chalks

TEXT-FIG. 1.6D. Facies distributions during Chalk deposition in the UK: Campanian–Early Maastrichtian.

comprising the upper Newhaven, Culver and Portsdown formations. These contain numerous layers of flint nodules, and a number of distinctive thin marl beds, both of which may be used in correlation. Important marker levels are narrow intervals with the echinoderms *Uintacrinus anglicus* Rasmussen, *Offaster pilula* (Lamarck), *Hagenowia blackmorei* Wright and Wright, and distinctive shape varieties of *Echinocorys*. Locally condensed horizons with hardgrounds and slumps occur in the Campanian, and these may contain an abundant fauna including ammonites.

The Campanian Chalk of southern England was truncated or completely removed by early Cenozoic erosion. It is only in Norfolk that uppermost Campanian and Maastrichtian chalks are preserved. The Maastrichtian is found only as glacially emplaced, tectonised masses on the coast to the east of Cromer, and to the east of Norwich. In the Northern Province (Yorkshire), the Lower Campanian Flamborough Formation comprises flint-free chalks with thin marls and, locally, finely preserved sponges. In Northern Ireland, the Campanian and lowermost Maastrichtian make up all but the lowest part of the White Limestone Formation, a hardened chalk with flints and rare hardgrounds.

FURTHER READING

Stratigraphy

BRISTOW, R., MORTIMORE, R. and WOOD, C. J. 1997. Lithostratigraphy for mapping the Chalk of southern England. *Proceedings of the Geologists' Association*, **108**, 293–317.

CHRISTENSEN, W. K. 1994. *Belemnitella* from the Upper Campanian and Lower Maastrichtian of Norfolk, England. *Special Papers in Palaeontology*, **51**, 1–80.

GALE, A. S. 1995. Cyclostratigraphy and correlation of the Cenomanian Stage in Western Europe. *Geological Society, London, Special Publication*, **85**, 177–197.

—— 1996. Correlation and sequence stratigraphy of the Turonian Chalk of southern England. *Geological Society, London, Special Publication*, **103**, 177–195.

—— and CLEEVELY, R. 1989. Arthur Rowe and the zones of the White Chalk of the English Coast. *Proceedings of the Geologists' Association*, **100**, 419–431.

MOGHADAM, H. V. and PAUL, C. R. C. 2000. Micropalaeontology of the Cenomanian at Chinnor, Oxfordshire, and comparison with the Dover–Folkestone succession. *Proceedings of the Geologists' Association*, **111**, 17–39.

MORTIMORE, R. N. 1986. Stratigraphy of the Upper Cretaceous White Chalk of Sussex. *Proceedings of the Geologists' Association*, **97**, 97–139.

ROWE, A. W. 1900. The zones of the White Chalk of the English Coast. I. Kent and Sussex. *Proceedings of the Geologists' Association*, **16**, 289–368.

Trace fossils

BROMLEY, R. G. 1970. Borings as trace fossils and *Entobia cretacea* Portlock as an example. *Geological Journal, Special Issue*, **3**, 49–90.
—— 1975. Trace fossils at omission surfaces. *In* FREY, R. W. (ed.). *The study of trace fossils*. Springer-Verlag, New York, 399–428.
—— 1984. Trace fossil preservation in flint in the European Chalk. *Journal of Paleontology*, **58**, 298–311.
—— 1986. Composite ichnofabrics and tiering of burrows. *Geological Magazine*, **123**, 59–65.

Sedimentology and palaeoceanography

GALE, A. S., YOUNG, J. R., CROWHURST, S. and SHACKLETON, N. J. 1999. Orbital tuning of Cenomanian marly chalk successions; towards a Milankovitch timescale for the Late Cretaceous. *Philosophical Transactions of the Royal Society, Series A*, **357**, 1815–1829.
—— SMITH, A. B., MONKS, N., YOUNG, J. R., HOWARD, A., WRAY, D. S. and HUGGETT, J. M. 2000. Marine biodiversity through the Late Cenomanian–Early Turonian; palaeoceanographic controls and sequence stratigraphic biases. *Journal of the Geological Society, London*, **157**, 745–757.
HÅKANSSON, E., BROMLEY, R. G. and PERCH-NIELSEN, K. 1974. Maastrichtian Chalk of north-west Europe – a pelagic shelf sediment. *In* HSÜ, K. J. and JENKYNS, H. C. (eds). Pelagic sediments on land and under the sea. *Special Publication of the International Association of Sedimentologists*, **1**, 211–233. [Nature of chalk sedimentation]
HANCOCK, J. M. 1975. The petrology of the Chalk. *Proceedings of the Geologists' Association, London*, **86**, 499–535. [Composition of chalks]
—— 1993. The formation and diagenesis of Chalk. *In* DOWNING, R. A., PRICE, M. and JONES, G. P. *The hydrogeology of the Chalk of North West Europe*. The Clarendon Press, Oxford, 14–34.
JENKYNS, H. C., GALE, A. S. and CORFIELD, R. 1994. The stable oxygen and carbon isotope stratigraphy of the English Chalk and the Italian Scaglia. *Geological Magazine*, **13**, 1–34. [Climate changes recorded in chalks]
MORTIMORE, R. N. 1987. Controls on Upper Cretaceous sedimentation in the South Downs, with particular reference to flint distribution. *In* SIEVEKING, G. DE C. and HART, M. B. (eds). *The scientific study of chalk and flint*. Cambridge University Press, Cambridge, 21–42.
—— and POMEROL, B. 1997. Upper Cretaceous tectonic phases and end Cretaceous inversion in the Chalk of the Anglo-Paris Basin. *Proceedings of the Geologists' Association*, **108**, 231–255.

Diagenesis

CLAYTON, C. J. 1986. The chemical environment of flint formation in Upper Cretaceous chalks. *In* SIEVEKING, G. DE C. and HART, M. B. (eds). *The scientific study of chalk and flint*. Cambridge University Press, Cambridge, 43–55.

KENNEDY, W. J. and GARRISON, R. E. 1975. Morpology and genesis of nodular chalks and hardgrounds in the Upper Cretaceous of southern England. *Sedimentology*, **22**, 311–386.

SCHOLLE, P. A. 1977. Chalk diagenesis and its relation to petroleum exploration: oil from chalk, a modern miracle? *Bulletin of the American Association of Petroleum Geologists*, **61**, 982–1009.

2. SPONGES

by RACHEL WOOD

Although sponges are common fossils in the Chalk they have received little attention and remain poorly known. The main systematic accounts are the works of Toulmin Smith (1847–48), in which he regarded the hexactinellid sponges as Bryozoa, Hinde's 'Catalogue of the Fossil Sponges' (1883) and various papers and an unfinished monograph by Reid (1954–64, 1958, 1961, 1962*a*, *b*).

Sponges belong to the phylum Porifera and are the most primitively organised of multicellular animals. They suspension-feed by pumping water through a system of canals and flagellated chambers. In most sponges, specialised cells secrete a mineral skeleton, generally in the form of spicules. It is this skeleton, as a whole or as isolated spicules, which becomes fossilised and provides the basis for the classification of Recent and fossil sponges. Spicules may be isolated in soft tissue or fused to form compound skeletons, and may be embedded in a massive calcareous skeleton. The latter are the most frequently preserved sponges in the fossil record. Sponges are found throughout the Chalk, but species are long-ranging and therefore of little stratigraphical value.

MORPHOLOGY

The Porifera have one basic cell type called a choanocyte. Choanocytes resemble individual flagellated protozoa (Text-fig. 2.1A). Sponges are aquatic and typically sessile, and feed via currents set up by the flagella in the choanocyte chambers. Sponge anatomy is extremely variable, but all are modifications of a single organisational type. This is an erect sac-like form, attached at the base, open at the top, and lined with choanocyte cells (Text-fig. 2.1c). Water enters the system through a series of small pores known as ostia, passes through the sponge wall into a central cavity, and is expelled through a common terminal opening known as an osculum (Text-fig. 2.1c). The flagella collect food particles, which are digested. There is no digestive cavity. Most sponges show complications of this basic structure by folding of the walls. The choanocyte cells protrude to form flagellated chambers (Text-fig. 2.1B). Canals develop for the separation of inhalant and exhalant circulation and these modify the flow of water to produce an aquiferous canal system.

osculum

A   B   C

ostia

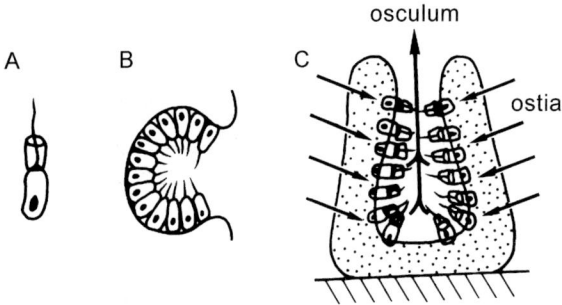

TEXT-FIG. 2.1. General features of the Porifera. A, choanocyte cell. B, choanocyte chamber. C, longitudinal section showing basic sponge organisation; arrows indicate direction of water circulation.

SYSTEMATICS OF THE PORIFERA

The classification of fossil sponges is based upon the chemical composition, geometric configuration, and positioning of the spicules within the sponge. Of the three classes of sponges described from the Chalk, two have siliceous spicules and one is characterised by calcareous spicules.

Class Demospongia (Text-fig. 2.2). Sponges with a skeleton composed of one- to four-rayed siliceous spicules, accompanied by organic spongin and/or collagen fibres. Spicules fall into two distinct size categories; megascleres and microscleres.

Class Calcarea (Text-fig. 2.2). Sponges with a skeleton of fused calcite spicules. The spicules can be one- to four-rayed and do not occur in the pronounced size categories of the demosponges and hexactinellids.

The tissues of the Demospongia and the Calcarea are histologically similar and some workers unite these two classes to form the Gelatinosa. The Gelatinosan wall consists of three layers: (1) an external dermal layer of flattened cells; (2) a gelatinous median strand; and (3) an inner gastral surface layer of choanocytes. In more advanced forms the choanocytes are restricted to flagellated chambers.

The canal system of the Gelatinosa consists of two sets of internal passages which open on opposite sides of the body wall. The exhalant passages may be lined with choanocyte chambers, or the chambers may discharge into the exhalant canals. The canal system of the Gelatinosa consists of originally external spaces, secondarily enclosed within the body wall.

Class Hexactinellida (Text-fig. 2.2). The hexactinellids, or Nuda, tend to be radial in symmetry and are usually vase-shaped. Most of the numerous spicule types are derived from the triaxon (three-rayed) or hexactine (six-rayed) in which three axes meet at right angles. Spicules are classified according to their position within the body. Microscleres and megascleres are present. The axial canal of hexactinellid sponge spicules is square in cross-section, as opposed to circular in the Gelatinosa.

The basic structure of the hexactinellid sponge wall consists of a choanocyte layer bounded on both sides by a network of filaments called trabeculae. These are modified at both external surfaces to form porous membranes, the dermal (external) and gastral (internal) bounding membranes. The area between the two membranes is known as the parenchyme. The dermal membrane is perforated by ostia and the gastral membrane by exhalant pores known as postica. Spicules are arranged into three categories, those lying beneath the dermal membrane, those near the gastral membrane and those in the parenchyme. The parenchymal skeleton may be fused to form an intricate framework.

The canal system is composed of flagellated chambers only and most hexactinellids develop convolutions of this simple chamber layer. The interspaces of the trabecular and choanocytal networks are filled with water. The canals are simply enlarged trabecular spaces, which form tubular passages.

These soft-tissue differences are not immediately recognisable in fossil material, but need to be considered when interpreting skeletal structures.

DISTRIBUTION

Chalk sponge faunas are predominantly siliceous, composed of lithistid demosponges, hexactinellids plus a few characteristic Calcarea. Although occurring in the south of England and Lincolnshire, they are more common on the Yorkshire coast. The appearance of the Hexactinellida seems to be associated with the onset of Chalk deposition. Immigration of hexactinellids occurred in most parts of Europe at the time of the Cenomanian transgression (Oakley 1937). In England, this is reflected by the sudden abundance of forms such as *Stauronema carteri* in the Lower Cenomanian (*M. mantelli* Zone) of the southern counties, in contrast to their paucity in the Albian Greensand. Hexactinellida are rare throughout the Cenomanian and basal Lower Turonian; above this they become more

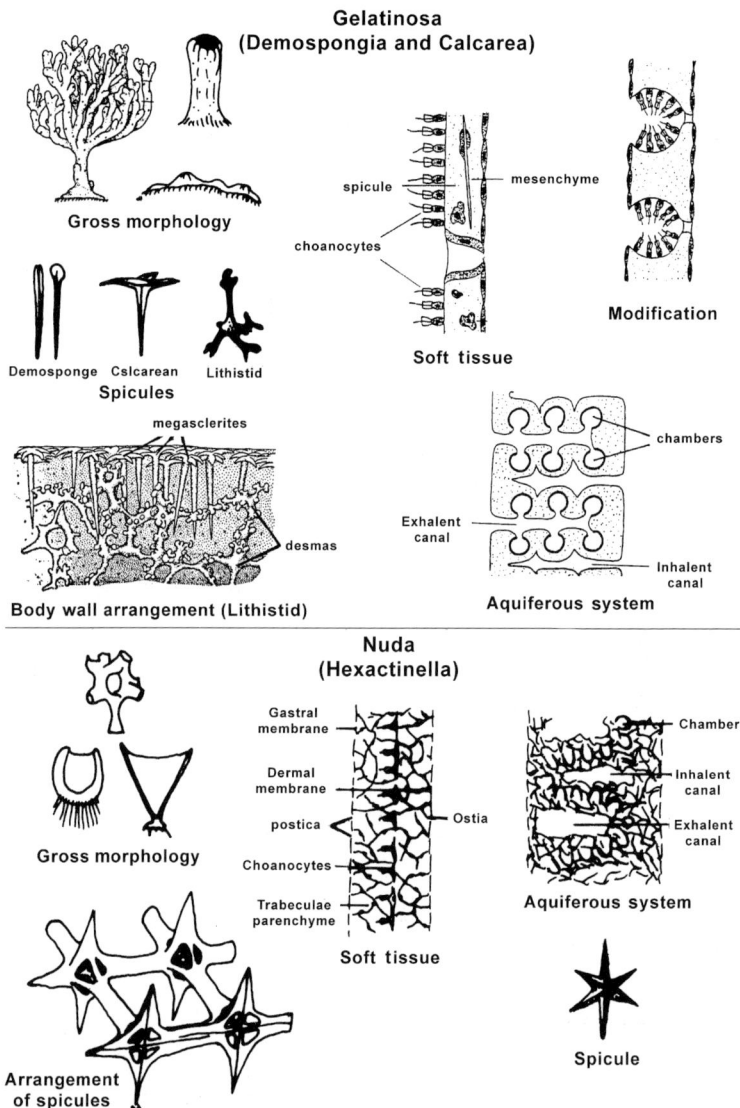

TEXT-FIG. 2.2. Morphology of the Gelatinosa and Nuda (after Reid 1954–54 and Hartman *et al.* 1979).

abundant and a new fauna develops. Hexactinellids dominate throughout the Middle and Upper Chalk together with some lithistids and the boring sponge *Cliona cretacea*.

DESCRIPTIONS

Spicule type and arrangement form the basis of fossil sponge classification at all taxonomic levels. Examination requires acid-dissolution of the spicules, thin sectioning of the specimens as well as specialist knowledge. There is not the space, nor is it appropriate in a 'field guidebook' to describe the details and rationale of spicule taxonomy. The following descriptions are working, rather than formal, diagnoses that will enable the reader to recognize forms in the field from their gross morphology and surface features. General characteristics of orders, including spicule form, are given where appropriate.

Phylum PORIFERA
Class HEXACTINELLIDA
Order LYCHNISCOSA

Remarks. Hexactinellids in which the parenchymal microscleres are hexasters (small, six-rayed and often with branched ends). The parenchymal megascleres unite to form a rigid framework. Each of the nodes is supported by 12 struts and is known as a lychnisc (Text-fig. 2.2). Only two families occur in Recent seas, but the group was common in the Mesozoic. They attached to a hard substratum or anchored to a soft bottom by basal tufts of spicules.

Family CALLODICTYIDAE

Genus POROCHONIA

Porochonia simplex (T. Smith)
Plate 1, figure 3

Description. Body funnel-shaped with or without a short initial stalk, supported by branching root-processes. Dermal skeletal surface with numerous small ostia, fine concentric ribbing and circular or oval apertures. Gastral skeletal surface with no distinct postica.
Occurrence. Turonian–Campanian; southern England.

Family VENTRICULITIDAE

Genus STAURONEMA

Stauronema carteri Sollas
Plate 3, figures 3–4

Description. Body saddle-, ear- or tongue-shaped or discoidal; no basal attachment. Dermal skeletal surface with many small ostia or none; the ostia are arranged without order. The surface is often prominently corrugated transversely or concentrically. The gastral skeletal surface has postica. Canalisation of the interior, if any, is in the form of irregular anastomosing passages.

Occurrence. Lower Cenomanian, *M. mantelli* Zone; southern England including the Isle of Wight.

Genus RHIZOPOTERION

Rhizopoterion cribosum (Phillips)
Plate 2, figure 4

Description. Body subcylindrical or cup-, funnel-, vase-, trumpet- or mushroom-shaped; sometimes with an initial stem, which is usually solid and supported by root processes. Dermal side of parenchymal framework is sometimes longitudinally ribbed, but never tuberculate. Ostia typically ovate, with their greater length arranged longitudinally, often separated in longitudinal series or depressed in furrows. Gastral side with rounded, ovate or irregular postica.

Remarks. *Rhizopoterion* can be distinguished from *Ventriculites* by the absence of external tuberculation on the skeletal surface.

Occurrence. Coniacian–Campanian; southern England.

Genus VENTRICULITES

Ventriculites chonoides (Mantell)
Plate 1, figures 4–5

Description. Body subcylindrical, funnel- or trumpet-shaped, or discoidal with a small initial funnel supported by branching root processes, typically slender and emitted from a small basal stalk. Dermal skeletal surface reticulate, and longitudinally or radially ribbed, tuberculate or showing an intermediate condition. Ostia usually depressed. Gastral side with round or ovate postica which alternate or are in longitudinal series. Root fibres formed from anastomosing filaments. Solid or showing a few longitudinal canals.

EXPLANATION OF PLATE 1

Fig. 1. *Paraplocia labyrinthica* (Mantell). Turonian, *T. lata* Zone, Dover, Kent; ×0·5.

Fig. 2. *Guettardiscyphia stellata* Michelin. 'Upper Chalk', horizon unknown, southern England; ×0·3.

Fig. 3. *Porochonia simplex* (T. Smith). Turonian, locality unspecified; ×0·4.

Figs 4–5. *Ventriculites chonoides* (Mantell). 'Upper Chalk', horizon unknown, southern England. 4, ×0·5; 5, ×0·3.

PLATE 1

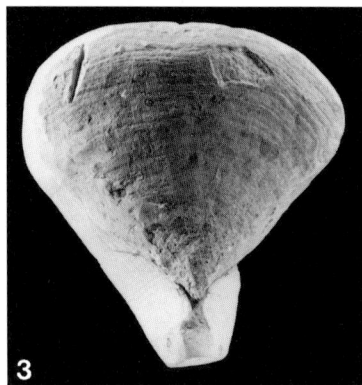

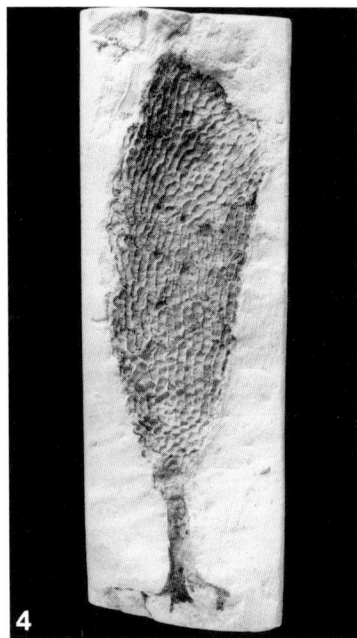

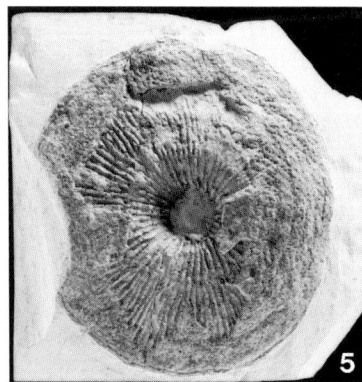

Remarks. *V. chonoides* is characteristically very variable in form.
Occurrence. Turonian–Campanian; especially common in the *S. plana* Zone; southern England.

Family DACTYLOCALYCIDAE

Genus PARAPLOCIA

Paraplocia labyrinthica (Mantell)
Plate 1, figure 1

Description. Body club- or cup-shaped, consisting of either dividing and anastomosing tubes or of an axial tube or funnel and open lateral tubes or solid lateral outgrowths, or of a central cluster of anastomosing tubes with solid peripheral outgrowths producing a reticulate surface. Circulatory canals in the form of a single system of dividing and anastomosing tubular passages which open through small rounded apertures in the skeletal surface, or elongate branching grooves. No apparent basal skeleton.
Remarks. The lack of a basal skeleton suggests that the basal mass of *P. labyrinthica* was embedded in the substratum and anchored by its weight.
Occurrence. Turonian–Campanian; especially in the *S. plana* Zone; southern England.

Order HEXACTINOSA
Family CRIBOSPONGIIDAE

Genus GUETTARDISCYPHIA

Guettardiscyphia stellata Michelin
Plate 1, figure 2

Description. Body stellate in plan, with axially-continuous compressed radial flanges, which originate as radial plications of a tubular stalk, or bilaterally compressed branches arising from an axial stalk. With or without a main osculum and with accessory oscula along the margins of the flanges. Dermal skeleton with alternating ostia. Gastral side with twice as many apertures, enlarged in longitudinal series.
Occurrence. Middle and Upper Cenomanian, Turonian (especially in the *S. plana* Zone) and Coniacian–Campanian; southern England.

Class CALCAREA
Subclass PHARETRONIDA

Remarks. A heterogeneous assemblage of calcareous sponges which possess 'tuning-fork'-shaped spicules.

Family PHARETROSPONGIIDAE

Genus PHARETROSPONGIA

Pharetrospongia strahani Sollas
Plate 3, figures 1–2

Description. Body formed of convoluted plates, sometimes becoming funnel-shaped or subcylindrical. Walls are 7–13 mm thick. No basal attachment. Outer skeletal surface is uneven, composed of fine reticulate tissues with small interspaces. The inner skeletal surface is smoother with coarser fibres and larger interspaces. No definite canals can be distinguished.
Occurrence. Coniacian–Campanian; Kent, Norwich (Norfolk), Warminster (Wiltshire) and Yorkshire.

Order MINCHINELLIDA

Genus POROSPHAERA

Porosphaera globularis (Phillips)
Plate 4, figure 3

Description. Roughly spherical, pea- or marble-shaped, ranging from 1–34 mm in diameter. Sometimes oval-, loaf- or cushion-shaped. For the most part free, with some growth layers partially covering the surface. Outer surface covered with small apertures and often with a large, open-ended canal through the centre. Shallow open grooves are occasionally seen on the surface.
Occurrence. Turonian (*I. labiatus* Zone) – Campanian; southern England to Yorkshire.

Porosphaera patelliformis Hinde
Plate 4, figure 6

Description. Limpet-shaped with peaked or rounded summits, base rounded or oval in outline. Usually deeply concave. Thin-walled. Slight development of radial canals.
Occurrence. Turonian–Campanian; southern England to Yorkshire.

Porosphaera arreta Hinde
Plate 4, figure 4

Description. Simple, conical pillar-shaped. Concave base with thin margins. Faint canal traces.

PLATE 2

Occurrence. Turonian–Campanian; rare in the Turonian of Devon, Lower Santonian of Yorkshire and Upper Santonian–Campanian of Kent and Sussex.

Class DEMOSPONGIA
Subclass TETRACTINOMORPHA
Order LITHISTIDA

Remarks. A polyphyletic assemblage of demosponges which share the presence of multi-branched siliceous spicules called desmas. The internal skeleton is composed of fused desmas which form a rigid, reticulate skeleton.

Suborder RHIZOMORINA
Family LEIODORELLIDAE

Genus STICHOPHYMA

Stichophyma tumidum (Hinde)
Plate 2, figure 3

Description. Simple elongate, club-shaped or cylindrical, usually widest near the summit and diminishing towards the basal end. The stem bears a series of horizontal swellings and constrictions. Very variable in size. Openings of vertical canals often project above the summit of the sponge. Lateral surfaces are covered with numerous circular ostia.
Occurrence. Coniacian–Campanian; southern England to Yorkshire.

Family KALIAPSIDAE

Genus LAOSCIADIA

Laosciadia plana (Phillips)
Plate 3, figure 6

Description. Expanded, mushroom-shaped. Upper surfaces plate-like, circular in outline often with a central depression immediately above the

---

EXPLANATION OF PLATE 2

Figs 1–2. *Pachinion scriptum* (Roemer). Santonian, Danes Dyke, Sewerby, Yorkshire; ×0·3.

Fig. 3. *Stichophyma tumidum* (Hinde). Coniacian or Santonian, Flamborough Head, Yorkshire; ×0·4.

Fig. 4. *Rhizopoterion cribosum* (Phillips). Campanian, *B. mucronata* Zone, Norwich, Norfolk; ×0·5.

PLATE 3

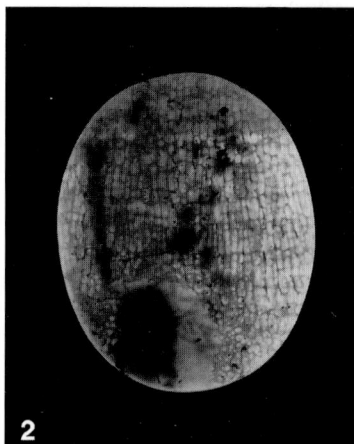

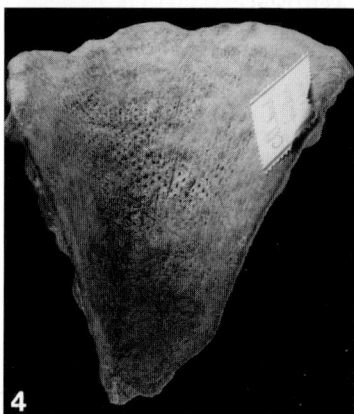

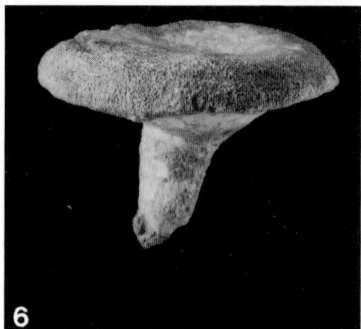

stalk. Surface may be flat or slightly convex. Inner surface is flat or concave. Stem is inversely conical and tapers to a blunt point. Upper surface is covered as far as the marginal edge with circular apertures.

Occurrence. Santonian–Campanian; Yorkshire.

Suborder MEGAMORINA
Family DORYDERMATIDAE

Genus DORYDERMA

Doryderma roemeri (Hinde)
Plate 3, figure 5

Description. Branching, with a thick stem and short lateral branches. Stems and branches transversed by numerous vertical canals 2 mm in diameter.

Remarks. Frequently found in flint nodules.

Occurrence. Coniacian–Campanian; Buckinghamshire, Wiltshire.

Suborder EUTAXICLADINA
Family GIGNOUXIIDAE

Genus PACHINION

Pachinion scriptum (Roemer)
Plate 2, figures 1–2

Description. Simple, cylindrical or inversely conical with the lower part gradually tapering to a cylindrical stem. The surface is smooth. The summit is rounded or depressed conical. The osculum is cylindrical and extends nearly the entire length of the sponge. No distinctive canals.

Occurrence. Coniacian–Campanian; southern England to Yorkshire.

Suborder TETRACLADINA
Family HALLIRHOIDAE

Genus SIPHONIA

EXPLANATION OF PLATE 3

Figs 1–2. *Pharetrospongia strahani* Sollas. Turonian, Pertwood, near Warminster, Wiltshire. 1, ×0·7; 2, ×0·1.

Figs 3–4. *Stauronema carteri* Sollas. Lower Cenomanian, *M. mantelli* Zone, Folkestone, Kent; ×0·5.

Fig. 5. *Doryderma roemeri* (Hinde). Lower Coniacian, Widnall Wood, Little Hampden, Buckinghamshire; ×0·7.

Fig. 6. *Laosciadia plana* (Phillips). Coniacian or Santonian, Flamborough Head, Yorkshire; ×0·4.

PLATE 4

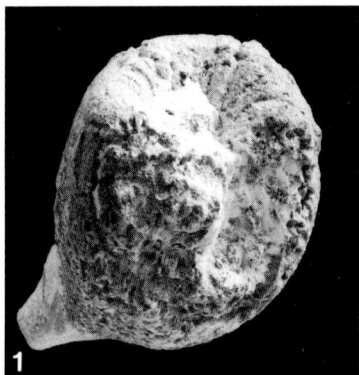

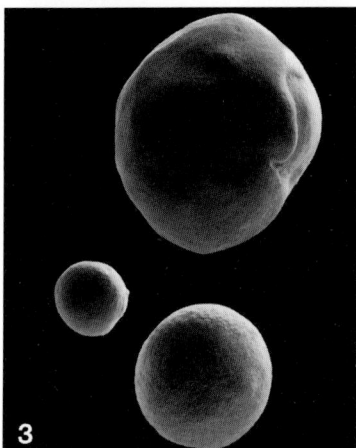
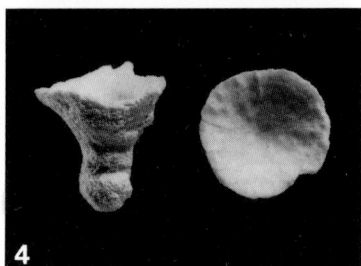

Siphonia koenigi (Mantell)
Plate 4, figures 1–2

Description. Pear-shaped or subspherical with a slender cylindrical stem. Cylindrical or funnel-shaped osculum with a very wide aperture. Margins are rounded. Strongly marked branching canals radiate from the margin of the osculum down the sides of the sponge.

Remarks. Most specimens are preserved in the interior of flint nodules which preserves the canal system but obliterates the spicular structure.

Occurrence. Coniacian–Campanian; most common as flint pebbles on Brighton beach; rare elsewhere but occurs throughout southern England and Yorkshire.

Order HADROMERIDA

Remarks. Demosponges with megascleres in the form of tylostyles (single, one-axis spicules with a basal boss) arranged in a radial pattern.

SPONGE ICHNOGENERA (TRACE FOSSILS)

Genus ENTOBIA

Entobia sp.
Plate 4, figure 5

Description. Solid spheroidal, ovate or depressed elongate siliceous bodies from 1·8–5·5 mm in diameter that are connected together by numerous stolons into small groups. Usually found in flints, partially filling the cavities formally occupied by belemnites, the tests of echinoderms or the shells of *Inoceramus*.

Remarks. These are casts in silica of the hollowed-out borings of the sponge *Cliona cretacea* (Portlock).

Occurrence. Upper Turonian–Maastrichtian; widespread throughout southern England.

EXPLANATION OF PLATE 4

Figs 1–2. *Siphonia koenigi* (Mantell). Coniacian or Santonian, Flamborough Head, Yorkshire. 1, ×0·5; 2, ×1.

Fig. 3. *Porosphaera globularis* (Phillips). Santonian, Thanet coast, Kent; ×1.

Fig. 4. *Porosphaera arreta* Hinde. Santonian, Ringwould, near Dover, Kent; ×1·3.

Fig. 5. *Entobia* sp. (Portlock). Santonian, Margate, Kent; ×1·3.

Fig. 6. *Porosphaera patelliformis* Hinde. 'Upper Chalk', horizon unknown, Sussex; ×1·3.

3. CORALS

by ANDREW. S. GALE

Small, solitary, scleractinian corals and octocorals are relatively common fossils in Upper Turonian–Maastrichtian chalks. All were originally attached to a hard substrate in life, and many smaller corals, including most octocorals, are found as epizoans on larger calcitic fossils. Larger solitary scleractinian corals are, however, found loose in the Chalk, either because they outgrew their substrate or because it later disintegrated or dissolved.

The foundation for Chalk coral taxonomy was established with the monograph by Edwards and Haime (1850–54) and Duncan (1869). Since then there has been very little published research on this group, with the exception of the octocorals which were revised by Voigt (1958). In particular, the caryophyllids, for which Duncan (1869) created so many new names, are badly in need of taxonomic revision.

Chalk scleractinian corals occur in two preservational styles, which reflect differences in original composition and probably also in taxonomy. The first is mouldic preservation, reflecting an originally aragonitic composition that has been lost during early diagenesis. Corals occur as hollow external and internal moulds in hardgrounds, and as crushed composite moulds in marly chalks of Cenomanian age. This material has, in general, been only poorly studied but is probably taxonomically separate from fossils in the next group.

The second preservational style is as three-dimensional calcite fossils, either loose within the chalk matrix or attached to hard substrates, most commonly on echinoid tests. This group is taxonomically diverse and includes the Caryophyllidae.

The fact that extant scleractinians are all composed of aragonite raises some interesting questions. Were some chalk scleractinian corals originally composed of calcite, and others of aragonite, or were they all composed originally of aragonite, but some converted to calcite on the sea floor?

DESCRIPTIONS

Subclass SCLERACTINIA
Family MICRABACIIDAE

Genus MICRABACIA

Micrabacia coronula (Goldfuss)
Plate 5, figures 7–8

Description. Small, button-like forms, circular in outline and 10 mm or less in diameter. The base is flat with concentric rings. Numerous finely denticulate septa extend down the vertical margin.
Remarks. Preserved as calcite.
Occurrence. Cenomanian–Turonian; common at some levels in Cenomanian chalks in southern and eastern England.

Family GUYNIIDAE

Genus ONCHOTROCHUS

Onchotrochus serpentinus Duncan
Plate 5, figure 4

Description. Elongate, solitary corallite, 2–4 mm in diameter; sinuous with angled bends. Surface smooth or with 4–6 longitudinal ridges. Septa few; no columella.
Remarks. Preserved as calcite.
Occurrence. Cenomanian–Maastrichtian; frequent to common throughout the English Chalk.

Family CARYOPHYLLIDAE

Genus PARASMILIA

Parasmilia centralis (Mantell)
Plate 5, figure 1

Description. Elongated, cylindrical-conical form, commonly sinuous, with flanged attachment area. Septal ridges weak and inconspicuous but more strongly developed on the outside of curves of the corallite. Columella spongy, comprising irregular blebs of calcite; septa ornate.
Occurrence. Turonian–Santonian; common and widespread in the English Chalk.

Parasmilia granulata Duncan
Plate 5, figures 11–12

Description. Conical, curved form. The outside of the corallite bears numerous low septal ridges which are finely granular. Septa bear granular ornament. Columella spongy.
Occurrence. Coniacian–Santonian; uncommon in Kent, Sussex, Hampshire, Wiltshire and the Isle of Wight.

Genus TROCHOSMILIA

Trochosmilia wiltshiri Duncan
Plate 5, figures 5–6

Description. Corallum symmetrically conical in form with prominent septal ridges on all but the earliest part. These, together with the irregular transverse ridges, form a raised reticulate sculpture on the surface of the corallite. The septa are smooth, and a columella is absent.

Occurrence. Upper Campanian; Norfolk, both around Norwich and on the coast west of Cromer.

Family OCULINIDAE

Genus DIBLASUS

Diblasus gravensis Lonsdale
Plate 5, figure 10

Description. Form irregular, roughly equidimensional, comprising a relatively small number of corallites which form low cylindrical projections from the smooth or weakly striated surface of the colony. Most corallites are nearly hollow, with only the mural bases of septa preserved.

Remarks. This is the only compound scleractinian coral to occur with any frequency in the Chalk.

Occurrence. Coniacian–Santonian; uncommon but widespread in southeast England.

EXPLANATION OF PLATE 5

Fig. 1. *Parasmilia centralis* (Mantell). Santonian, Canterbury, Kent; lateral view of corallite; ×2·2.

Figs 2–3. *Moltkia* sp. 2. Lower Campanian, Salisbury, Wiltshire; attachment base of root-like form; 3, Lower Santonian, Kent; sheet-like form attached to echinoid; both ×1·6.

Fig. 4. *Onchotrochus serpentinus* Duncan. Cenomanian, Eastbourne, Sussex; lateral view of corallite; ×2·5.

Figs 5–6. *Trochosmilia wiltshiri* Duncan. Upper Campanian, Norwich, Norfolk. 5, lateral view; ×1·5; 6, calical view; ×2·3.

Figs 7–8. *Micrabacia coronula* (Goldfuss). Cenomanian, Eastbourne, Sussex. 7, calical view; 8, lateral view; both ×2·3.

Fig. 9. *Epiphaxum auloporoides* Duncan. Lower Santonian, Northfleet, Kent; colony; calcified stolon attached to the echinoid *Conulus albogalerus*; ×3·5.

Fig. 10. *Diblasus gravensis* Lonsdale. Upper Coniacian or Lower Santonian, Gravesend, Kent; colony; ×1·5.

Figs 11–12. *Parasmilia granulata* Duncan. Lower Santonian, Ramsgate, Kent. 11, calical view; 12, lateral view; both ×2·5.

PLATE 5

Subclass OCTOCORALLIA
Family CLAVULARIIDAE

Genus EPIPHAXUM

Epiphaxum auloporoides Duncan
Plate 5, figure 9

Description. Calcified stolon which attaches to hard substrates and has irregularly meandering and branching habit. Stolon (1–2 mm across) composed of 5–8 fused rods. At irregular intervals and at branching points smooth, concave-lipped cavities are present from which uncalcified polyps would have arisen.

Occurrence. Upper Turonian–Campanian; widely distributed and fairly common in the English Chalk.

Family ISIDIDAE

Genus MOLTKIA

Moltkia sp.
Plate 5, figures 2–3

Description. Calcified attached base, comprising either (1) a root-like, multi-branching structure with central irregular cavity, or (2) an irregularly lobate structure with fine, sinuous, radially arranged striae.

Remarks. Usually preserved attached to hard substrates, most commonly echinoid tests. Found free in chalk where original substrate has been lost by decomposition or early diagenesis.

Occurrence. Cenomanian–Maastrichtian; widespread but not common in the English Chalk.

4. SERPULIDS

by ANDREW S. GALE

Calcite tubes constructed by serpulid polychaete worms are common but rather inconspicuous fossils in the Chalk. They are most often found attached to hard substrates (including fossils, intraclasts and more rarely hardground surfaces) but also occur loose within the sediment, either because they were originally free-living or because the original attachment site has been lost through sea-floor or early diagenetic dissolution.

Many new species of serpulid from the English Chalk were described by J. Sowerby (1815–18) and J. de C. Sowerby (1826–45), and the zonal fossil collectors of the late nineteenth and early twentieth centuries collected extensive material of the group. In particular, Arthur Rowe provided notes on the taxonomy and distribution of species in the 'Zones of the White Chalk' series (1900–08), and left extensive annotations on labelled specimens in his collection. However, the English material still lacks any published taxonomic study since the original Sowerby descriptions. In contrast, German, Danish, Polish and Russian workers have taken considerable interest in the serpulids of the Chalk, and important taxonomic works by Regenhardt (1961), Jaeger (1983) and others have done much to standardize the taxonomy of this group. Some of these works involved partial revision of Sowerby's original English material, so it is at least possible to apply updated names to Chalk serpulids.

Fossil serpulids are classified on the structure, coiling, and external sculpture of the tube, and less commonly the opercula present in some forms.

DESCRIPTIONS

Class POLYCHAETA
Order SEDENTARIA
Family SERPULIDAE
Subfamily FILOGRANINAE

Genus CYCLOSERPULA

Remarks. Tubes circular to oval in cross-section, externally smooth, with simple aperture.

Cycloserpula gordialis (Schlotheim)
Plate 6, figure 14

Description. Solitary or gregarious smooth tubes with highly variable growth pattern; planispiral, meandering, or tightly recurved in consecutive layers, or irregular, or combination of these. Cross-section oval or circular, up to 1 mm in diameter.

Remarks. Usually attached to hard substrates (e.g. shells, belemnites, intraclasts) but sometimes found free.

Occurrence. Lower Jurassic–?Eocene; common throughout the Chalk.

Cycloserpula plexus (J. de C. Sowerby)
Plate 6, figure 1

Description. Gregarious, forming a complex tangle of smooth, uniform tubes 1–2 mm in diameter; forms irregular masses with bulbous or elongated shape, 1–5 cm in maximum dimension.

Occurrence. Cenomanian–Campanian; uncommon but widespread.

Subfamily SERPULINAE

Genus ROTULARIA

Rotularia (*Praerotularia*) *umbonata* (J. Sowerby)
Plate 6, figure 13

Description. Posterior tube forms low trochospiral coil, anterior part forms gently curved projection. Umbilicus deep and open; tube circular in cross-section and thick-walled. Surface bears fine comarginal growth lines and irregular flanges.

Occurrence. Cenomanian; widespread in marly chalk facies of England.

Genus PROLISERPULA

Remarks. Tube gently tapering, curved or meandering; attached. Periodic annular swellings mark successive positions of peristomes. Longitudinal ridges sometimes present.

Proliserpula ampullacea (J. de C. Sowerby)
Plate 6, figures 11–12

Description. Tubes attached at least in early stages. Posterior part of tube coiled planispirally or looped, anterior part forms straight or weakly curved projection. Outer surface with fine corrugations and irregularly spaced annular swellings. Solitary.

Occurrence. Cenomanian–Maastrichtian; common in the English Chalk facies.

Proliserpula avita (J. de C. Sowerby)
Plate 6, figure 8

Description. Tubes attached, meandering and looped, 1–3 mm in diameter. Upper surface with single low central ridge. Irregular periodic annular swellings.
Remarks. Gregarious; found densely encrusting the bivalve *Mytiloides*.
Occurrence. Lower Turonian; abundant at a single level across southern England.

Proliserpula obtusa (J. de C. Sowerby)
Plate 6, figure 15

Description. Tubes attached, 3–5 mm in diameter, meandering or in loose spirals, with strong cockscomb-like median ridge, and two less well-developed lateral ridges. Termination of median ridge projects over peristome.
Remarks. Gregarious, often found densely encrusting exterior of large inoceramid bivalves.
Occurrence. Coniacian–Santonian; Norfolk, Kent, Sussex.

Genus ORTHOCONORCA

Orthoconorca turbinella (J. Sowerby)
Plate 6, figure 3

Description. Small (5–10 mm high), tightly coiled, high trochospiral form, variably elongated. Surface smooth; simple circular peristome.
Occurrence. Coniacian–Maastrichtian; common in southern England.

Genus GLANDIFERA

Glandifera rustica (J. Sowerby)
Plate 6, figures 6–7

Description. Tube initially rather strongly curved and quadrangular, then straight or weakly curved, gently tapering and subquadrate. Outer surface smooth, with irregular, incomplete, transverse constrictions and dimples. Peristome smoothly rounded and circular.
Occurrence. Cenomanian; apparently restricted to a single marl bed in the *C. inerme* Zone; widespread in Kent, Sussex and the Isle of Wight.

PLATE 6

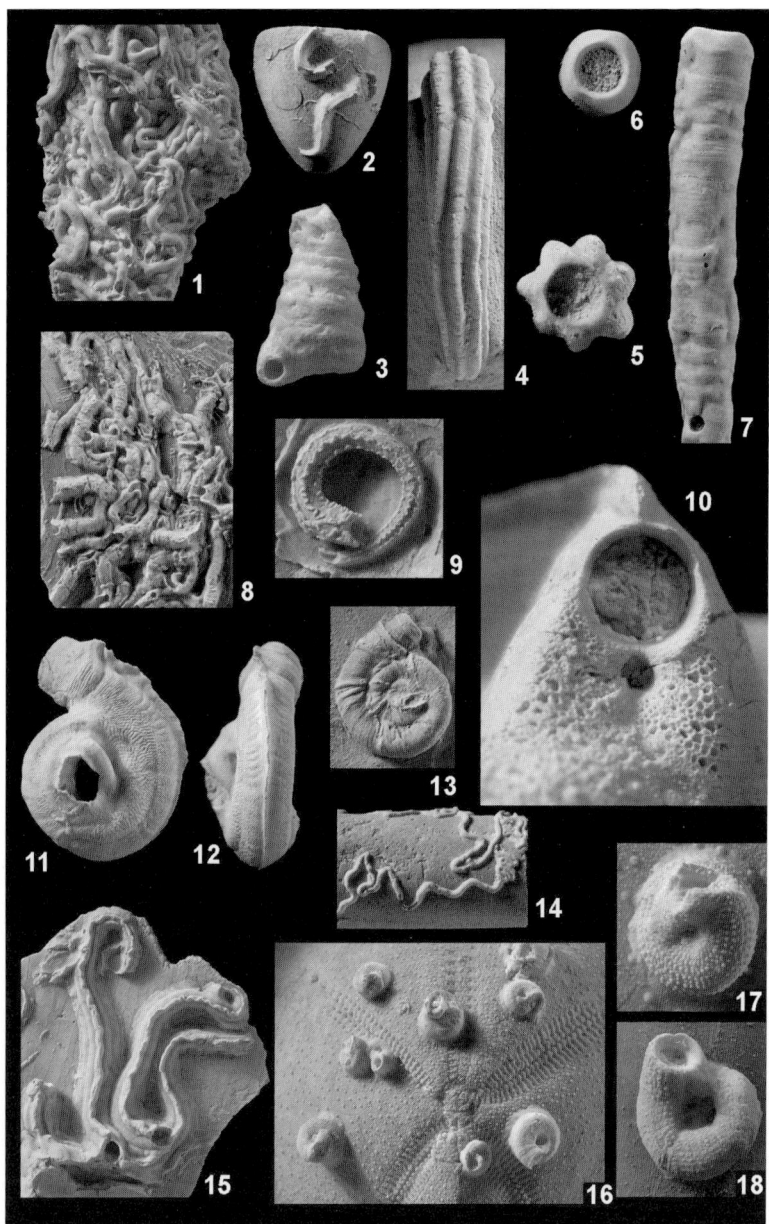

Genus VERMILIOPSIS

Vermiliopsis fluctuata (J. Sowerby)
Plate 6, figure 9

Description. Attached, open spiral or curved form with 4–5 longitudinal cockscomb-like ridges along length of tube. Peristome circular to oval in outline.
Occurrence. Upper Campanian–Lower Maastrichtian; uncommon, Norfolk.

Genus SCLEROSTYLA

Sclerostyla macropus (J. de C. Sowerby)
Plate 6, figures 2, 10

Description. Attached, sinuous tube tapering quite rapidly; triangular in cross-section with high, sharp, median ridge. Tube wall septate, peristome

---

EXPLANATION OF PLATE 6

Fig. 1. *Cycloserpula plexus* (J. de C. Sowerby). Upper Santonian, Thanet, Kent; mass of interwoven tubes; ×1.

Figs 2, 10. *Sclerostyla macropus* (J. de C. Sowerby). Lower Santonian, Northfleet, Kent. 2, two specimens attached to the sea urchin *Conulus albogalerus*; ×0·9. 10, apertural view of the larger of the two specimens; ×7.

Fig. 3. *Orthoconorca turbinella* (J. Sowerby). Upper Campanian, Studland, Dorset; lateral view of tube; ×3.

Figs 4–5. *Hepteris difformis* (Lamarck). Upper Cenomanian, Eastbourne, Sussex. 4, lateral view; 5, apertural view; both ×2·2.

Figs 6–7. *Glandifera rustica* (J. Sowerby). Middle Cenomanian, between Dover and Folkestone, Kent. 6, lateral view; 7, apertural view; both ×1·5.

Fig. 8. *Proliserpula avita* (J. de C. Sowerby). Lower Turonian, Dover, Kent; mass of tubes growing on the outer shell surface of the bivalve; ×2·5.

Fig. 9. *Vermiliopsis fluctuata* (J. Sowerby). Upper Campanian, Norwich, Norfolk; tube attached to oyster; ×2·5.

Figs 11–12. *Proliserpula ampullacea* (J. de C. Sowerby). Upper Campanian, Norwich, Norfolk; ×1·2.

Fig. 13. *Rotularia* (*Praerotularia*) *umbonata* (J. Sowerby). Cenomanian, Folkestone, Kent; lateral view of tube; ×1.

Fig. 14. *Cycloserpula gordialis* (Schlotheim). Upper Santonian, Brighton, Sussex; tube attached to belemnite; ×1.

Fig. 15. *Proliserpula obtusa* (J. de C. Sowerby). ?Coniacian, Dover, Kent; tubes growing on surface of inoceramid shell fragment; ×1.

Figs 16–18. *Neomicrorbis granulatus* (J. Sowerby). 16, group of tubes attached to the sea urchin *Micraster* from the Upper Coniacian–Lower Santonian, Kent; ×1·5. 17, detail of one of these; ×5. 18, finely granulated variety from the Upper Campanian, Norwich, Norfolk; ×5.

circular. Swollen area of porous vesicular tissue developed between peristome and substrate.

Occurrence. Upper Turonian–Maastrichtian; widespread but uncommon in the English Chalk.

Genus HEPTERIS

Hepteris difformis (Lamarck)
Plate 6, figures 4–5

Description. Tube evenly curved and tapering rapidly. Surface bears seven longitudinal ridges and intervening grooves which may twist around the length of the tube. Peristome circular.

Occurrence. Upper Cenomanian, *M. geslinianum* Zone; widespread but uncommon.

Subfamily SPIRORBINAE

Genus NEOMICRORBIS

Neomicrorbis granulatus (J. Sowerby)
Plate 6, figures 16–18

Description. Small (3–7 mm in diameter), attached, low trochospiral form with open umbilicus. Surface with closely spaced growth lines and variably developed fine granular sculpture. Conical operculum known but never found in life position.

Occurrence. Cenomanian–Maastrichtian: widespread throughout England.

5. BRYOZOANS

by PAUL D. TAYLOR

The bryozoan fauna of the British Chalk has considerable evolutionary importance and is very rich in terms of species diversity and abundance of specimens. Nevertheless, the great majority of species are in acute need of revision. Few species have been studied using modern methods, such as scanning electron microscopy, and early descriptions of bryozoans by Reuss (1846), Lonsdale (*in* Dixon 1850), d'Orbigny (1850, 1851–54; partly revised by Pergens 1890 and Canu 1900), Gregory (1899, 1907, 1909), Brydone (1906, 1909, 1910, 1911, 1912, 1913, 1914, 1916, 1917, 1918, 1929, 1930, 1936) and Lang (1914*a*, *b*, 1915, 1916, 1919*a*, *b*, 1921, 1922) have yet to be superseded. Only a small number of publications connected with British Chalk bryozoans have appeared during the last fifty years (Thomas and Larwood 1960; Larwood 1962, 1973, 1985; Medd 1965, 1966*a*, *b*, 1972, 1979; Whittlesea 1983, 1985, 1991, 1996*a–c*; Taylor 1988, 1994; McKinney and Taylor 1997), although Voigt and co-workers have published additional papers on continental European bryozoans which are of significance (see Voigt 1979, 1981, 1983, 1991; Håkansson and Voigt 1996). Specific and even generic determination of Chalk bryozoans is hampered by inadequate original descriptions and uncertain genus concepts. Furthermore, systematic problems inherent in the early descriptions by d'Orbigny and others have too often been compounded by subsequent misinterpretations of species. Therefore, the names applied to the small selection of common or distinctive species selected for description here must be regarded as tentative pending full revision. Just as species identities remain uncertain, so establishing species stratigraphical ranges is a challenge yet to be met for the great majority of British Chalk bryozoans.

MORPHOLOGY

Bryozoans are colonial metazoans in which the colony consists of numerous small zooids budded asexually. Primitively each zooid is capable of suspension feeding and all other life functions. However, in most species specialization of function results in zooidal polymorphism. The majority of bryozoans have mineralized skeletons of calcite; species with skeletons of aragonite or mixed mineralogy are of negligible importance in the Cretaceous. This fossilizable calcareous component of each bryozoan zooid is termed the zooecium, and that of the colony the zoarium.

Two orders of bryozoans with mineralized skeletons occur in the Chalk: cyclostomes and cheilostomes (Text-fig. 5.1). Feeding zooids (autozooids) are typically box-shaped in cheilostomes but tubular in cyclostomes. They have a distal opening (orifice in cheilostomes, aperture in cyclostomes) through which a crown of tentacles is protruded and retracted during life. In cheilostomes the orifice is closed by a hinged operculum, typically uncalcified, and is set within a membranous region (opesia; pl. opesiae) on the frontal surface of the zooid. Opesiae tend to form extensive elliptical to semicircular openings in anascan cheilostomes, but in ascophoran cheilostomes they are less extensive, almost coextensive with the orifice, and vary in shape from semicircular to sinusuate to circular with a subsidiary proximal opening (spiramen). Cyclostome apertures have a circular, elliptical or polygonal shape in most species. They lack an operculum in all cases except for one aberrant family (Eleidae) which is characterised by semielliptical apertures closed by calcified opercula. A tubular peristome commonly develops around the aperture in cyclostomes, and a similar structure is occasionally found surrounding the orifice of ascophoran cheilostomes. Old zooids that have ceased to feed may be sealed by closure plates (cheilostomes) or terminal diaphragms (cyclostomes).

Most species of Cretaceous cyclostomes have frontal exterior walls pierced by minute holes termed pseudopores. The frontal surface of cheilostome zooids may have a calcified frontal wall (or shield) composed of a combination of the following skeletal elements: (1) costae: hollow, overarching spines (characteristic of cribrimorph ascophorans); (2) gymnocyst: a smooth, exterior wall covered by cuticle during life; (3) cryptocyst: smooth or pustulose interior wall, usually depressed, and developing beneath a secretory epithelium during life. Gymnocyst commonly forms the proximal and outer margins of the frontal wall, with cryptocyst or costae situated on its inner side. Erect, articulated spines may also occur in cheilostomes, especially around the orifice, leaving slightly raised circular holes (spine bases) after the spines become disarticulated.

Non-feeding polymorphs, known as heterozooids, can be important in bryozoan taxonomy. Chalk cyclostomes may develop the following varieties of heterozooids: (1) kenozooids: space-filling zooids typically smaller than autozooids and often without an aperture (Text-fig. 5.1); (2) gonozooids: enlarged brooding zooids equipped with an opening (ooecio-opore) for the release of the larvae (Text-fig. 5.1); (3) eleozooids: these presumed defensive zooids, found only in melicerititids, have an aperture and operculum that is enlarged relative to the size of the zooid (Pl. 8, fig. 7). Cheilostomes in the Chalk may have the following heterozooids: (1) kenozooids; and (2) avicularia (Text-fig. 5.1), probable defensive zooids, often small but with a relatively enlarged operculum closing against a pointed or rounded rostrum, and either budded within the normal sequence of zooids (interzooidal and vicarious avicularia) or on the frontal surface of the colony (adventitious avicularia). Larvae are brooded in ovicells before their release in most cheilostomes. These globular structures are located immediately distal of the orifice of a maternal zooid which, unlike the brooding gonozooid of a cyclostome, usually retains the ability to feed.

Zooids of the same type of polymorph may vary within a colony as a function of the developmental stage of the zooid (ontogeny), that of the colony (astogeny), and local conditions (microenvironment). Zooids often

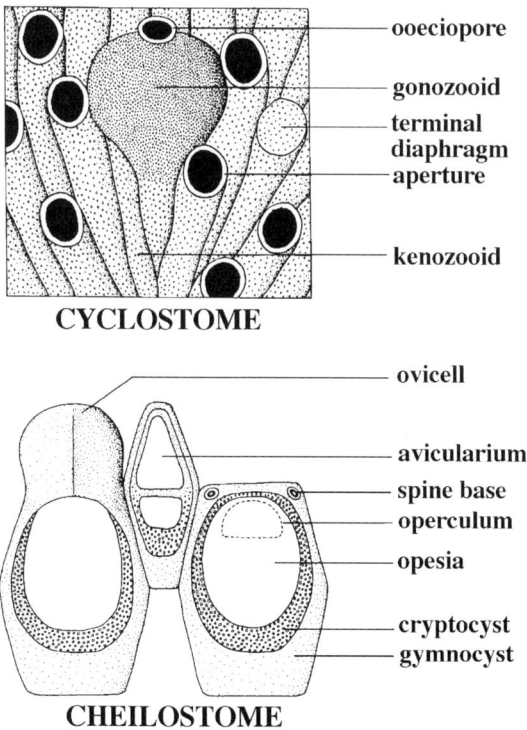

TEXT-FIG. 5.1. Morphological features in idealised cyclostome and cheilostome bryozoans from the Chalk.

accumulate calcification and become more complex during their ontogeny. Astogenetic variation of zooid morphology is most apparent close to the origin of the colony in a 'primary zone of astogenetic change' through which zooids typically become larger and more complex. The founding zooid of the colony, termed the ancestrula, tends to be especially small and simple. Cheilostome ancestrulae are ovoidal in outline and often have opesiae encircled by spine bases, whereas the ancestrulae of cyclostomes have a hemispherical protoecium located at the proximal end. Micro-environmental variations occur, for example, when colonies encounter obstacles impeding their growth.

The enormous variety of colony-forms found in the Bryozoa is largely a result of variations in budding pattern and relative rate of budding. Although colony-form is often a good clue to taxonomic identity, plasticity in growth ensures that colonies of the same species are seldom identical in size and shape. Many Chalk species have entirely encrusting colonies. These vary from runner-like (e.g. *Voigtopora*, Pl. 7, fig. 1) to ribbon-like (e.g. *Proboscinopora*, Pl. 7, fig. 2) with bifurcating branches one or a few zooids wide, to sheet-like (e.g. *Castanopora*, Pl. 7, fig. 3), often roughly circular in plan view and sometimes becoming multilamellar by self overgrowth. Growth upwards from the encrusting base of the colony gives an erect colony-form comprising cylindrical branches ('vinculariiform') or flattened branches or fronds ('adeoniform' and 'eschariform' respectively). Colonies having cylindrical branches (e.g. '*Biflustra*', Pl. 9, fig. 7) can be solid or hollow and may bifurcate in a single plane or three-dimensionally to give respectively planar (e.g. *Homoeosolen*, Pl. 7. fig. 6) and bushy shapes. Frondose species from the Chalk generally have strap-like bifoliate fronds consisting of two layers of zooids back-to-back (e.g. *Porina*, Pl. 7, fig. 5). Some bifoliate species construct reticulate colonies by a regular pattern of frond bifurcation and anastomosis. A few erect species in the Chalk have fungiform colonies with a narrow stalk supporting an expanded head (e.g. *Bicavea*, Pl. 8, figs 8–9). All of the colony-forms mentioned above are sessile and live firmly anchored to an immobile substratum. However, free-living bryozoans (see below) can also be found in the Chalk. These include cap-shaped 'lunulitiform' colonies of *Lunulites* (Pl. 9, fig. 10), and spindle-shaped colonies of *Volviflustrellaria* (Pl. 9, fig. 6).

PALAEOECOLOGY

Bryozoan palaeoecology has been reviewed by Smith (1995) and aspects of functional morphology by McKinney and Jackson (1989) and Taylor

(1999). Like most present-day bryozoans, species from the Chalk were evidently stenohaline suspension-feeders which required well-oxygenated waters with adequate circulation and replenishment of phytoplanktonic food supplies. Larvae of most bryozoan species have a short free-swimming stage before they locate a suitable firm substratum, settle and metamorphose into an ancestrula, which will found a new colony. Exceptions are the Chalk cheilostomes *Herpetopora* and *Pyripora*, which can be inferred to have possessed long-lived, planktotrophic larvae enabling more widespread dispersal.

Substrata colonized by bryozoans in the Chalk sea included echinoid tests, bivalves, brachiopods, sponges, belemnite guards, crinoid ossicles and sedimentary hardgrounds. Bryozoans are especially abundant on the tests of irregular echinoids, notably *Echinocorys*. Several bryozoan species may encrust a single echinoid test, but individual colonies tend to be less than a few centimetres in diameter and total coverage by bryozoans and other encrusters rarely exceeds half the surface area of the test. Nevertheless, there is often clear evidence of competition for living space in the occurrence of overgrowths between colonies, including reciprocal overgrowths that furnish indisputable evidence of interactions between living competitors. Cheilostomes usually dominated in competition for space with cyclostomes (McKinney 1995). Bryozoan colonies also competed for substratum space with sponges, serpulid worms, and cemented bivalves and brachiopods. Shells of the epifaunal bivalve *Inoceramus*, seemingly plentiful as potential substrata for bryozoan encrustation, are less often encrusted than echinoid tests. Cemented bases of erect species commonly occur on the same substrata as entirely encrusting species. Fragments of erect colonies with narrow cylindrical branches (vinculari-iform) are a dominant element of many Chalk faunas and can be abundant in the coarser fractions of sediment washings, including the soft chalk filling cavities in flints ('flint meals'). Vinculariiform colonies live today mainly in water too deep for significant wave action and with minimal current activity, although this was not necessarily so in the Cretaceous (e.g. Pitt and Taylor 1990). Some erect colonies from the Chalk have hollow tubular bases and were evidently attached during life to the stems of soft-bodied organisms (?hydrozoans, gorgonians, plants) which held them aloft above the sea-bed. Free-living lunulitiform bryozoans can be common in parts of the Chalk, especially the Maastrichtian (e.g. Håkansson and Voigt 1996). Judging by Recent analogues, colonies had marginal avicularia with hair-like setae that acted to support them a little above the sediment surface. Modern lunulitiform bryozoans are able to live on mobile sediments, and some species can right themselves when turned upside-down and resurface when buried. Spindle-shaped or ball-shaped colonies of Chalk *Volviflustrellaria* represent a second free-living morphology. In *Volviflustrellaria* the colony seems to have been rolled periodically (Kvatchko 1997), perhaps as a result of currents or the activities of vagile infauna or epifauna.

DISTRIBUTION

The great majority of British Chalk bryozoans in museum collections are from southern England. Bryozoans can be particularly abundant in the Coniacian and Santonian Chalk of the North Downs in the Chatham and Dover regions, and South Downs along the Sussex coast, and in the Campanian Chalk of Hampshire and Norfolk. Relatively few have been recorded from the Chalk of Yorkshire and Northern Ireland, which is probably explained, at least partly, by the difficulty of extracting specimens from these typically hard lithologies. They are less common in the Turonian of southern England and relatively rare in the Cenomanian, although marginal facies of the Cenomanian (e.g. Wilmington Sands, Glauconitic Marl) may contain distinctive and diverse bryozoan faunas.

DESCRIPTIONS

Order CYCLOSTOMATA
Suborder TUBULOPORINA
Family STOMATOPORIDAE

Genus VOIGTOPORA

Voigtopora dixoni (Vine)
Plate 7, figure 1; Plate 8, figure 3

Description. Colony encrusting, comprising branching chains of pseudo-uniserially arranged zooids often extending across large areas of substrate. Branch ramification is lateral, with new branches arising from the sides of existing branches at more-or-less 90 degrees. Zooids large, their apparent width exaggerated because each branch section comprises the distal part of one zooid flanked on either side by narrow proximal parts of the next zooid in series. Zooidal apertures 0·15 mm in diameter, small relative to branch width (0·45–0·60 mm), and with a slight peristome. Ancestrula short and wide with a narrow protoecium, budding two daughter zooids growing in opposite directions. Gonozooids unknown, presumed to be either absent or budded atop autozooidal peristomes and not fossilized.

Remarks. Uniserial or pseudo-uniserial cyclostomes are common in the Chalk and have been traditionally assigned to *Stomatopora*. Very frequent as an encruster of *Echinocorys* tests.

Occurrence. Coniacian, *M. cortestudinarium* Zone, to Campanian, *G. quadrata* Zone; widespread in southern England.

Genus PROBOSCINOPORA

Proboscinopora toucasiana (d'Orbigny)
Plate 7, figure 2; Plate 8, figure 4

Description. Colony encrusting, comprising multiserial branches, generally 3–7 zooids wide (0·35–1·00 mm), dividing irregularly and anastomosing occasionally. Branches low and smooth-surfaced except for slight furrows at lateral zooidal boundaries. Autozooids slender (0·6–0·9 mm long by 0·15 mm wide), apertures small (0·07–0·09 mm), circular or transverse, with a slight peristome. Ancestrula with a large protoecium (0·20 mm wide). Kenozooids occur along sloping branch margins. Gonozooids unknown, presumed to be either absent or budded atop autozooidal peristomes and broken off.

Remarks. Many species with probosciniform colonies occur in the Chalk. They vary in zooidal dimensions, branching patterns, branch convexity, and presence/absence of gonozooids.

Occurrence. Coniacian, *M. cortestudinarium* Zone, to Santonian, *M. coranguinum* Zone; ?Campanian, *O. pilula* Zone; southern England; abundant in the Chatham area.

Family PLAGIOECIIDAE

Genus PLAGIOECIA

Plagioecia sp.
Plate 8, figure 2

Description. Colony encrusting, a thin multiserial sheet of zooids, usually about 1 cm in diameter, more or less circular in outline, and bounded by a peripheral growing edge with a wide distal fringe of basal lamina. Autozooids small and slender, their frontal walls flat proximally but slightly convex distally, about 0·60 mm long by 0·15 mm wide. Apertures small, longitudinally elongate (diameter 0·10 by 0·07 mm), arranged in a quincunx, and with peristomes inclined at 45 degrees. Gonozooids large, with a bulbous distal portion elongated transversely (*c.* 0·55–0·70 mm long by 1·6–2·2 mm wide) and penetrated by autozooidal peristomes. Ooeciopore subcircular, about 0·06 mm in diameter, located in the centre of the bulbous gonozooidal frontal wall.

PLATE 7

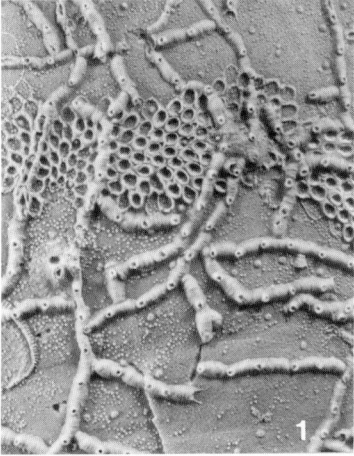

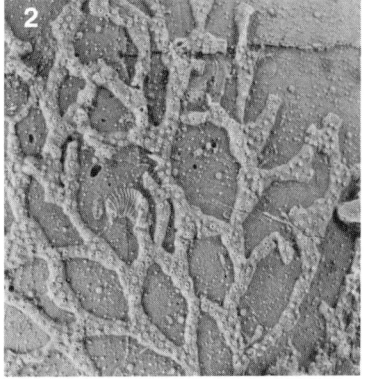

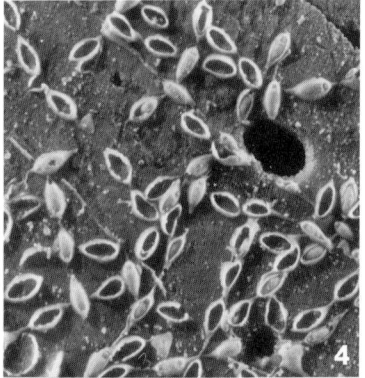

Remarks. This and similar species have traditionally been referred to the genus *Berenicea* but should properly be assigned to one of a number of other genera (e.g. *Plagioecia, Hyporosopora, Microeciella, Mesonopora*) distinguished mainly on the basis of gonozooidal morphology (Taylor and Sequieros 1982; see also McKinney and Taylor 1997). Unfortunately, not every colony possesses gonozooids. When gonozooids are not available, the informal name '*Berenicea*' is best used.

Occurrence. Stratigraphical range unknown; the figured specimen is from the Coniacian, *M. cortestudinarium* Zone; the genus *Plagioecia* ranges from Jurassic to Recent; Kent, and probably other southern counties.

Family *incertae sedis*

Genus 'ENTALOPHORA'

'*Entalophora*' sp.
Plate 8, figure 1

Description. Colony erect with bifurcating cylindrical branches, highly variable in diameter (0·2–2·8 mm, commonly *c*. 0·5 mm in white chalk facies). Zooids with slender frontal walls (0·6–4·3 mm long), zooidal boundaries poorly-defined, apertures about 0·15 mm in diameter, longitudinally elliptical when worn, subcircular when peristomes are preserved. Gonozooids unknown.

Remarks. This and similar species of vinculariiform tubuloporinids are commonly referred to *Entalophora*, although the type species of *Entalophora* (*E. cellarioides* Lamouroux) differs in having a narrow axial canal within the branches. The figured specimen resembles '*Entalophora*'

EXPLANATION OF PLATE 7

Colony-form in Chalk cyclostomes and cheilostomes.

Fig. 1. *Voigtopora dixoni* (Vine). Campanian, West Harnham, Salisbury, Wiltshire; runner-like encrusting colony; ×3·6.

Fig. 2. *Proboscinopora toucasiana* (d'Orbigny). Coniacian, Chatham, Kent; ribbon-like encrusting colony; ×2·9.

Fig. 3. *Castanopora magnifica* (d'Orbigny). Campanian, Norwich, Norfolk; sheet-like encrusting colony; ×2·3.

Fig. 4. *Herpetopora anglica* Lang. Turonian, Lewes, Sussex; runner-like encrusting colony; ×7.

Fig. 5. *Porina* sp. Campanian, Norwich; bifoliate erect branches arising from encrusting base; ×2·7.

Fig. 6. *Homoeosolen ramulosus* Lonsdale. Coniacian or Santonian, Gravesend, Kent; planar branching colony; ×1.

raripora described by d'Orbigny (1850) from the Coniacian of Fécamp which, however, has larger zooids.

Occurrence. Stratigraphical range unknown; the figured specimen is from the Coniacian; Kent; probably distributed throughout southern England.

Family ELEIDAE

Genus MELICERITITES

Meliceritites dollfusi Pergens
Plate 8, figures 6–7

Description. Colony erect with bifurcating cylindrical branches 1·5–3·0 mm in diameter, sometimes developing lamellar overgrowths. Autozooids large with subhexagonal frontal walls (*c.* 0·6 by 0·4 mm), large semi-elliptical apertures (0·2 mm long) borne on short but distinct peristomes, and opercula with sparse, irregularly-arranged pores. Eleozooids moderately abundant, with long (up to 0·75 mm) parallel-sided or slightly spatulate rostra. Gonozooids rare, their distal frontal walls ovoid in outline, slightly longer than wide (*c.* 1·8 by 1·6 mm); ooeciopore transversely elliptical, 0·15 mm wide.

Remarks. This and the following two species belong to an aberrant family of cyclostomes, usually referred to as melicerititids. They are remarkably convergent with cheilostomes with which they share feeding zooids equipped with opercula and mandibulate polymorphs.

Occurrence. Coniacian–Campanian; common in the *M. cortestudinarium, M. coranguinum* and *G. quadrata* zones; southern England.

Meliceritites durobrivensis (Gregory)
Plate 8, figure 5

Description. Colony erect with bifurcating cylindrical branches, 0·75–1·5 mm in diameter. Autozooids generally with rhomboidal frontal walls, rounded distally, and about 0·45 mm long by 0·30 mm wide; apertures semi-elliptical, 0·18 mm long, large relative to the frontal wall and with a prominent rim; opercula convex, a crescent of slit-like pores visible in worn examples. Eleozooids smaller than autozooids, their apertures ⊥-shaped and closed by a small semi-elliptical operculum which is seldom preserved. Secondary eleozooids, formed by reparative budding within old autozooids, are common and have a ⊥-shaped aperture set within a pseudoporous terminal diaphragm. Gonozooids very rare, extremely bulbous, longitudinally elliptical in outline shape, with a transversely elongate ooeciopore (*c.* 0·09 by 0·16 mm).

Remarks. A very characteristic species of Upper Cretaceous white chalk facies, e.g., in the *M. cortestudinarium* Zone of Chatham.

Occurrence. Turonian, *S. plana* Zone, to Campanian; widespread in southern England.

Genus REPTOMULTELEA

Reptomultelea sarissata Gregory

Text-figure 5.2A

Description. Colony encrusting, unilamellar or multilamellar, sometimes growing freely from an initial substratum. Autozooids generally with rhomboidal frontal walls, pointed distally, large, averaging 0·92 mm long by 0·38 mm wide; apertures shaped like an 'ogee' arch drawn out distally almost to a point, very large, averaging 0·40 mm long by 0·24 mm wide; opercula convex, often preserved *in situ*. Eleozooids common, larger than autozooids, about 1·4 mm long by 0·5 mm wide, their apertures longitudinally elongate, parallel-sided or slightly spatulate. Secondary eleozooids formed by reparative budding within old eleozooids may occur. Gonozooids rare, inverted pear-shaped, with an ooeciopore about 0·15 mm in diameter.

Remarks. Although there are numerous species of the encrusting melicerititid genus *Reptomultelea* in marginal facies of Late Cretaceous age, only three have been recorded from the British Chalk (Taylor 1994). *R. filiozati* is distinguished from *R. dixoni* (Lang) and *R. auris* Taylor by the peculiar shape of the apertures, whereas *R. auris* can be easily recognized by the presence of small kenozooids on either side of the autozooidal apertures.

Occurrence. Coniacian, ?basal Santonian; southern England.

Suborder CANCELLATA

Family PETALOPORIDAE

Genus PETALOPORA

Petalopora pulchella (Roemer)

Plate 8, figure 10

Description. Colony erect with bifurcating cylindrical branches about 1·0–2·5 mm in diameter. Branch surfaces have low longitudinal ridges defining series of autozooids with small (diameter 0·06–0·08 mm) circular apertures, sometimes slightly raised, widely-spaced and arranged in zones. A polygonal reticulum of small, crater-shaped cancelli, each with a central perforation, occupies the space between autozooidal apertures, normally with three rows of cancelli between adjacent longitudinal ridges. Brood chambers are rare, bulbous, with a frontal surface of cancelli and a large transverse ooeciopore often situated centrally.

Remarks. *P. pulchella* with three rows of cancelli between adjacent longitudinal ridges is distinguished from *P. costata* d'Orbigny, which usually has two rows of cancelli and narrower branches.

Occurrence. Stratigraphical range uncertain; Gregory (1899) gave the range as Turonian–Maastrichtian, but no other occurrences in rocks younger than Campanian are known; southern England.

Family CYTIDIDAE

Genus BICAVEA

Bicavea rotaformis Gregory
Plate 8, figures 8–9

Description. Colony erect, fungiform, a discoidal head or capitulum arising from a narrow stalk. Capitulum about 3 mm in diameter, with a concave lower surface where the stalk is attached, and an upper surface bearing approximately 7–10 upstanding radial fascicles protruding outwards at the edge of the capitulum. Autozooidal apertures polygonal, about 0·09 mm in diameter, restricted to the curved distal (outer) ends of the fascicles. Elongate autozooidal walls form striae on lateral edges of fascicles. Cancelli with thick-walled apertures, highly variable in diameter (often *c.* 0·08 mm), occupying the centre of the upper surface of the capitulum and the spoke-like areas between fascicles. Brood chamber

---

EXPLANATION OF PLATE 8

Chalk cyclostomes; all scanning electron micrographs, and all specimens uncoated apart from that in Fig. 5; magnifications approximate.

Fig. 1. '*Entalophora*' sp. Coniacian, Chatham, Kent; ×20.

Fig. 2. *Plagioecia* sp. Coniacian, Chatham, Kent; part of colony with a band of gonozooids; ×17.

Fig. 3. *Voigtopora dixoni* (Vine). Santonian, Dartford, Kent; ×20.

Fig. 4. *Proboscinopora toucasiana* (d'Orbigny). Coniacian, Chatham, Kent; ×31.

Fig. 5. *Meliceritites durobrivensis* (Gregory). Coniacian, Seaford Head, Sussex; three autozooids, one preserving the operculum (top left); ×45.

Fig. 6. *Meliceritites dollfusi* Pergens. Santonian, Northfleet, Kent; autozooids; ×30.

Fig. 7. *Meliceritites dollfusi* Pergens. Santonian, Thanet Coast, Kent; two eleozooids with an autozooid between; ×26.

Figs 8–9. *Bicavea rotaformis* Gregory. Turonian, Isle of Wight. 8, side view of colony; 9, upper surface of colony; both ×11.

Fig. 10. *Petalopora pulchella* (Roemer). Upper Chalk, England; ×23.

PLATE 8

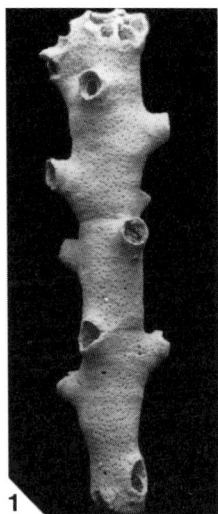

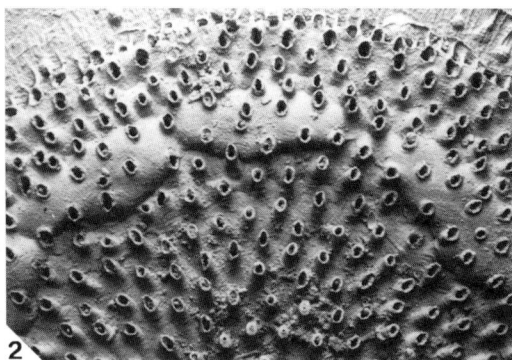

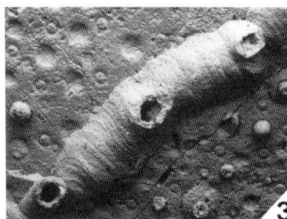

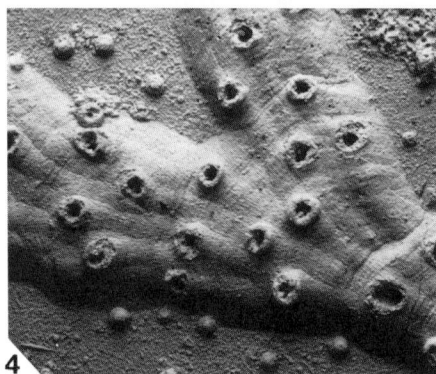

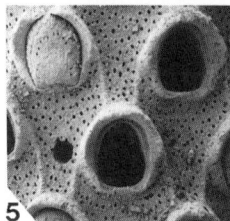

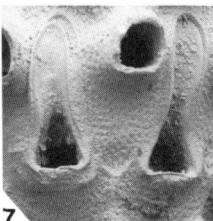

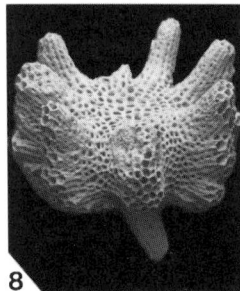

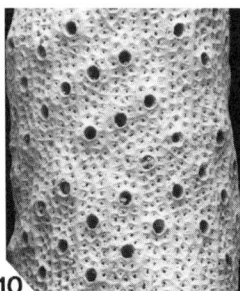

undescribed.

Remarks. Fungiform (mushroom-shaped) colonies also occur in the Chalk genus *Trochiliopora* (see Thomas 1939) which, however, lacks a concave lower surface to the capitulum and has fascicles projecting only slightly from the capitulum.

Occurrence. Turonian, *S. plana* Zone; southern England; common on the Isle of Wight.

Genus HOMOEOSOLEN

Homoeosolen ramulosus Lonsdale
Plate 7, figure 6

Description. Colony erect, planar with a dendritic pattern of irregularly bifurcating branches that do not anastomose. Branch diameter variable (<1–3 mm), many branches tapering almost to a point. Reverse surfaces of branches gently convex, formed of calcified exterior wall through which traces of interzooidal walls may be visible; obverse surfaces strongly convex, bearing contiguous, elongate apertures of zooids which intersect the branch surface acutely. Autozooidal apertures about 0·12 mm wide are interspersed with smaller kenozooidal apertures, the latter particularly abundant at branch bifurcations. Gonozooids open on branch frontal surfaces, possess globose frontal walls (about 1 mm wide) and have large ooeciopore.

Remarks. *H. gamblei* Gregory differs in having a pinnate colony-form with a more regular branching pattern.

Occurrence. Turonian–Maastrichtian; found especially in the *S. plana–M. cortestudinarium* zones of the Chalk; southern England.

Order CHEILOSTOMATA
Suborder ANASCA
Family ELECTRIDAE

Genus HERPETOPORA

Herpetopora anglica Lang
Plate 7, figure 4; Plate 9, figure 1

Description. Colony encrusting, with fragile branching chains of uniserially-arranged zooids; new branches originate as disto-lateral buds orientated at 60–90 degrees to parent branches. Autozooids elongate pyriform, highly variable in length (0·65 – >5 mm), proximal gymnocysts (caudae) becoming progressively longer in successive zooids of each branch, less variable in width (0·24–0·47 mm). Opesiae large, oval, variable in size, 0·25–0·70 mm long by 0·14–0·27 mm wide, without a cryptocyst. Reparative budding produces zooids with two or more concentric mural rims, and sealed zooids have closure plates bearing a crescentic impression of the operculum. Kenozooids narrower than autozooids, with small opesiae, commonly budded at the ends of branches. No ovicells or avicularia.

Remarks. *Herpetopora* differs from *Pyripora* in lacking a cryptocyst and showing a continuous astogenetic increase in zooid length along each branch. A second Chalk species of *Herpetopora*, *H. laxata* (d'Orbigny), occurs mainly in the Campanian and Maastrichtian and is distinguished by its broader opesiae (Thomas and Larwood 1960) and a two-phase, stepped astogenetic gradient along each branch (Taylor 1988).

Occurrence. Turonian, *T. lata* Zone, to Campanian, *O. pilula* Zone; southern England.

Genus PYRIPORA

Pyripora magna Larwood
Text-figure 5.2B

Description. Colony encrusting, with zooids arranged uniserially or in irregular multiserial patches. Autozooids large, elongate-pyriform, 1·20–1·75 mm long by 0·53–0·83 mm wide, the proximal gymnocyst accounting for about half of the total zooidal length; opesiae oval, 1·3–1·7 times longer than wide; cryptocyst smooth, widest proximally where it forms a modest shelf, pinching out distally. Pore windows visible around margins of zooids and small pore chambers seen in abraded zooids. Kenozooids infrequent, small. No ovicells or avicularia.

Remarks. Although other British Chalk species have previously been assigned to *Pyripora*, most are now regarded as belonging to *Herpetopora*, and *P. magna* may be the only true species of *Pyripora* present.

Occurrence. Campanian, *B. mucronata* Zone; Norfolk.

Family CALLOPORIDAE

Genus WILBERTOPORA

Wilbertopora woodwardi (Brydone)
Plate 9, figure 2

Description. Colony encrusting, a multiserial sheet of zooids. Autozooids large (*c.* 1 mm long by 0·7 mm wide), rhomboidal with an extensive elliptical to pear-shaped opesia, a broad, finely pustulose cryptocyst and a small proximal gymnocyst. Ovicells rounded-rectangular in shape,

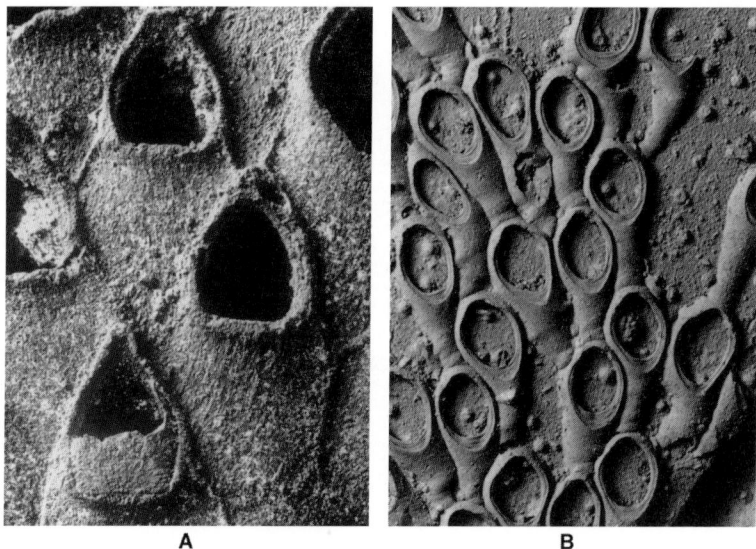

A   **B**

TEXT-FIG. 5.2. Scanning electron micrographs of Chalk bryozoans. A, *Reptomul-telea sarissata* Gregory; three autozooids, the lower one with a broken operculum; Coniacian, Chatham, Kent; ×7·5. B, *Pyripora magna* Larwood; note irregular arrangement of autozooids; Campanian, Harford Bridges, Norwich, Norfolk; ×13.

sometimes with a median suture line. Interzooidal avicularia narrow, about 0·78 mm long by 0·39 mm wide, with an extensive cryptocyst proximally, a complete pivotal bar, and an acuminate rostrum which may be raised slightly. Kenozooids fill irregular spaces, and have a small central opening, a narrow cryptocyst and a broad gymnocyst.

Remarks. Like most other 'membraniporimorphs' from the Chalk, this species was originally assigned to *Membranipora*. The Recent type species of *Membranipora*, *M. membranacea*, lives as an algal epiphyte, is feebly calcified and aragonitic, has a twinned ancestrula, and lacks ovicells. None of these features applies to species of so-called *Membranipora* from the Chalk. Some Cretaceous species have already been reassigned to other genera such as *Ellisina* (see Medd 1979), but much further work is needed to establish the true generic identities of the enormous number of membraniporimorphs present in the Chalk. Consequently, the generic assignments of this and the following two species are tentative.

Occurrence. Turonian, *S. plana* Zone, to Campanian, *G. quadrata* Zone; southern England.

Genus PYRIPORELLA

Pyriporella? *sagittaria* (Brydone)
Plate 9, figure 3

Description. Colony encrusting, a multiserial sheet of zooids. Autozooids elongate elliptical (*c*. 0·7 mm long) with an oval opesia, a proximal gymnocyst usually obscured by adventitious avicularia or ovicells, a broad pustulose cryptocyst, and a variable number of spine bases, including a distal pair, especially numerous in early zooids and in zooids containing reparative buds. Ovicells slightly longer than broad, about 0·17 mm wide. Interzooidal avicularia roughly twice as long (0·47 mm) as wide, with a broad, crescent-shaped proximal cryptocyst and a hammer-head shaped rostrum ending in a bowed transverse wall; complete pivotal bars not observed. Adventitious avicularia generally developed in pairs at the distal ends of autozooids, one on either side of the ovicell, of a similar shape to the interzooidal avicularia but about half their length and directed proximally relative to colony growth direction. Probable incompletely developed adventitious avicularia common.

Remarks. Provisionally assigned to the little-known *Pyriporella* because of the abundant interzooidal avicularia, this species probably requires a new genus.

Occurrence. Turonian, *S. plana* Zone, to Campanian, *B. mucronata* Zone; southern and eastern England.

Genus BIAVICULIGERA

Biaviculigera? *furina* (Brydone)
Plate 9, figure 4

Description. Colony encrusting, a multiserial sheet of zooids. Autozooids rounded rhomboidal, elongate, about 0·4–0·5 mm long by 0·25–0·35 mm wide, with an elongate oval opesia and a narrow pustulose cryptocyst bordered by small spine bases. Reparatively budded autozooids with concentric mural rims common. Ovicells slightly longer than wide (*c*. 0·14 mm), covering the proximal gymnocyst of the next distal autozooid in series. Interzooidal avicularia of two types: one is large and rhomboidal with a small triangular rostrum (*c*. 0·12 mm wide), complete pivotal bar, and extensive distal gymnocyst, surrounded by a very extensive smooth gymnocyst; the second is small and hour-glass shaped.

Occurrence. Campanian; characteristic of the *B. mucronata* Zone; southern and eastern England, including the Isle of Wight ('Lower *mucronata* Chalk'), the Sponge Beds at Trimingham, and the Weybourne Chalk of the Norfolk Coast.

Genus 'BIFLUSTRA'

'Biflustra' argus d'Orbigny
Plate 9, figure 7

Description. Colony erect, comprising narrow (*c.* 1 mm wide), cylindrical branches (vinculariiform) bifurcating at intervals, and quadrate, pentagonal or hexagonal in cross-section according to the number (4–6) of longitudinal rows of zooids present. Autozooids rhomboidal, slender, about 1·0–1·2 mm long by 0·4–0·6 mm wide, with an elongate elliptical opesia (0·6–0·7 mm long by 0·3 mm wide) and a slightly depressed cryptocystal frontal wall bordered by numerous minute spine bases (barely visible with a light microscope). Concentric mural rims, indicating reparative budding, and perforate closure plates are common. Interzooidal avicularia developed mainly at branch bifurcations, moderately small (*c.* 0·7 mm long by 0·4 mm wide), with a large opesia and slightly projecting rostrum. Ovicells of uncertain occurrence.

Remarks. The abundant vinculariiform anascan species of the Chalk are generally assigned to *Vincularia*. However, they are quite distinct from true *Vincularia* which has articulated colonies with asymmetrical zooids. *'Biflustra'* is used here informally for this species acknowledging that true *Biflustra* lacks avicularia.

Occurrence. Stratigraphic range unknown; southern and eastern England; fairly abundant in the Campanian of Weybourne, Norfolk, according to Brydone (1930).

Family ONYCHOCELLIDAE

Genus ONYCHOCELLA

Onychocella inelegans (Lonsdale)
Plate 9, figure 9

Description. Colony usually erect, with flattened bifoliate branches (*c.* 3–4 mm wide) which divide at intervals in the plane of the branch; encrusting sheet-like colonies (?bases of immature erect colonies) may also occur. Autozooids variable in shape, often broad and rhomboidal (*c.* 0·45–0·75 mm long by 0·45 mm wide), with a semi-elliptical opesia (*c.* 0·12 mm long by 0·15 mm wide) indented by opesiules at the two proximal corners, and an extensive cryptocystal frontal wall which is slightly depressed. Interzooidal avicularia elongate (*c.* 0·45 mm long by 0·15 mm wide), with a long curved rostrum and a tulip-shaped opesia which has a short slit in the proximal margin. Ovicells inconspicuous, immersed in the frontal wall of the succeeding zooid.

Remarks. Bifoliate fragments of this and similar species are among the most commonly encountered bryozoans in the Chalk and are often found on the surfaces of flints.

Occurrence. Turonian, *S. plana* Zone, to Maastrichtian; southern and eastern England.

Family LUNULITIDAE

Genus LUNULITES

Lunulites tenax Brydone
Plate 9, figure 10

Description. Colony free-living, small (generally <1 cm in diameter), circular in plan view, with a gently convex upper surface bearing zooidal apertures, a concave lower surface ornamented by radial furrows, and a fluted circumferential growing edge. Autozooids with quadrate opesia (*c.* 0·15–0·20 mm) and a depressed cryptocystal frontal wall; a strong astogenetic gradient of zooidal-size increase occurs outwards from the central ancestrula, and early autozooids have opesiae with a median proximal sinus. Ovicells developed only in zooids near the colony margin, globose and prominent with a large semicircular opening. Interzooidal avicularia numerous, arranged in radial rows between rows of autozooids, deeply sunken, elongate, about 0·45 mm long by 0·15 mm wide, with a rounded rostrum and a median proximal sinus in the opesia.

Remarks. *Lunulites* is the sole genus of lunulitiform bryozoans represented in the British Chalk and usually first appears in the Campanian. Brydone (1929) distinguished several British species, including some from the pre-Campanian, which require confirmation.

Occurrence. Campanian, upper *G. quadrata* and lower *B. mucronata* zones; Hampshire.

Genus VOLVIFLUSTRELLARIA

Volviflustrellaria taverensis (Brydone)
Plate 9, figure 6

Description. Colony free-living, small (<2 cm in diameter), spheroidal or spindle-shaped; a single growing edge extends between the two opposite poles of the colony and buds zooids in a continuous spiral, multilamellar overgrowth. Autozooids subrectangular, about 0·75 mm long by 0·45–0·60 mm wide, with a rounded-quadrate opesia (*c.* 0·30 mm long by 0·25–0·30 mm wide) and a finely pustulose cryptocystal frontal wall, depressed and occupying about one-half of the frontal area. Ovicells inconspicuous,

immersed in the frontal wall of the next distal zooid. Interzooidal avicularia abundant, deeply sunken between rows of autozooids, long and narrow (*c*. 0·75 mm long by 0·12 mm wide), with a short rounded rostrum.
Occurrence. Campanian, *B. mucronata* Zone; possibly also *G. quadrata* Zone; eastern England and Northern Ireland; abundant in the basal *mucronata* Chalk of Norfolk.

Family MICROPORIDAE

Genus STICHOMICROPORA

Stichomicropora marginula (Brydone)
Plate 9, figure 8

Description. Colony encrusting, comprising a multiserial sheet of zooids, generally with a stepped growing edge. Autozooids rhomboidal, broad, about 0·6–0·7 mm long by 0·45–0·65 mm wide, with a raised shelf surrounding a cryptocystal frontal wall, which is convex and pierced by a pair of slit-shaped opesiules situated close to the edge and about mid-length; opesia small, semi-elliptical, transversely elongate (about 0·07 mm long by 0·15 mm wide), and with an interior distal shelf and about six orificial spine bases. Ovicell remains are visible as slight depressions in the widened proximal shelf of the zooid distal to the maternal zooid, and are bordered distally by a crescent of eight spine bases and laterally by the raised edges of the adjacent zooids. Avicularia unknown.
Remarks. The unusual ovicell of *Stichomicropora*, consisting of a cage formed by multiple spines, is very delicate and seldom survives fossilisation; normally only a crescent of spine bases remains.
Occurrence. Coniacian, *M. cortestudinarium* Zone; Kent and Sussex.

Suborder ASCOPHORA
Family PELMATOPORIDAE

Genus PELMATOPORA

Pelmatopora gregoryi (Brydone)
Plate 9, figure 11

Description. Colony encrusting, comprising a multiserial sheet of zooid. Autozooids elongate-rhomboidal, almost parallel-sided, 0·75–1·2 mm long by 0·4–0·5 mm wide, surrounded by a wide zone of lacunate interzooidal calcification; frontal shield flat, with 11–20 well-defined costae each with two pelmata and a pelmatidium, two paired rows of intercostal spaces and a wide apertural bar. Orifice semi-elliptical, longer than wide (0·15–0·18 mm long by 0·21 mm wide), overarched by a pair of distal oral spines in well-preserved zooids. Ovicell immersed in the next distal zooid. Adventitious avicularia small, stalked, usually paired, one near each proximolateral corner of the orifice; rostrum pointed.

Remarks. Larwood (1962) distinguished 17 named species of *Pelmato-pora* from the British Chalk and provided a useful key to their identification.

Occurrence. Santonian, *M. testudinarius* Zone, to Campanian, *G. quadrata* Zone; Sussex, Hampshire and Wiltshire.

Genus CASTANOPORA

Castanopora magnifica (d'Orbigny)
Plate 7, figure 3; Plate 9, figure 12

Description. Colony encrusting, comprising a multiserial sheet of zooids. Autozooids large, rounded-rhomboidal, 0·75–1·50 mm long by 0·45–1·20 mm wide; frontal shield oval, gently convex, with 14–24 costae each bearing 4–8 lateral costal fusions and pelmatidia; apertural bar prominent; orifice semi-elliptical, 0·15–0·33 mm long by 0·15–0·30 mm wide, with four large oral spines, the lateral spine bases wider than the median ones. Ovicells rare, large, globular, and bearing two converging ridges. Adventitious avicularia variable in occurrence, small, 0·17–0·25 mm long by 0·09–0·18 mm wide, proximally directed, with a pivotal bar and a rounded rostrum.

Remarks. Larwood (1962) provided a key to species of *Castanopora*.

Occurrence. Campanian, *B. mucronata* Zone, to Maastrichtian; Norfolk.

Family PORINIDAE

Genus PORINA

Porina sp.
Plate 7, figure 5; Plate 9, figure 5

Description. Colony erect, comprising bifoliate branches, 1–2 mm wide, thicker and more rounded near the base of the colony, and bifurcating at intervals in the plane of the branch. Autozooids with a circular orifice (0·11–0·14 mm in diameter), a slight peristome, and a cryptocystal frontal wall, finely crenulate and pierced by large irregular pores and a spiramen (diameter *c.* 0·02 mm) which projects slightly and is located medially, about midway along the length of the frontal wall. Thick secondary calcification obscures zooidal boundaries and occludes pores and eventually the orifices of old zooids. Ovicell globose, overarching the distal part of the fertile autozooidal orifice, and with a finely reticulate surface.

PLATE 9

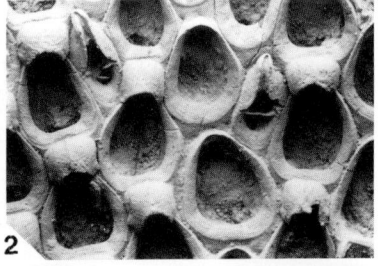

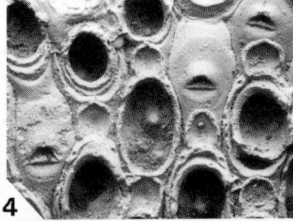

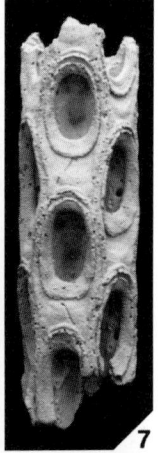

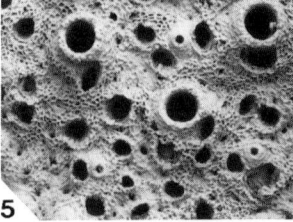

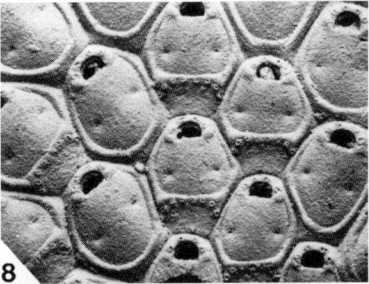

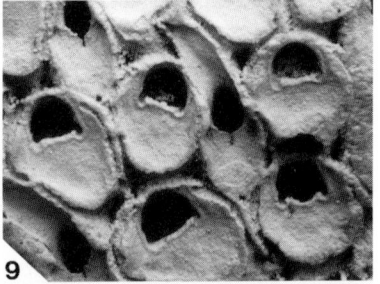

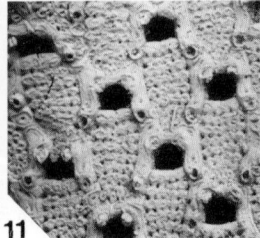

Remarks. Porinid species are difficult to identify because secondary calcification obscures taxonomically useful features in all but the most distal branches. Several of the species described by Brydone (1930) may well be based on such ontogenetic variation. *Porina* first appears in Britain in the lowermost Campanian but does not become common until the *B. mucronata* Zone (Brydone 1930).

Occurrence. Stratigraphic range unknown; described specimens come from the Campanian, *B. mucronata* Zone; Norfolk, and possibly elsewhere.

Acknowledgements. I am grateful to Mike Nowicki and Paul Whittlesea for advice on stratigraphical ranges, Andrea Burgess-Faulkner for SEM assistance, and Prof. E. Voigt for encouragement and freely availing himself of his unique knowledge of Cretaceous bryozoans. Prof. Voigt and the late Dr G. P. Larwood kindly read the manuscript for the first edition.

EXPLANATION OF PLATE 9

Chalk cheilostomes; all scanning electron micrographs of uncoated specimens; magnifications approximate.

Fig. 1. *Herpetopora anglica* Lang. Turonian, Lewes, Sussex; autozooid with a closure plate and two disto-laterally budded autozooids with open opesiae; ×26.

Fig. 2. *Wilbertopora woodwardi* (Brydone). Campanian, West Meon Station, south-east of New Alresford, Hampshire; autozooids (some with ovicells) and two interzooidal avicularia; ×17.

Fig. 3. *Pyriporella? sagittaria* (Brydone). Campanian, Ropley, east of New Alresford, Hampshire; autozooids (most with ovicells) and large interzooidal and small adventitious avicularia; ×23.

Fig. 4. *Biaviculigera? furina* (Brydone). Campanian, near Weybourne, Norfolk; autozooids (with broken ovicells and intramural reparative buds) and large interzooidal avicularia; ×32.

Fig. 5. *Porina* sp. Campanian, Norwich, Norfolk; distal (young) part of colony showing orifices of three autozooids, large frontal pores and spiramen; ×28.

Fig. 6. *Volviflustrellaria taverensis* (Brydone). Campanian, Norwich, Norfolk; entire spindle-shaped colony; ×7.5.

Fig. 7. '*Biflustra*' *argus* d'Orbigny. Campanian, Norwich, Norfolk; ×19.

Fig. 8. *Stichomicropora marginula* (Brydone). Coniacian, Dover, Kent; autozooids, some with ovicells represented by crescents of spine bases; ×30.

Fig. 9. *Onychocella inelegans* (Lonsdale). Coniacian or Santonian, Bromley, Kent; autozooids and curved avicularia; ×25.

Fig. 10. *Lunulites tenax* Brydone. Campanian, East Harnham, Salisbury, Wiltshire; upper colony surface; ×3·3.

Fig. 11. *Pelmatopora gregoryi* (Brydone). Santonian, Sussex; ×16.

Fig. 12. *Castanopora magnifica* (d'Orbigny). Campanian, Norwich, Norfolk; autozooids and broken ovicells; ×14.

6. BRACHIOPODS

by ELLIS OWEN

During the nineteenth century the British Mesozoic brachiopod faunas were known chiefly from the works of James Sowerby (1815–18) and later, his son James de Carl Sowerby (1826–29). Excellent though their descriptions and illustrations were for the time, the works lacked any useful stratigraphical detail. This is not surprising since the science of stratigraphy was only just becoming established as a result of the pioneer work of William Smith. By the time Thomas Davidson (1852–54, 1874) had produced his excellent monographs of British Brachiopoda, published by the Palaeontographical Society, London, William Smith's concept of British stratigraphical succession had been widely accepted and was thus used by Davidson.

In the volumes on Cretaceous Brachiopoda (1852–54, 1874), Davidson described many British endemic species for the first time and incorporated monographical studies of the more ubiquitous species described by authors from France (d'Archiac 1847; d'Orbigny 1848–51) and Germany (Roemer 1840; Schloenbach 1867), which occur in this country.

Accounts of more diverse species within the Lower Cretaceous were popular with many authors of the time but it was not until several decades later that a work of any significance was published on British Chalk brachiopods. Sahni (1929) produced a monograph on British Chalk Terebratulidae that remained the major work of reference for several years. The Chalk Rhynchonellidae were the subject of a monograph by Pettitt (1949, 1954); this dealt largely with Turonian and Senonian species.

Since the publication of the works of Sahni and Pettitt, Owen (1962, 1970, 1977, 1988) has revised Cretaceous genera with representative species occurring in the Chalk and, more recently, has written on the systematics, ecology and distribution of some Cenomanian genera and species.

DESCRIPTIONS

For information on terminology and classification, see Moore (1965).

Phylum BRACHIOPODA
Class INARTICULATA
Superfamily CRANIACEA

Family CRANIIDAE

Genus ANCISTROCRANIA

Ancistrocrania parisiensis (Defrance)
Plate 11, figure 2

Description. One of the few inarticulate brachiopods possessing an entirely calcareous shell. Attachment formed by cementation to hard surfaces, often on the test of echinoids or large mollusc shells. The pedicle valve, which forms the attachment, is seen as a shallow subcircular concavity with steep, thickened shell margins. The apical area of the shell has well-formed adductor muscle scars and faint vascular markings. The brachial valve is often seen with a low, short, median septum that separates two slender, posterolaterally-directed brachial processes. The valve margins are not thickened as in the pedicle valve.

Occurrence. Upper Turonian–Campanian; Devon, Dorset, Isle of Wight, Sussex, Kent, Norfolk, Yorkshire, and Northern Ireland.

Genus ISOCRANIA

Isocrania paucicostata (Bosquet)
Plate 11, figure 1

Description. Attached to hard surfaces by apical region of the pedicle valve only. Both valves are almost circular in outline and are conical; they are ornamented by 10–14 strong costae, with some intercalations, originating from the apex of each valve. The muscle scars are well developed in each valve.

Occurrence. Santonian–Campanian; Wiltshire, Hampshire, Kent, and Norfolk.

Class ARTICULATA

Family WELLERELLIDAE

Genus ORBIRHYNCHIA

Orbirhynchia mantelliana (J. de C. Sowerby)
Plate 10, figure 4

Description. Small, biconvex, broadly oval rhynchonellid with 16–18 deeply incised, rounded, radiating costae on each valve; with four on the almost imperceptible dorsal fold and three or four in the sulcus of the ventral valve. The anterior commissure is uniplicate with a marked and extensive trapezoidal linguiform extension. The pedicle umbo is short, with a small circular foramen and fairly well-defined beak-ridges. The deltidial plates are conjunct but poorly exposed. Internal characters, as seen in transverse serial sections, show a thickened shell and short, slightly converging dental lamellae in the pedicle valve: the brachial valve has no median septum but displays the typical falciform crura.

Occurrence. Middle Cenomanian, *A. rhotomagense* Zone; Devon, Dorset, Isle of Wight, Sussex, Kent, Cambridgeshire, Norfolk, Lincolnshire and Yorkshire.

Orbirhynchia wiesti (Quenstedt)
Plate 10, figure 6

Description. Shell evenly biconvex, subcircular in general outline with oval anterior and lateral profiles. Ornament of 25–28 fine, subangular to rounded, radiating costae: 7–8 on low, almost imperceptible, dorsal fold and a corresponding number in a low arcuate anterior sulcus with a fairly extensive linguiform extension. Growth lines are faint but there are usually two or three prominent growth rings at about one-half to two-thirds the length of the shell. The umbo is short, with a small circular foramen and indistinct beak-ridges.

Occurrence. Upper Cenomanian; appears to be confined to the more chalky facies of Devon and Dorset.

Orbirhynchia dispansa Pettitt
Plate 13, figure 7

Description. Moderately biconvex shell, subcircular in general outline with oval anterior and lateral profiles. The umbo is massive, erect, with a sharp beak and small, circular to pyriform foramen. The conjunct deltidial plates are sometimes thickened or slightly produced into a labial extension. Shell ornament consists of 18–22 strong, rounded costae not deeply incised but with shallow intervening sulci. The anterior commissure is low arcuate and occasionally asymmetrical.

Occurrence. Upper Turonian, *S. plana* Zone; southern England.

EXPLANATION OF PLATE 10

Fig. 1. *Cyclothyris difformis* (Valenciennes, *in* Lamarck). Lower Cenomanian, South Devon; ×1.
Fig. 2. *Orbirhynchia parkinsoni* Owen. Lower Cenomanian, Cambridge; ×1.
Fig. 3. *Burrirhynchia devoniana* Owen. Lower Cenomanian, South Devon; ×1·5.
Fig. 4. *Orbirhynchia mantelliana* (J. de C. Sowerby). Middle Cenomanian, Kent; ×1·5.
Fig. 5. *Orbirhynchia multicostata* Pettitt. Upper Cenomanian, Sussex; ×1·5.
Fig. 6. *Orbirhynchia wiesti* (Quenstedt). Upper Cenomanian, South Devon; ×1·5. a, dorsal view; b, lateral view; c, anterior view.

PLATE 10

1a

1b

2a

1c

2c

3a

3b

4a

2b

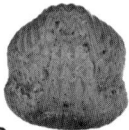

3c

5b

4b

6a

5a

4c

5c

6c

6b

Orbirhynchia cuvieri (d'Orbigny)
Plate 11, figure 5

Description. Resembling *O. wiesti* illustrated here (Pl. 10, fig. 6) but differing from that species in its more circular general outline and in shell ornament, having 30–34 subangular to rounded radiating costae. Its pedicle valve is also less acutely convex and slightly flattened anteriorly. The linguiform extension is short and arcuate.

Occurrence. Turonian; Surrey, Kent, Hertfordshire, Buckinghamshire, Cambridgeshire, Berkshire and possibly Yorkshire.

Orbirhynchia multicostata Pettitt
Plate 10, figure 5

Description. Broadly oval to subpentagonal in general outline with elongate-oval lateral and anterior profiles. Umbo short with small circular foramen and indistinct beak-ridges. Shell evenly biconvex with ornament of approximately 40 fine, rounded costae; 12 on very low median fold and 11 in broad arcuate sulcus.

Occurrence. Upper Cenomanian, *M. geslinianum* and *N. juddii* zones; Sussex, Surrey, Bedfordshire, Buckinghamshire and Cambridgeshire.

Orbirhynchia parkinsoni Owen
Plate 10, figure 2

Description. Shell large, biconvex, subcircular in general outline with elongate-oval lateral and anterior profiles. Umbo short, massive, with moderate to large circular foramen and well-defined beak-ridges. Deltidial plates are conjunct and well exposed. Shell ornament consists of 16–20 strong, angular, and deeply incised radiating costae. Median dorsal fold poorly developed. Low arcuate anterior commissure often asymmetrical. Linguiform extension provides an extensive shallow sulcus in the pedicle valve.

Occurrence. Upper Albian–Lower Cenomanian; confined to the Cambridge Greensand and uppermost bed of the Red Rock at Hunstanton, Norfolk.

Family RHYNCHONELLIDAE
Subfamily CYCLOTHYRIDINAE

Genus CYCLOTHYRIS

Cyclothyris difformis (Valenciennes, *in* Lamarck)
Plate 10, figure 1

Description. Medium-sized *Cyclothyris*. Biconvex, broadly oval to sub-triangular in general outline. Dorsal fold low, indistinct. Anterior commissure broadly arcuate with well-marked uniplication, often asymmetrical. The massive umbo is slightly incurved and truncated by a large circular foramen. Deltidial plates are conjunct and extend forwards almost completely encircling the foramen. The beak-ridges are distinct and border an extensive interarea. The shell surface is ornamented by 40–45 rounded radiating costae, with nine on the fold and ten or 11 in the sulcus. Two, sometimes three, step-like growth-marks occur at about one-third and two-thirds of the length of the shell.

Occurrence. Lower Cenomanian; Devon, Dorset, Wiltshire and the Isle of Wight; chiefly confined to consolidated limestone and chalky facies.

Genus BURRIRHYNCHIA

Burrirhynchia devoniana Owen
Plate 10, figure 3

Description. Medium to small subquadrate shell, acutely biconvex in outline. Umbo short, massive with sharp, suberect beak. Ornament of approximately 24 rounded but deeply incised costae with five or six costae on well-defined dorsal fold and an equivalent number in shallow extensive ventral sulcus. Median fold low, sulcus shallow. Anterior commissure with high arcuate linguiform extension. Concentric growth lines faint, becoming more prominent marginally.

Occurrence. Lower Cenomanian, *M. mantelli* Zone; south Devon.

Genus CRETIRHYNCHIA

Cretirhynchia plicatilis (J. Sowerby)
Plate 13, figure 8

Description. Shell transversely oval to subpentagonal in general outline with oval anterior contour; subcircular to oval in lateral profile. Although acutely biconvex, the brachial valve is more highly developed with an almost semicircular lateral profile. The umbo is small and the beak suberect with a small circular foramen. Beak-ridges well marked but not sharply defined, bordering a moderately broad but not extensive interarea. Shell ornament consists of about 65–70 rounded costae, with about 11 on the indistinct brachial fold and ten in the pedicle sulcus. The anterior commissure, sometimes flattened, shows a broad shallow sulcus with an extensive trapezoidal linguiform extension. Growth lines are more prominent towards the shell margins.

PLATE 11

1a

1b

2a

3b

4b

3a

6a

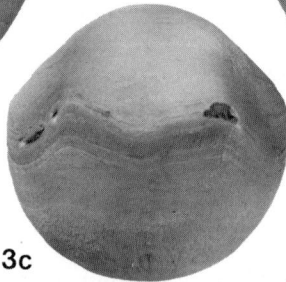
3c

4c

5a

6b

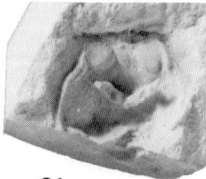
2b

4a

5b

6c

7a

7b

7c

5c

Occurrence. Upper Coniacian–Upper Santonian; Sussex, Kent and Norfolk; common.

Cretirhynchia arcuata Pettitt

Plate 13, figure 6

Description. Medium-sized biconvex *Cretirhynchia*, subcircular to subtriangular in outline. Anterior contour oval to lenticular, lateral profile elongate-oval. A low median fold is developed anteriorly on the dorsal valve and a corresponding shallow sulcus is present on the anterior part of the ventral valve. The umbo is slightly produced and incurved. Beak-ridges are well defined and the flat interareas fairly extensive. The deltidial plates are well exposed. The shell surface is smooth in the early stages of growth but short, rounded, subangular costae develop along the anterolateral commissure in later stages of growth.
Occurrence. Upper Campanian, *B. mucronata* Zone; Norfolk and northern Hampshire.

Superfamily TEREBRATULACEA

Family TEREBRATULIDAE

Subfamily SELLITHYRIDINAE

Genus OVATATHYRIS

Ovatathyris ovata (J. Sowerby)

Plate 16, figure 4

Description. Medium-sized elongate-oval, sulco-carinate terebratulid. Umbo short, massive, truncated by large, circular foramen. Beak erect;

<hr>

EXPLANATION OF PLATE 11

Fig. 1. *Isocrania paucicostata* (Bosquet). Santonian, Sussex: a, dorsal exterior; b, internal view of dorsal valve; both ×3.

Fig. 2. *Ancistrocrania parisiensis* (Defrance). Horizon unknown, Kent: a, internal view of ventral valve; b, internal view of dorsal valve; ×1·5.

Fig. 3. *Concinnithyris obesa* (J. de C. Sowerby). Upper Cenomanian, Wiltshire; ×1.

Fig. 4. *Terebratulina striatula* (Mantell). Horizon unknown, Sussex; ×1.

Fig. 5. *Orbirhynchia cuvieri* (d'Orbigny). Turonian, Hertfordshire; ×1.

Fig. 6. *Arenaciarcula beaumonti* (d'Archiac). Lower Cenomanian, South Devon; ×1·5.

Fig. 7. *Kingena elegans* Owen. Turonian, Lincolnshire; ×1.

Figs 3–7: a, dorsal view; b, lateral view; c, anterior view.

beak-ridges clearly defined; permesothyrid. Shell ornamented by step-like concentric growth-lines, more prominent on the anterior half of the shell. Shell surface covered with short, almost hair-like spinules, often poorly preserved. Dorsal valve with shallow sulcus that broadens and deepens anteriorly. Ventral valve with clearly defined carina in umbonal area that diminishes anteriorly. Anterior commissure sulcate to sulco-carinate.

Occurrence. Lower Cenomanian, *M. mantelli* Zone; Wiltshire, Dorset and South Devon.

Ovatathyris potternensis Owen
Plate 15, figs 3–4

Description. Like *O. ovata* (J. Sowerby) but lacking a carina in the ventral valve, and in having uniplicate to strongly biplicate anterior commissure, and numerous well-marked concentric growth-lines.

Occurrence. Upper Albian–Lower Cenomanian, *M. mantelli* Zone; Wiltshire, Dorset and the Isle of Wight.

Genus BOUBEITHYRIS

Boubeithyris diploplicata Owen
Plate 16, figs 5–6

Description. Medium-sized elongate-oval to subquadrate, acutely biplicate, biconvex terebratulid. Umbo short, massive with suberect beak truncated by a large circular foramen. Beak-ridges distinct; permeso-thyrid. Interarea flat and short; delthyrium poorly exposed. The concentric growth-lines are well marked, particularly at the shell margins. The strongly biplicate anterior commissure is sometimes obscured by gross gerontic thickening.

Remarks. The species has been widely misquoted as *Terebratula biplicata* Sowerby.

Occurrence. Lower Cenomanian; South Devon, Dorset, Wiltshire, Isle of Wight and Sussex.

EXPLANATION OF PLATE 12

Fig. 1. *Concinnithyris subundata* (J. Sowerby). Cenomanian, Wiltshire; ×1·5.
Fig. 2. *Gibbithyris semiglobosa* (J. Sowerby). Turonian, Sussex; ×1.
Fig. 3. *Gibbithyris media* Sahni. Turonian, Wiltshire; ×1.
Fig. 4. *Gemmarcula canaliculata* (Roemer). Lower Cenomanian, South Devon: a, dorsal view; b, anterior view; ×1.
Figs 1–3: a, dorsal view; b, lateral view; c, anterior view.

PLATE 12

1a

4a

2a

1b

2b

1c

4b

2c

3a

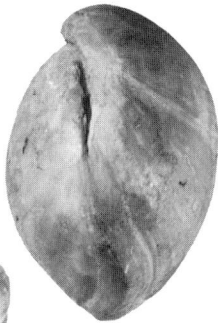

3b

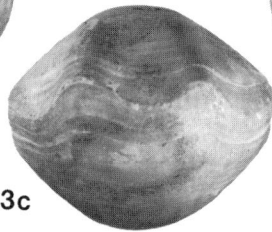

3c

Subfamily RECTITHYRIDINAE

Genus RECTITHYRIS

Rectithyris wrightorum Owen
Plate 14, figure 3

Description. Large oval, evenly biconvex, rectimarginate terebratulid. Umbo short and erect; truncated by large circular foramen. Fairly distinct mesothyrid beak-ridges bordering a short, flat interarea. Delthyrium not exposed. Lateral profile elongate-oval. Anterior commissure elliptical in outline. Shell surface smooth with numerous faint concentric growth-lines.
Occurrence. Lower Cenomanian, *M. mantelli* Zone; Devon, rare.

Genus TROPEOTHYRIS

Tropeothyris vectis Owen
Plate 14, figure 2

Description. Shell elongate-oval to subpentagonal in general outline. Lateral profile oval; lateral commissure oblique. Anterior commissure oval in outline, biplicate. Umbo short, beak suberect and truncated by large circular foramen. Beak-ridges rounded; mesothyrid. Maximum width of shell occurs just anterior to mid-line.
Occurrence. Lower Cenomanian, *N. carcitanense* Zone; Sussex, Isle of Wight and Kent.

Genus MOUTONITHYRIS

Moutonithyris dutempleana (d'Orbigny)
Plate 14, figure 1

Description. Large oval to elongate-pentagonal biplicate terebratulid. Umbo erect and massive. Foramen circular, marginate to labiate. Beak-ridges distinct, permesothyrid. Although biconvex, the dorsal valve of the shell is less acutely convex than the ventral valve and often exhibits a flattened umbonal area. The anterior commissure is incipient to strongly biplicate, some specimens having only a very slight median dorsal sulcation within a well-defined uniplication.
Occurrence. Upper Albian–Lower Cenomanian; Kent, Cambridgeshire, Norfolk and Yorkshire.

Moutonithyris obtusa (J. Sowerby)
Plate 15, figure 1

Description. Medium-sized subquadrate, uniplicate, obtuse terebratulid. Umbo short, suberect and massive, truncated by a large circular foramen with well-developed pedicle collar. Beak-ridges distinct, permesothyrid. Anterior commissure oval in general outline; showing wide uniplication and numerous marginal growth-lines which tend to thicken the shell margin in gerontic specimens. Shell surface ornamented by numerous concentric growth-lines.

Remarks. This species has previously been assigned to the genus *Ornatothyris* because of the density of concentric growth-lines on both valves and the uniplicate anterior margin. Transverse serial sections, however, confirm that the species belongs to the genus *Moutonithyris*.

Occurrence. Lower Cenomanian; confined to the Cambridge Greensand and the top bed of the Red Rock of Hunstanton, Norfolk.

Subfamily GIBBITHYRIDINAE

Genus GIBBITHYRIS

Gibbithyris semiglobosa (J. Sowerby)
Plate 12, figure 2

Description. Medium-sized subcircular to oval, evenly biconvex terebratulid. Umbo short and slightly incurved; foramen small and circular. Beak-ridges fairly well-defined; permesothyrid. Anterior commissure biplicate; oval to subcircular in profile. Lateral profile distinctly oval.

Occurrence. Turonian; Wiltshire, Sussex, Surrey, Kent, Hertfordshire and Cambridgeshire.

Gibbithyris media Sahni
Plate 12, figure 3

Description. Considerably larger than *G. semiglobosa* with a more massive umbo and larger foramen. The general outline is more pentagonal than in either *G. semiglobosa* or *G. merensis*, with which it has been compared, and the anterior part of the shell shows a somewhat earlier and stronger development of the characteristic biplication. The lateral profile shows almost equal biconvexity.

Occurrence. Turonian; Wiltshire and Kent.

Gibbithyris merensis Sahni
Plate 13, figure 1

Description. Medium-sized, oval, acutely biconvex, biplicate *Gibbithyris*. Umbo massive; beak slightly incurved; foramen small and circular. Beakridges indistinct; epithyrid. Symphytium completely obscured by the incurvature of the beak. Anterior commissure strongly biplicate, subcircular in profile. Lateral profile broadly oval.

PLATE 13

1a

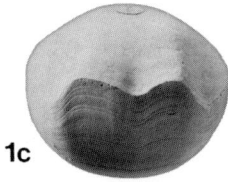

1c

1b

4

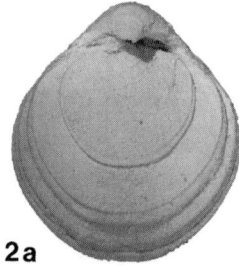

2a

2b

3

5a

8a

6a

6b

5b

8c

6c

7b

8b

7a

7c

5c

Occurrence. Upper Turonian, *S. plana* Zone, to Upper Coniacian Lower Santonian, *M. cortestudinarium* Zone; Wiltshire, Surrey and Kent.

Genus CONCINNITHYRIS

Concinnithyris subundata (J. Sowerby)
Plate 12, figure 1

Description. Large, uniformly oval terebratulid, almost evenly biconvex, uniplicate. Lateral profile elongate-oval, anterior profile oval. Umbo short, massive, suberect. Foramen large and circular. Symphytium not exposed. Beak-ridges indistinct; permesothyrid. Concentric growth-lines more prominent marginally.
Occurrence. Cenomanian; Wiltshire, Dorset, Sussex and Kent; probably also Cambridgeshire and Yorkshire.

Concinnithyris obesa (J. de C. Sowerby)
Plate 11, figure 3

Description. Large, acutely biconvex, elongate-oval, biplicate terebratulid. Umbo short, massive and suberect. Foramen large and circular. Symphytium poorly exposed. Beak-ridges smooth and indistinct; permesothyrid. Morphological variation includes less acutely developed anterior biplication, labiate foramen and more constricted anterior part of shell. The lateral profile is oval and the anterior profile is distinctly subcircular at all stages of growth.
Occurrence. Middle and Upper Cenomanian; Wiltshire, Dorset, Sussex and the Isle of Wight.

EXPLANATION OF PLATE 13

Fig. 1. *Gibbithyris merensis* Sahni. Upper Turonian, Wiltshire; ×1.
Fig. 2. *Magas pumilus* J. Sowerby. Horizon unknown, Hampshire: a, dorsal view; b, anterior view; ×4.
Fig. 3. *Terebratulina lata* (R. Etheridge). Turonian, *T. lata* Zone, South Devon; ×4.
Fig. 4. *Carneithyris carnea* (J. Sowerby). Campanian, Norfolk; ×1.
Fig. 5. *Capillithyris squamosa* (Mantell). Lower Cenomanian, Kent; ×2.
Fig. 6. *Cretirhynchia arcuata* Pettitt. Campanian, Norfolk; ×1·5.
Fig. 7. *Orbirhynchia dispansa* Pettitt. Upper Turonian, Surrey; ×1·5.
Fig. 8. *Cretirhynchia plicatilis* (J. Sowerby). Coniacian–Santonian, Wiltshire; ×1.
Figs 1, 5–8: a, dorsal view; b, lateral view; c, anterior view.

Genus ORNATOTHYRIS

Ornatothyris sulcifera (Morris)
Plate 15, figure 2

Description. Small- to medium-sized, plano-convex to biconvex tere-bratulid. Umbo short and massive with large, circular to labiate foramen. Symphytium poorly exposed. Beak-ridges rounded; permesothyrid. The anterior commissure is distinctly uniplicate with a subcircular general profile. Shell ornament consists of numerous concentric growth-lines occurring as a raised pattern of rugae. Lateral profile evenly biconvex.
Occurrence. Upper Cenomanian; comparatively common at Fulbourn, Cambridge but may also occur as a variant at South Cave, Humberside, and in Lincolnshire.

Subfamily CARNEITHYRIDINAE

Genus CARNEITHYRIS

Carneithyris carnea (J. Sowerby)
Plate 13, figure 4

Description. Medium-sized oval to subcircular, evenly biconvex terebratulid. Umbo short and massive; beak slightly incurved; foramen small and circular. Symphytium poorly exposed. Beak-ridges rounded; mesothyrid. Lateral profile elongate-oval; anterior commissure recti-marginate, elliptical in general profile.
Occurrence. Upper Campanian, *B. mucronata* Zone; Norfolk.

Subfamily CAPILLITHYRIDINAE

Genus CAPILLITHYRIS

Capillithyris squamosa (Mantell)
Plate 13, figure 5

Description. Small terebratulid, broadly oval to subpentagonal in general outline. Umbo short, slightly produced and suberect. Foramen large and

---

EXPLANATION OF PLATE 14

Fig. 1. *Moutonithyris dutempleana* (d'Orbigny). Lower Cenomanian, Cambridge; ×1.
Fig. 2. *Tropeothyris vectis* Owen. Lower Cenomanian, Isle of Wight; ×1·5.
Fig. 3. *Rectithyris wrightorum* Owen. Lower Cenomanian, South Devon; ×1.
Fig. 4. *Terebratulina etheridgei* Owen. Cenomanian, Cambridgeshire; ×1.
  a, dorsal view; b, lateral view; c, anterior view.

PLATE 14

1a

4

2a

1b

2b

1c

2c

3b

3a

3c

circular. Beak-ridges distinct; permesothyrid. Lateral profile evenly biconvex. Anterior commissure rectimarginate to sulcate. Ornament of numerous well-defined concentric growth-lines that tend to become more step-like towards anterior. This ornament is sometimes crossed by fine longitudinal striae, radiating from the umbonal areas.

Occurrence. Lower–Middle Cenomanian; confined to the more marly chalks of southern and south-eastern England.

Family CANCELLOTHYRIDIDAE
Subfamily CANCELLOTHYRIDINAE

Genus TEREBRATULINA

Terebratulina etheridgei Owen
Plate 14, figure 4

Description. Subtriangular in general outline. Umbo slightly produced; foramen comparatively large and circular. Beak-ridges distinct, bordering a flat, triangular interarea. Symphytium well exposed. Lateral profile plano-convex to weakly biconvex. Anterior commissure rectimarginate. Shell ornament consisting of 30–35 rounded costellae with frequent bifurcation and marginal intercalation.

Remarks. This species has previously been referred to as *Terebratulina triangularis* Etheridge.

Occurrence. Upper Albian–Lower Cenomanian; confined to the Cambridge Greensand of the Cambridge district.

Terebratulina striatula (Mantell)
Plate 11, figure 4

Description. Medium-sized terebratulinid that is elongate-oval to pentagonal in general outline. Umbo produced, foramen small, symphytium poorly exposed. Beak-ridges rounded or smooth, mesothyrid. Lateral profile biconvex. Anterior commissure incipiently uniplicate but with shallow sulcus originating from a point just anterior to the ventral umbo which broadens anteriorly. Shell ornamented by numerous fine, rounded costellae and crossed by 8–10 concentric growth-lines that are more apparent at the anterior marginal areas.

Occurrence. Turonian–Campanian; most localities in south-east and south-west England and Hertfordshire, Cambridgeshire, Norfolk and Yorkshire.

Terebratulina lata (R. Etheridge)
Plate 13, figure 3

Description. Plano-convex to biconvex terebratulinid. Small and sub-circular in general outline. Umbo short and foramen small. Hinge-line slightly extended to form a broadly based triangular interarea. Beak-ridges are distinct and mesothyrid. Anterior commissure rectimarginate. Shell ornament consists of approximately 45–50 smooth, rounded costellae with frequent intercalations and bifurcations. Concentric growth-lines well defined, typically numbering four or five.

Occurrence. Turonian, *T. lata* Zone; chiefly in southern England but also occurs in Hertfordshire, Cambridgeshire, Berkshire and Norfolk; it may not be the same species as *T. lata* from Lincolnshire and Yorkshire.

Superfamily TEREBRATELLACEA
Family KINGENIDAE
Subfamily KINGENINAE

Genus KINGENA

Kingena elegans Owen
Plate 11, figure 7

Description. A small, neat-looking, elongate-pentagonal, biconvex kingenid. Umbo suberect, slightly produced, symphytium well exposed. Foramen large and subcircular. Beak-ridges well defined, permesothyrid. Anterior commissure sulcate. A flattened area, corresponding to the sulcus in the brachial or dorsal valve, is bordered by faint radiating carinae on the anterior margin of the ventral valve.

Occurrence. Turonian, *T. lata* and *S. plana* zones; Hertfordshire, Norfolk, Lincolnshire and Yorkshire.

Family DALLINIDAE
Subfamily GEMMARCULINAE

Genus GEMMARCULA

Gemmarcula canaliculata (Roemer)
Plate 12, figure 4

Description. Small, biconvex, subtriangular dallinid. Umbo massive; truncated by large circular foramen; suberect. Deltidial plates conjunct. Hinge-line extended, forming a fairly extensive triangular, flat interarea. Both valves are ornamented by approximately 16 strong, rounded, radiating costae. These are crossed by seven or eight well-defined growth-lines which become more evident towards the shell-margins. The anterior commissure is parasulcate, a deep anterior sulcus in the ventral valve

PLATE 15

1a

1c

1b

2a

2c

2b

3a

3b

3c

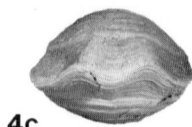

4a

4b

4c

corresponding with a marked or well-developed median fold in the dorsal valve. A more extensive hinge-line, more circular general outline and more clearly defined costae can be found as intraspecific variation.

Occurrence. Early Cenomanian, *M. mantelli* Zone; Devon.

Subfamily MAGADINAE

Genus MAGAS

Magas pumilus J. Sowerby
Plate 13, figure 2

Description. Small, smooth, plano-convex to biconvex terebratellid. Subcircular in general outline. Umbo slightly produced, beak incurved, beak-ridges distinct, epithyrid. Hinge-line slightly extended, forming a flat, narrow, triangular interarea. Deltidial plates are just visible. Anterior commissure incipiently sulcate. Concentric growth-lines often more prominent marginally.

Occurrence. Campanian–Maastrichtian; Wiltshire, Hampshire, Isle of Wight, Dorset, Kent, Sussex and Norfolk.

Subfamily *incertae sedis*

Genus ARENACIARCULA

Arenaciarcula beaumonti (d'Archiac)
Plate 11, figure 6

Description. Elongate-oval, evenly biconvex, costate terebratellid. Umbo produced, suberect; foramen small and circular. Beak-ridges are well defined, permesothyrid and border a short, flat, triangular interarea. Triangular deltidial plates are well exposed. The umbonal region of the brachial or dorsal valve is characteristically inflated. Shell ornament consists of 15–18 rounded but well-incised radiating costae, which seldom bifurcate. The anterior commissure is rectimarginate and subcircular to oval in outline.

Occurrence. Lower Cenomanian; Devon and Wiltshire.

EXPLANATION OF PLATE 15

Fig. 1. *Moutonithyris obtusa* (J. Sowerby). Lower Cenomanian, Cambridgeshire; ×1·5.

Fig. 2. *Ornatothyris sulcifera* (Morris). Upper Cenomanian, Cambridgeshire; ×1.

Figs 3–4. *Ovatathyris potternensis* Owen. Lower Cenomanian, Potterne, Wiltshire; ×1·5.

a, dorsal view; b, lateral view; c, anterior view.

PLATE 16

2a

2b

3a

3b 2c

1

3c

4b

4a

6a

5b

4c

6c

5c

5a

6b

Family TEREBRATELLIDAE
Subfamily TRIGONOSEMINAE

Genus TEREBRIROSTRA

Terebrirostra lyra (J. Sowerby)
Plate 16, figure 1

Description. Large, elongate-subpentagonal, biconvex terebratellid. Umbo extremely produced, forming an erect, subtriangular rostrum. The foramen is large and subcircular in outline. Conjunct deltidial plates are very well exposed. Shell ornament consists of approximately 24 strong, rounded, radiating costae originating from the umbones and frequently bifurcating. Concentric growth-lines are prominent or step-like and vary in number from four to six. The anterior commissure is rectimarginate and sub-quadrate in outline. Marginal thickening of the shell is common in older individuals.
Occurrence. Lower Cenomanian; Devon, Wiltshire and Dorset.

Genus DERETA

Dereta pectita (J. Sowerby)
Plate 16, figure 2

Description. Medium-sized, biconvex, costate terebratellid. Subcircular in general outline. Umbo short, massive, suberect. Foramen large and circular. Beak-ridges permesothyrid, defining a broad, triangular inter-area. The deltidial plates are conjunct and well exposed. The hinge-line is slightly extended. The shell ornament consists of 45–48 rounded costae, crossed by 5–7 concentric growth-lines, which become more prominent towards the anterior section of the shell. The lateral profile shows the even biconvexity of the shell, but some individuals show a greater inflation of the dorsal umbo. The anterior commissure is rectimarginate.
Occurrence. Lower Cenomanian; Wiltshire, Isle of Wight and Dorset.

EXPLANATION OF PLATE 16

Fig. 1. *Terebrirostra lyra* (J. Sowerby). Lower Cenomanian, Wiltshire; ×1·5.
Fig. 2. *Dereta pectita* (J. Sowerby). Lower Cenomanian, Dorset; ×1.
Fig. 3. *Dereta incerta* (Davidson). Lower Cenomanian, Somerset; ×3.
Fig. 4. *Ovatathyris ovata* (J. Sowerby). Lower Cenomanian, Dorset; ×1·5.
Figs 5–6. *Boubeithyris diploplicata* Owen. Lower Cenomanian, Wiltshire; ×1·5.
   a, dorsal view; b, lateral view; c, anterior view.

Dereta incerta (Davidson)
Plate 16, figure 3

Description. Small, almost circular, evenly biconvex *Dereta*. The umbo is slightly produced and suberect with a small, circular foramen. The permesothyrid beak-ridges are sharp, clearly defining the comparatively broad, triangular interarea and exposed conjunct deltidial plates. The hinge-line is slightly extended, as in *D. pectita,* and the umbones very slightly inflated. Shell ornament consists of 20–28 rounded, radiating costae given to frequent bifurcation. Concentric growth-lines are faint, except for one or two well developed at about two-thirds of the shell-length.

Occurrence. Lower Cenomanian; Somerset.

7. INTRODUCTION TO MOLLUSCS
AND BIVALVES

by R. J. CLEEVELY *and* N. J. MORRIS

The fossil Mollusca in Chalk faunas have attracted interest from the earliest days of palaeontology and have been described by a long succession of authors extending from the early works of Mantell and the Sowerbys to the more detailed accounts of the present-day. Many of the commoner Chalk fossils were first illustrated and described in the earlier classical literature dealing with British and European Cretaceous faunas, e.g. James and J. de Carle Sowerby (1812–46), Mantell (1822) for the former and Geinitz (1849–50, 1875), Goldfuss (1833–44) and d'Orbigny (1844–47) for the continent. In recent times the relevant publications of the Geological Survey have provided other limited illustrations of the fauna (e.g. Worssam and Taylor 1969).

Unfortunately, the information associated with many of the better preserved, or unique, specimens needed to illustrate this field guide, is inadequate. Consequently, although this has been corrected in some cases, a more accurate understanding of the stratigraphical occurrence and geographical distribution of Chalk Mollusca can only be obtained by referring to the detailed studies of particular horizons made by specialists. The most important of these, giving faunal lists and stratigraphic sections, are Carter and Hart (1977), Kennedy (1969, 1970) and Wright and Kennedy (1984) for Cenomanian localities; Bromley and Gale (1982) for the Upper Turonian Chalk Rock; Kennedy and Garrison (1975*b*), Peake and Hancock (1961), Bailey *et al.* (1983, 1984), Mortimore (1986), and Robinson (1986) for Coniacian–Maastrichtian chalks.

The preservation of the Mollusca is governed by the nature of the calcium carbonate forming their hard parts. Aragonite is a component of many molluscan shells, but under normal conditions of deposition it is metastable and tends to revert to calcite (see Introduction). Kennedy (1969) and Carter (1972) have summarized the principal ways in which the remains of aragonite fossils have been preserved. There are three broad categories of preservation.

1. Normal preservation, in which the original aragonite shell is either replaced by chalky calcite, or the shell outline is marked by a thin film

of marcasite or pyrites, and is usually oxidized to limonite; such preservation is nearly always associated with hardground horizons.
2. Pebble preservation, resulting from rolling, phosphatisation and/or glauconitisation, with patches of altered shell sometimes being retained. Often the fossil is partly obscured by adherent phosphatised sediment.
3. Oyster-cast preservation, where external moulds of the shells of gastropods, ammonites and bivalves occur on the attachment areas of cemented or encrusting organisms, e.g. particularly *Pycnodonte* and *Exogyra*; other indications may also be given by the xenomorphic sculpture of these cemented bivalves.

It must also be remembered that apart from the exigencies of preservation, and the nature of Chalk sedimentation and deposition, molluscan faunas are also naturally extremely variable. Many of the bivalve and gastropod species of the British Cretaceous are only found in the more diverse and inshore faunas of south-west England. We have occasionally used such fossils to illustrate certain taxa, but have adhered to a relatively strict use of the term 'Chalk fossils'. The Late Cretaceous was a time of considerable diversity in many molluscan families. In particular the Arcoidea developed epifaunal and infaunal groups, but diversity was only high among inshore faunas and remained low in offshore faunas. Consequently, taxa belonging to this superfamily are hardly dealt with in this guide, because examples from the Chalk are scarce and not worth illustrating. It is for this reason that members of many other inshore, or infaunal families, such as the Nuculoidea and Trigonioidea, are not described in this section, although they are commonly found at localities in south-west England in which the sandier facies of the Cretaceous are exposed.

BIVALVES

For a general description of the morphological features of bivalves and an account of their classification, see the appropriate sections in Moore (1969) and Murray (1985); for another glossary of conchology, see Arnold (1965); and for further illustrations, see Cox (1962). Henry Woods (1899–1913) provided illustrations and descriptions of most species occurring in the British Chalk in his monograph of British Cretaceous 'Lamellibranchia'; for revisions of many genera belonging to the Pectinacea, see under Dhondt in the references. Carter (1972) attempted a functional analysis of several of the common bivalve members of the Chalk fauna, while the broad approach of Stanley (1970) on the relationship of shell form and life habits provides more general information. Savazzi has also published a number of papers on the functional shell morphology of burrowing bivalves.

Dhondt (1989) emphasized that there is a particularly marked difference in the faunas from 'northern' (=Temperate) regions and those from Tethys, seen in the families Pectinidae, Limidae, Ostreoidea and inoceramids. Several genera and endemic species belonging to these families appear to have thrived in the white chalks of Coniacian–Maastrichtian age.

DESCRIPTIONS

Order ARCOIDA
Superfamily ARCOIDEA
Family ARCIDAE
Subfamily ARCINAE

Genus EONAVICULA

?*Eonavicula* sp.
Plate 17, figures 11–12

Description. Subtrapezoidal to rectangular inaequilateral shell. Anterior margin rounded and shorter than sloping posterior margin; byssal sinus present; ventral margin not parallel to dorsal margin. Taxodont hinge with both anterior and posterior teeth sloping to a point well below the umbo; posterior series larger. Ornament of radial ribs and growth rugae which form a reticulate network; ribs crossed by fine growth lines; strong carina from umbo to postero-ventral margin. A relatively wide denticulate inner shell margin.

Remarks. Usually preserved as internal and external moulds. *Arca strehlensis* Geinitz (Turonian) is more rectangular in shape, and has more regular dentition; *Barbatia geinitzi* Reuss (Turonian) is a smaller, less elongate shell with much finer radial ornament.

Occurrence. Turonian, S. *plana* Zone; south-east England, Berkshire, Hampshire, Surrey and Kent.

Order MYTILOIDA
Superfamily PINNOIDEA
Family PINNIDAE

Genus PINNA

Pinna decussata Goldfuss
Plate 17, figure 3

Description. Moderately elongate triangular shell; laterally compressed with a rhombic to lenticular cross-section dependent on age. Dorsal margin straight, ventral margin slightly curved. Each valve divided equally into a dorsal flattened part, with 7–9 strong, rounded ribs separated by broad shallow areas, linear ridges crossing both ribs and interspaces at regular intervals; and a ventral part (5–7 uniform ribs) which has several stronger ridges or folds at its most ventral margin.

Remarks. Less elongate than *P. cretacea* (Schlotheim) and its ventral ridges are more curved; has a larger apical angle than *P. robinaldina* d'Orbigny (Aptian) in which the ventral ribs gradually decrease in size. Most specimens from the Chalk are poorly preserved; those from Norfolk (Upper Campanian) have rather broader and more rounded ribs.

Occurrence. Lower Cenomanian, *M. mantelli* Zone, to Upper Campanian, *B. mucronata* Zone; Hampshire, Norfolk and south-east England.

Genus STEGOCONCHA

Stegoconcha sp.
Plate 17, figure 1

EXPLANATION OF PLATE 17

Fig. 1. *Stegoconcha* sp. Cenomanian, Dover, Kent; *c.* ×0·35.

Fig. 2. *Myoconcha cretacea* d'Orbigny. Cenomanian, Maiden Newton, Wiltshire; ×1.

Fig. 3. *Pinna decussata* Goldfuss. Cenomanian, Wilmington, Devon; ×0·5.

Fig. 4. '*Septifer*' *lineatus* J. de C. Sowerby. Campanian, *G. quadrata* Zone, East Harnham, near Salisbury, Wiltshire; ×1·5.

Fig. 5. *Ctenoides divaricata* (Dujardin). Senonian, Houghton Pit, Arundel, Sussex; left valve; ×0·75.

Fig. 6. *Ctenoides rapa* (d'Orbigny). Upper Albian, Chardstock, Devon; left valve; ×0·75.

Fig. 7. '*Unicardium*' *ringmeriense* (Mantell). Lower Cenomanian, Ringmer, Sussex; left valve; *c.* ×0·5.

Figs 8–9. '*Beguina*' *turoniensis* (Woods). 8, Turonian, *S. plana* Zone, Aston Hill, Oxfordshire; right valve; ×4. 9, Lower Cenomanian, *S. varians* Zone, Spring Barn, near Lewes, Sussex; right valve; ×1·5.

Fig. 10. ?*Ludbrookia cottaldina* (d'Orbigny). Lower Cenomanian, Peter's Pit, Burham, Kent; left valve; ×2.

Figs 11–12. ?*Eonavicula* sp. Turonian, *S. plana* Zone, Sparsholt, Hampshire. 11, fragment of internal mould of left valve showing part of hinge; ×2·5. 12, cast of external mould of left valve showing ornament; ×3.

Fig. 13. *Jouannetia rotundata* (J. de C. Sowerby). Turonian, ?Kent; latex cast from external mould of adult left valve showing anterior part of composite shell; *c.* ×2.

Fig. 14. *Durania mortoni* (Mantell). Cenomanian, Folkestone, Kent; fragment of attached valve showing siphonal bands/pillars and honeycomb structure; ×0·5.

PLATE 17

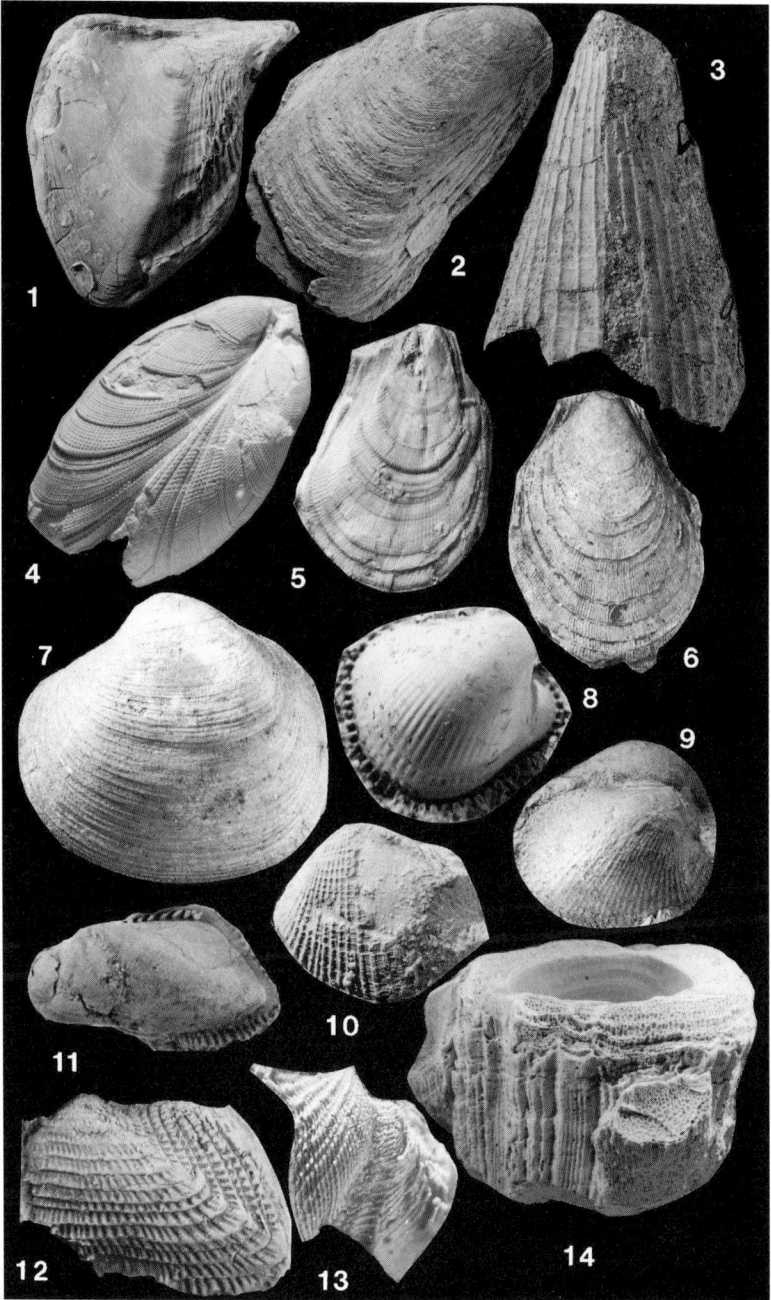

Description. Large, equivalve, ham-shaped, moderately thick shell. Most inflated along rounded ridge running from umbo to postero-ventral corner. Height just exceeds length; anterior smaller than posterior part of shell. Beak at anterior end of straight hinge margin, which is produced and pointed anteriorly; anterior margin rounded. Posterior wing not differentiated; merged into the rounded posterior ventral margin. Ornament of left valve consists of irregular radial ribs on anterior side of ridge, strongest near the margin; fine concentric ribs occur over the entire valve.
Occurrence. Middle Cenomanian, *A. rhotomagense* Zone, to Campanian, *B. quadrata* Zone; relatively scarce; known only from Kent and Wiltshire.

Superfamily MYTILOIDEA
Family MYTILIDAE

Genus SEPTIFER

'*Septifer*' *lineatus* J. de C. Sowerby
Plate 17, figure 4

Description. Ovate-oblong, slightly curved, thin, equivalve shell (other forms occur); regularly convex from umbo to posterior extremity, but compressed perpendicular to its greatest length. Dorsal margin slightly convex; antero-ventral margin slightly concave and face flattened. Umbones small, curved and terminal. Ornament of numerous fine, slightly irregular, radial ribs, except for an oval area beneath umbo crossed by concentric growth lines. Growth lines fairly strong at intervals.
Occurrence. Coniacian–Campanian; most horizons of Chalk in the counties around London; internal casts frequently found.

Superfamily PTERIOIDEA
Family BAKEVELLIIDAE

Genus PSEUDOPTERA

Pseudoptera caerulescens (Nilsson)
Plate 22, figure 9

Description. Small- to medium-sized, oblique, rhomboidal–triangular, inequivalve shell. Umbo at anterior extremity and scarcely protruding; its anterior ear small and not separated from valve; the posterior auricle (=pterioid wing) large, triangular and continuous with postero-ventral margin. Left valve convex with a strong oblique anterior tooth and an elongate longitudinal posterior tooth. A double flexure produces a prominent ridge from the umbo to the posterior margin (not present in juvenile specimens). Right valve not known. Ornament of weak radial ribs/threads, which may bear variably spaced spiny or scaly projections, but these are not regularly distributed over the valve.

Remarks. The palaeoecological significance of shell torsion and thickening in species of *Pseudoptera* was discussed by Savazzi (1984); he inferred that the ontogenetic changes in shell shape reflect a change in life habits, and also noted that the sculpture of the shell is only developed on the uppermost left valve.

Occurrence. Campanian, *G. quadrata* and *B. mucronata* zones; Wiltshire and Norfolk.

Superfamily INOCERAMOIDEA

Family INOCERAMIDAE

Remarks. The Inoceramidae are possibly the biostratigraphically most important family of non-cephalopod molluscs in the Chalk. Their cosmopolitan biogeographic distribution, arising from a long pelagic larval stage, and the considerable variation in shell morphology make them an excellent biostratigraphic tool (Dhondt 1983). However, the homogeneity of the family causes difficulty in establishing adequate differentiating characteristics, which together with their commonly poor preservation has resulted in taxonomic difficulties (see Kauffman *in* Herm *et al.* 1979).

Woods (1910 *in* 1899–1913) provided detailed descriptions of most Chalk inoceramids. Owing to inadequate stratigraphic details and the different standards of taxonomic interpretation (his species concept is now thought to be too broad) these taxa are no longer acceptable. Unfortunately, his illustrations alone have been used as the basis of species interpretation by other workers, resulting in numerous differing concepts of particular species.

The morphological characters of inoceramid shells have been illustrated by Seitz (1961), Tröger (1967), Tröger and Christensen (1991) and Walaszcyk and Cobban (2000). Important features for distinguishing species are: (1) shell and valve form; (2) the convexity and inclination of the shell; (3) the nature of the posterior and anterior auricles; (4) the position, projection and obliquity of the beak/umbo; (5) the ornament: its components and variation; and (6) the hinge and ligament pits. Unfortunately, inoceramids are seldom sufficiently well preserved in the Chalk to reveal many of these features.

There is considerable ecophenotypic and ontogenetic variation within species, as well as probable dimorphism (Seitz 1965). Consequently, inoceramids display a complex evolutionary mosaic of external shell characters involving all taxa, with extensive homeomorphy and parallelism. Kauffman (*in* Kauffman *et al.* 1977, and Herm *et al.* 1979) has argued that the more conservative internal shell features (e.g. musculature and ligamental structure) are more useful in establishing true relationships. These are, however, seldom sufficiently well preserved in Chalk specimens. Other authors have resorted to different methods for establishing subdivisions within the inoceramids. These include the use of the nomenclatorially inadequate generic/subgeneric names of Heinz (1932), informal species groups based on morphological characters (Dobrov and Pavlova 1959), systematic groups at subgeneric level (e.g. Seitz 1961; Sornay 1966–73), or evolutionary lineages (Kauffman *et al.* 1977).

It is not possible to illustrate, or even list, all the inoceramids that may be found in the Chalk. Text-figure 7.1 provides the stratigraphic ranges of some of the more important taxa. We have endeavoured to place British species in genera, but have recognised the practicality of using species groups for rapidly evolving lineages. It is very likely that the morphology of specimens of *Inoceramus* will not conform to those described here because, apart from the very large number of species that can occur, there are also 'linking forms' which have the characters of several taxa.

Genus INOCERAMUS

Inoceramus lamarcki Parkinson
Plate 19, figure 5; Text-figure 7.2

Description. Medium- to large-sized, virtually equivalve, moderately inflated inoceramid; its anterior margin perpendicular to its thick hinge line, with a prominent, posteriorly projecting auricle that is almost wing-like and has deep auricular and posterior umbonal sulci; strongly convex valves. Ornament consists of a series of large concentric folds and lacks the mid-ventral flattening of the growth line trace.

Remarks. *I. cuvieri* (J. Sowerby) (Middle Turonian, *T. lata* Zone) is similar but has a smooth shell, is less inflated, and is distinguished by its broad, low, wave-like rugae (ribs), sharply defined anterior margin and lack of postero-ventral sulcus. It is usually much larger. There are two common varieties in the British Chalk: *I. lamarcki lamarcki* and *I. lamarcki stumckei* (Text-fig. 7.2), the latter being larger and distinguished by its prominent, angular and terrace-like rugae, and by its lack of convexity near the margin of the shell.

Occurrence. Turonian, *T. lata* and *S. plana* zones; widespread in England from Yorkshire to Kent.

Inoceramus atlanticus Heinz
Plate 18, figure 5

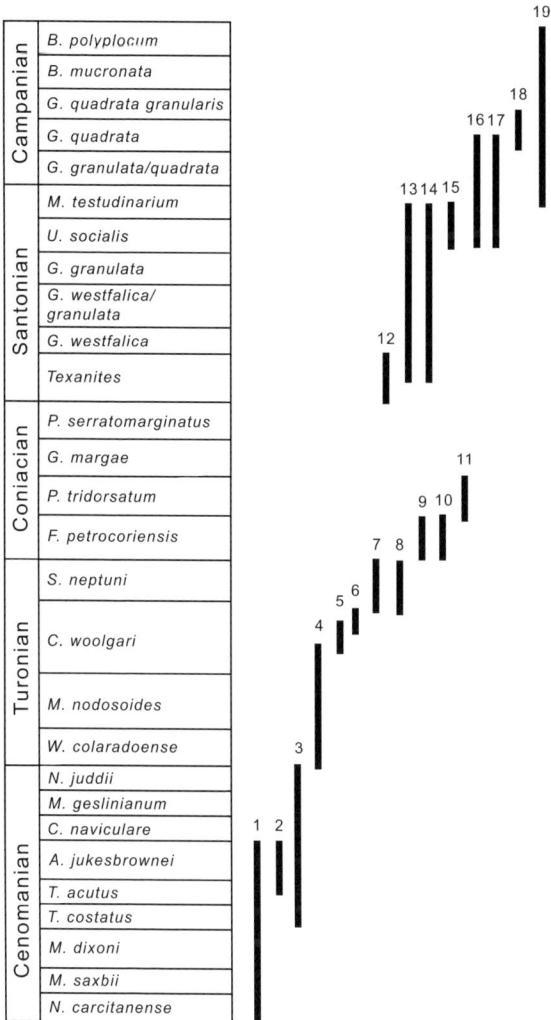

TEXT-FIG. 7.1. Range chart showing the distribution of inoceramid taxa described here (after Tröger 1989). Species as follows: 1, *Inoceramus crippsii* Mantell; 2, *I. atlanticus* Heinz; 3, *I. pictus* Sowerby; 4, *Mytiloides labiatus* (Schlotheim); 5, *I. apicalis* Woods; 6, *I. lamarcki* Parkinson; 7, *I. undulatus* Mantell; 8, *Cremnoceramus websteri* (Woods); 9, *C. inconstans* (Woods); 10, *C.* cf. *deformis* (Meek); 11, *Voluticeramus involutus* (J. de C. Sowerby); 12, *Sphenoceramus pachtii* (Archanguelsky); 13, *Cladoceramus digitatus* (J. Sowerby); 14, *Cordiceramus cordiformis* (J. de C. Sowerby); 15, *Sphenoceramus pinniformis* (Willett); 16, *S. patootensis* (de Loriol); 17, *Platyceramus rhomboides* Seitz; 18, *S. steenstrupi* (de Loriol); 19, *Cataceramus balticus* (Böhm).

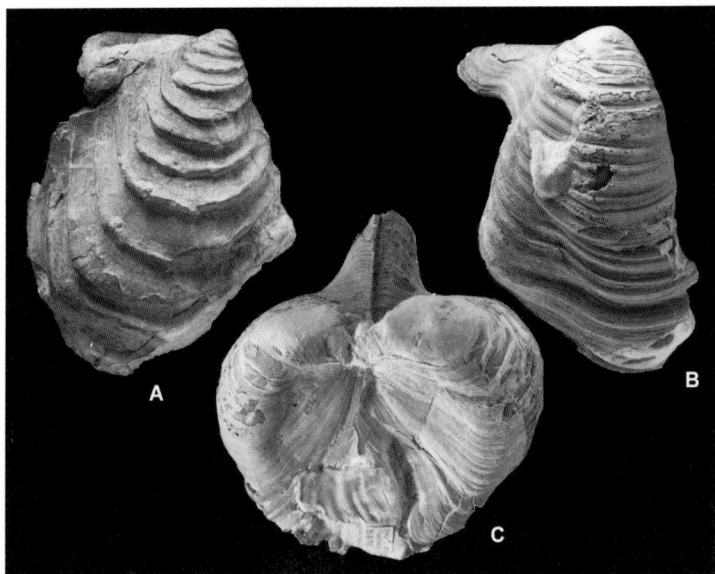

TEXT-FIG. 7.2. A, *I. lamarcki stumckei* Heinz, Chalk, locality and horizon unknown; right valve; ×0·125. B–C, *I. lamarcki lamarcki* Parkinson, Turonian, *T. lata* Zone, Essex. B, right valve showing wing-like projection; C, view of hinge of both valves; both ×0·5.

Description. Subquadrangular, thin, low to moderately convex shell; antero-dorsal margin shorter than in *I. crippsii*; terminal rounded beak and umbo project slightly above a straight hinge-line. Ornamented by well-spaced, sharply defined, sub-regular rugae; the growth lines may be very fine and numerous, or few and equivalent in size to the rugae.
Remarks. *I. crippsii* Mantell (Early Cenomanian) is similar but has more closely-spaced rugae, and is smaller and less quadrangular in curvature. This may eventually prove to be a synonym of *I. reachensis* Etheridge.
Occurrence. Middle Cenomanian, *H. subglobosus* Zone; throughout southern England; also found in Northern Ireland.

Inoceramus crippsii Mantell
Plate 18, figure 2

Description. Sub-equivalve, thin, low to moderately convex shell; its terminal rounded beaks and umbo do not project significantly. Anterior margin slopes back normal to the hinge. Ornament of well-spaced, alternating, sub-regular rugae and raised growth lines or lamellae.

Remarks. Distinguished by its thick ligamental plates, open ligamental areas that face centrodorsally, and much coarser resilifers. It retains the primitive internal rib between the posterior auricle and the disc. The *crippsii* lineage has been questionably related (Kauffmann *et al.* 1977) to the genus *Mytiloides* on the basis of similar shell morphology and ornament, as well as musculature. The lack of a posterior auricle, or auricular sulcus, separate it from *I. cuvieri* and *I. lamarcki*. *I. crippsii crippsii* Mantell and *I. c. hoppenstedtensis* Tröger are important Early and Middle Cenomanian representatives occurring throughout Europe. They are distinguished from *I. crippsii* by the relatively circular arrangement of ribs in the former and the narrower, more elongated pattern of ribs in the latter.
Occurrence. Cenomanian, *M. mantelli* Zone; widespread in southern England, Lincolnshire and Cambridgeshire.

Inoceramus pictus J. de C. Sowerby
Plate 18, figure 7

Description. Suberect, moderately high, inflated, equivalve shells; height exceeds length; umbos/beaks incurve and project; weak to prominent folds and sulci occur posteriorly. Subtriangular posterior auricle separated from disc by a sulcus. Large hinge plate with wide resilifers. Ornament of close, evenly spaced ribs and subequal fine to coarse growth lines, flat lamellae and variable rugae.
Remarks. Members of this species typically have evenly spaced and evenly developed growth lines. *I. pictus pictus* is characterised by its broad outline, moderate convexity and ornament composed of fine, raised growth lines forming regular concentric ribs and numerous small subequal rugae causing larger irregular wrinkles. It is distinguished from the more coarsely ornamented *I. pedalionoides* Nagao and Matsumoto (Turonian), which has far fewer subequal rugae, and *I. apicalis* Woods (Turonian) which is much narrower, has smaller posterior auricles, even finer concentric growth lines, much weaker rugae, and a less projecting umbo.
Occurrence. Middle–Upper Cenomanian, *H. subglobosus* and *C. naviculare* zones; widespread throughout south-east England.

Inoceramus apicalis Woods
Text-figure 7.3c

Description. Small, ovoid species with a narrow, moderately convex umbo and with fine, closely packed, rounded concentric ribs in regular groups of 4–8. The hinge line forms an angle of less than 90 degrees with the anterior margin.

PLATE 18

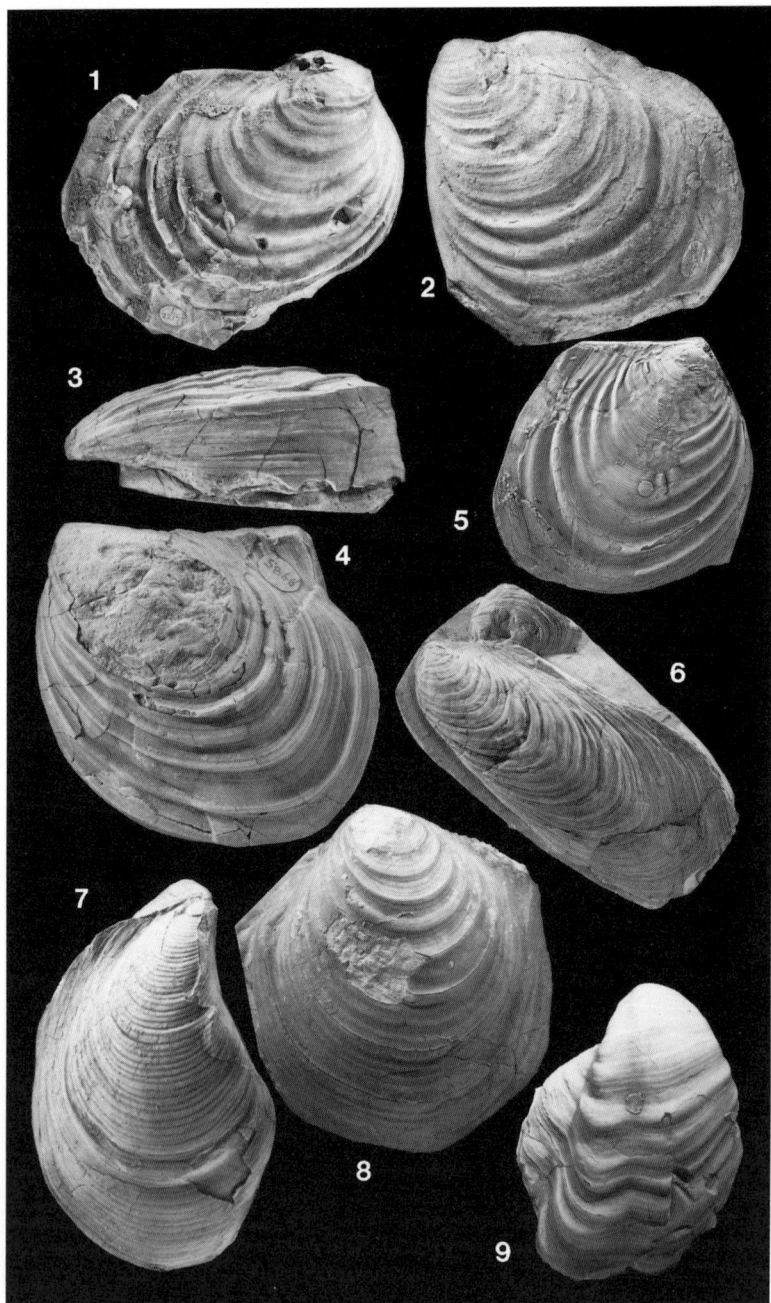

Occurrence. Lower Turonian, *I. labiatus* Zone; south-east England.

Inoceramus undulatus Mantell
Text-figure 7.3D

Description. Small, very convex species with the posterior ear well separated. It has strong, rounded, and regularly close-packed concentric ribs that are finer on the posterior ear and disappear on the anterior surface. The angle between the hinge line and the anterior margin is usually greater than 90 degrees.

Occurrence. Upper Turonian, *S. plana* Zone; south-east England.

Subgenus INOCERAMUS (CATACERAMUS)

Inoceramus (Cataceramus) balticus J. Böhm
Plate 18, figure 1

Description. Medium to large, rectangular, sub-equivalve, very inequilaterally rounded, moderately convex shell, becoming more convex with age as the shell margins grew at an angle. Length greater than height, this difference increasing with age. Long, straight hinge line. Anterior and ventral margins rounded, posterior margin forms an obtuse angle with hinge. Umbones broadly rounded, possibly almost terminal, but usually eroded. Ornament of strong, sharp, narrow, asymmetrically curved, almost elliptical, concentric ribs and broad concave interspaces showing closely packed growth lines.

EXPLANATION OF PLATE 18

Fig. 1. *Inoceramus (Cataceramus) balticus* J. Böhm. Campanian, Worbarrow Bay, Dorset; flint cast of right valve; ×0·25.

Fig. 2. *Inoceramus crippsii* Mantell. Lower Cenomanian, locality unknown; internal cast of left valve; ×1.

Figs 3–4. *Cremnoceramus inconstans* Woods. Locality and horizon unknown. 3, *c*, ×0·5; 4, left valve; *c*. ×0·5.

Fig. 5. *Inoceramus atlanticus* Heinz. Middle Cenomanian, *H. subglobosus* Zone, Bluebell Hill, Burham, Kent; right valve; ×0·5.

Fig. 6. *Mytiloides labiatus* (Schlotheim). Early Turonian, *I. labiatus* Zone, locality unknown; left valve; ×0·25.

Fig. 7. *Inoceramus pictus* J. de C. Sowerby. Middle Cenomanian, *H. subglobosus* Zone, Burham, Kent; right valve; ×0·5.

Fig. 8. *Cremnoceramus* cf. *deformis* (Meek). Coniacian, *M. cortestudinarium* Zone, Chatham, Kent; left valve; ×0·5.

Fig. 9. *Cordiceramus cordiformis* (J. de C. Sowerby). Santonian, *M. coranguinum* Zone, Micheldever, Hampshire; right valve; *c*. ×0·5.

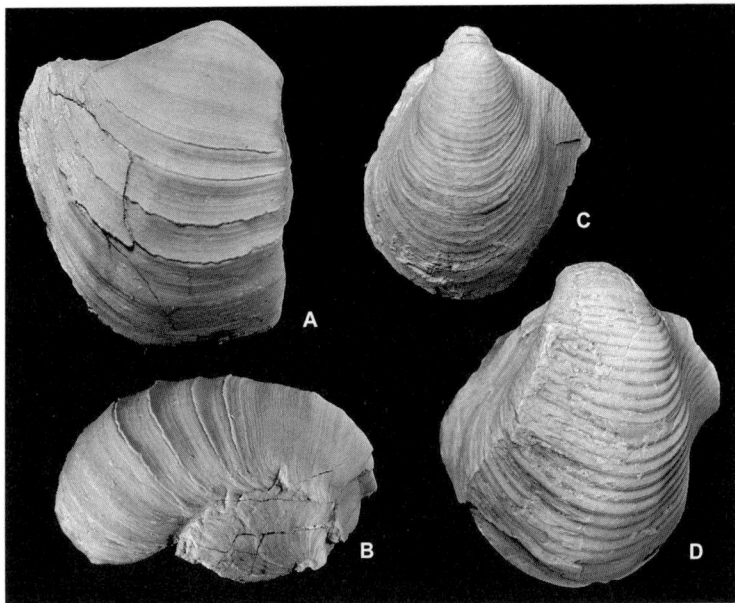

TEXT-FIG. 7.3. *Inoceramus* spp. A–B, *Cremnoceramus websteri* (Woods), Turonian, *S. plana* Zone, Borstal, Kent; right valve; ×1. C, *Inoceramus apicalis* Woods. Turonian, *I. labiatus* Zone; Peter's Pit, Burham, Kent; left valve; ×1. D, *Inoceramus undulatus* Mantell, Turonian, *S. plana* Zone, Lewes, Sussex; left valve; ×1.

Occurrence. Campanian, *O. pilula–B. mucronata* zones; throughout southern England, including Devon and Dorset.

Genus MYTILOIDES

Mytiloides labiatus (Schlotheim)
Plate 18, figure 6

Description. Moderately to slightly convex, obliquely inclined, thin, elongate, mytiliform, nearly equivalve shell. Beak terminal and pointed or slightly inflated; often crushed. Straight, moderately long hinge, seemingly not extended anterior of the beaks but continued along anterior margin. Ornament of alternating, regularly spaced, concentric rugae and fine, slightly raised, growth lines.

Remarks. The nomenclature of the various species and forms belonging to the Turonian *M. labiatus* lineage has continually been misconstrued. *M. labiatus sensu* Schlotheim [=*mytiloides* (Mantell)] is recognised as having regular concentric ornament and by being more highly inclined and less curved than the more inflated *M. labiatus sensu* Seitz, which is ornamented by irregular crowded rugae and numerous flat-lying lamellae. The hinge beneath the umbo in the former is very angular, but quite rounded in the labiatoid form.

Occurrence. Lower Turonian, *I. labiatus* Zone; widespread throughout England.

Genus CREMNOCERAMUS

Cremnoceramus inconstans (Woods, *non* Mantell)
Plate 18, figures 3–4

Description. A medium-sized, variably shaped, subrounded to subrhombic, frequently inequivalve shell, its convexity dependent upon the persistence of the flatter early stages and the occurrence of the characteristic abrupt change in direction of the adult shell. Umbones broadly rounded and not prominent. Early ornament consists of regularly spaced, concentric, asymmetric folds (=crests) and finer growth lines; adult/ gerontic shell relatively smooth and steep.

Remarks. *C. inconstans* could be described as a polyphyletic potpourri of taxa (e.g. see Kauffman *in* Herm *et al.* 1979), but specimens can be distinguished by their relatively long hinge line and the virtual right angle this forms with the anterior margin. Adult specimens typically have a threefold ornament pattern, ending with the sharp-angled gerontic shell that causes the characteristic break in slope.

Occurrence. Lower Coniacian; widespread throughout Europe in a wide variety of facies.

Cremnoceramus websteri (Woods, *non* Mantell)
Text-figure 7.3A–B

Description. Small, convex species with conspicuous but not well-differentiated, posterior ear. The hinge line is approximately at right angles to the anterior margin. Distinguished by its low, well-spaced, concentric ribs which bear noticeable lamellae along their crests.

Occurrence. Upper Turonian, *S. plana* Zone; south-east England.

Cremnoceramus cf. *deformis* (Meek)
Plate 18, figure 8

Description. Moderately large, subglobose, irregularly subovate to subrhombic inflated shell; inequivalve; short, thick hinge-line; beaks broad and not very prominent. Ornamented by regular concentric rugae and growth lines, which show a trace of characteristic ventral flattening.

114   *Fossils of the Chalk*

Remarks. *C. deformis* (Meek) appears to be intermediate between *C. inconstans* (Woods) and *C. schloenbachii* (Böhm), and may indicate a gradualistic change. It has also been suggested that the supposedly successive species *rotundatus–erectus–deformis–schloenbachii* were contemporaneous ecomorphs occupying different positions on the continental slope.
Occurrence. Coniacian, *M. cortestudinarium* Zone; southern England, particularly Sussex.

Genus SPHENOCERAMUS

Sphenoceramus patootensis (de Loriol)
Plate 19, figure 8

Description. Very inequilateral, moderately inflated, oblique shell divided into a V-shaped, inflated anterior part and a compressed, sharply distinct, postero-dorsal wing. Much higher than long; strongly pointed anterior beaks; long, straight anterior margin. Shallow sulcus behind beaks extending to postero-ventral margin. Ornament of small, closely spaced ribs and large, more widely and irregularly spaced rugae. Typically the ribs split the growth lines.
Remarks. Distinguished from many other *Sphenoceramus* species by its complete lack of radial ornament. However, the wide range of variation that occurs has led to it being described under a number of other names.
Occurrence. Campanian, *G. quadrata* Zone; common in northern Europe and North America from Middle Santonian to Lower Campanian. Frequently found as internal moulds at localities in Yorkshire.

Sphenoceramus pachtii (Archanguelsky)
Plate 19, figure 1

Description. Relatively broad, small, thin-shelled inoceramid with rounded anterior margin. Ornament consists of two orders of concentric ribbing; wide, stepped rugae and much finer ribs; a sulcus and ridge occur postero-dorsally; radial ribbing mainly on central part of shell.
Remarks. Distinguished by its wider shell, stepped ornament and rounded anterior margin. A closely related species, *S. cardissoides* (Goldfuss) (Santonian–Campanian) is differentiated by its very narrow elongate shell, with a steep, straight, anterior margin, prominent umbo and a small central area of reticulate ornament. Seitz (1965) utilised ratios of rib width and length, together with rib-hinge angles, to separate forms within the *pachtii/cardissoides* group.
Occurrence. Santonian; only common in the Lower Santonian; Yorkshire, Essex and possibly Kent.

Sphenoceramus pinniformis (Willett)
Plate 19, figures 3–4

Description. Medium to very large with shell much higher than long; moderately convex; large posterior wing. Right valve ornamented by distinctive strong, rounded, radial ribs, which may have a median furrow and also bear much finer parallel ribs. Several strong, widely spaced, concentric folds give a tuberculate effect to the ribs. Left valve has a regular series of evenly spaced, concentric ribs and widely spaced rugae that are much stronger than the radial ribs. The ribs are restricted to the central part of the shell and contribute to the more reticulate-tuberculate appearance of this valve.
Occurrence. Santonian, *M. coranguinum* Zone, to Campanian, *G. quadrata* Zone; Yorkshire and south-east England.

Sphenoceramus steenstrupi (de Loriol)
Plate 19, figure 9

Description. Very inequilateral, oblique shell much higher than long; dorsal part moderately convex, ventral part only slightly convex. Small, almost terminal umbones; antero-dorsal area flattened, nearly smooth. Hinge line at an acute angle to anterior margin. Ornament consists of numerous concentric ribs, but at a short distance from the umbo these are cut by strong radial furrows, which produce the distinctive rows of tubercles that characterise this species.
Remarks. Has a more regular reticulate pattern than *S. cardissoides.*
Occurrence. Lower Campanian, *G. quadrata* Zone; Yorkshire, Sussex and Wiltshire.

Genus VOLVICERAMUS

Volviceramus involutus (J. de C. Sowerby)
Plate 19, figure 2

Description. Medium to large, highly inequivalve, inequilateral shell with curved beaks directed anteriorly. Right valve slightly convex to nearly flat, length greater than height, margins rounded. Extensive hinge line with numerous rectangular to oblong ligament pits. Ornamented by strong asymmetrical concentric folds that are separated by broad, concave interspaces; radial striae may occur. Left valve much larger than right valve; very inflated and spiral; its prosogyrate umbo curves inwards. Shell relatively smooth with the growth stages providing the only ornament.
Remarks. Involute inoceramid distinguished by its very inequivalve and inequilateral shells. *V. koeneni* Müller (Coniacian), a very rare coiled form

PLATE 19

in which both valves are spiral and coil inwards, is less inequivalve than *V. involutus*. *V. koeneni* is the key inoceramid of the lower Middle Coniacian and rarely occurs at other horizons.

Occurrence. Coniacian–Santonian; throughout southern England and East Anglia; there are floods of *V.* cf. *involutus* at three levels within the Coniacian of Sussex.

Genus CLADOCERAMUS

Cladoceramus digitatus (J. Sowerby)
Plate 19, figure 7

Description. Large, radially corrugate inoceramid with sub-equivalve, ovate shell; much higher than long. Hinge line forms obtuse angle with anterior margin. Ornament consists of concentric and radial ribs; the early growth stages only have concentric ribs, but broad V- and U-shaped divaricating folds develop on later part of shell. These folds/ribs diverge from a median line and are separated by broad interspaces; tubercles, or knolls, may develop at the intersections.

Remarks. Two species, *C. digitatus* and *C. undulatoplicatus* (Roemer), have been distinguished on differences in ribbing, with the former being supposedly much stronger and with fewer ribs on the posterior part than on the anterior part of the shell. Clear differentiation is, however, impossible.

EXPLANATION OF PLATE 19

Fig. 1. *Sphenoceramus pachtii* (Archanguelsky). Santonian, locality unknown; left valve; *c.* ×0·75.

Fig. 2. *Volviceramus involutus* (J. de C. Sowerby). Tertiary flint gravel, presumably derived from Coniacian or Santonian, Charing, Kent; internal mould; *c.* ×0·4.

Fig. 3. *Sphenoceramus pinniformis* (Willett). Upper Santonian, *U. socialis* Zone, Beverley Barracks Pit, East Yorkshire; left valve; ×0·2.

Fig. 4. *Sphenoceramus pinniformis* (Willett). Upper Campanian, *G. quadrata* Zone, Brighton, Sussex; right valve; ×0·2.

Fig. 5. *Inoceramus lamarcki* Parkinson. Middle Turonian, *T. lata* Zone, Great Chesterford, Essex; ×0·5.

Fig. 6. ?*Platyceramus rhomboides* Seitz. Coniacian–Santonian, *M. coranguinum* Zone, Grays, Essex; left valve (note ornament on early part of shell); ×0·15.

Fig. 7. *Cladoceramus digitatus* (J. Sowerby). Lower Santonian, Arreton Down, Isle of Wight; left valve; ×0·2.

Fig. 8. *Sphenoceramus patootensis* (de Loriol). Upper Campanian, *G. quadrata* Zone, East Leys, Yorkshire; right valve; ×0·25.

Fig. 9. *Sphenoceramus steenstrupi* (de Loriol). Upper Campanian, *G. quadrata* Zone, ?Brighton, Sussex; right valve; ×0·2.

Occurrence. Santonian; Kent, Sussex and Haldon, Devon; especially abundant at certain levels in the *M. cortestudinarium* Zone of Sussex.

Genus CORDICERAMUS

Cordiceramus cordiformis (J. de C. Sowerby)
Plate 18, figure 9

Description. Small- to medium-sized, rounded, equivalve, highly inflated shell, with a posterior sulcus. Ornament consists of broad, rounded concentric folds, which bend upwards wherever they cross the two radial sulci; these folds are less distinct on the anterior and posterior margins. There are numerous closely spaced growth lines.
Occurrence. Coniacian, *M. cortestudinarium* Zone, to Campanian, *O. pilula* Zone; widespread: Yorkshire to Devon, Hampshire, Kent and Sussex.

Genus PLATYCERAMUS

Remarks. *Platyceramus* is relatively thin-shelled and grows to a considerable size. Consequently, in chalk sediments the shells are easily deformed. It is thought to be related to *Cladoceramus* because several of its species have a tendency to develop a radially diverging ornament in their later stages. *Sphenoceramus* can be distinguished in that its radially diverging folds form the only well-developed ornament of its older, larger valves.

?Platyceramus rhomboides Seitz
Plate 19, figure 6

Description. Large to very large, flattish, equivalved shell; early growth stages with concentric folds, later stages inclined to be smooth.
Remarks. The form of the type species, *P. mantelli*, is in doubt and does not appear to conform to the present interpretation of the genus. Consequently this species is questionably assigned to *Platyceramus*. *Inoceramus cycloides* Wegner (Santonian and Campanian) differs in that the posterior part of its concentric folds become more angled towards the posterior ventral margin.
Occurrence. Upper Coniacian–Lower Santonian, *M. coranguinum* Zone; Essex and Sussex.

Superfamily AVICULOPECTINOIDEA
Family OXYTOMIDAE

Genus HYPOXYTOMA

Hypoxytoma tenuicostata (Roemer)
Plate 22, figure 12

Description. Small, obliquely oval shell, longer than high. Left valve moderately convex, with evenly convex margins but postero-dorsal concave. Ears large, anterior ear not clearly delimited, posterior ear long and its dorsal part wing-like and pointed. Ornament of closely spaced, subequal ribs separated by flat interspaces. Right valve much smaller, flattened but slightly convex in centre; smooth with concentric growth lines; anterior ear has a deep sinus but this is less deep than in *Oxytoma*.

Remarks. Distinguished from *Oxytoma* by the smaller size of its right valve sinus and by the left valve ornament, which consists of only two orders of ribs. Its finer and more numerous ribs and smaller size serve to distinguish this species from the Lower Cretaceous *O. pectinata* J. de C. Sowerby. *O. dubia* (Etheridge) from the Middle Cenomanian is inadequately known.

Occurrence. Campanian, *G. quadrata* Zone; Wiltshire.

Superfamily PECTINOIDEA
Family PECTINIDAE
Subfamily CHLAMYDINAE

Genus CHLAMYS

Chlamys? elongata (Lamarck)
Plate 20, figure 9; Text-figure 7.4D

Description. Ornamented by tripartite, often divided ribs that are crossed by concentric striae; these ribs are covered with small scales. The valves are slightly elongated, prosocline and flattened; the auricles are large and very unequal.

Remarks. Differs from other species in having a rather irregular rib distribution and a very high number of small scales on all ribs. Ribs vary with age and their position on the valve.

Occurrence. Albian–Cenomanian; widespread throughout southern England.

Chlamys? subacuta (Lamarck)
Plate 21, figure 6; Text-figure 7.4C

Description. Valves rather flattened and covered with 19–28 well-developed, undivided ribs and numerous concentric growth lines. Spines or scales occur at the rib tops, especially on the left valve. Marginal striae

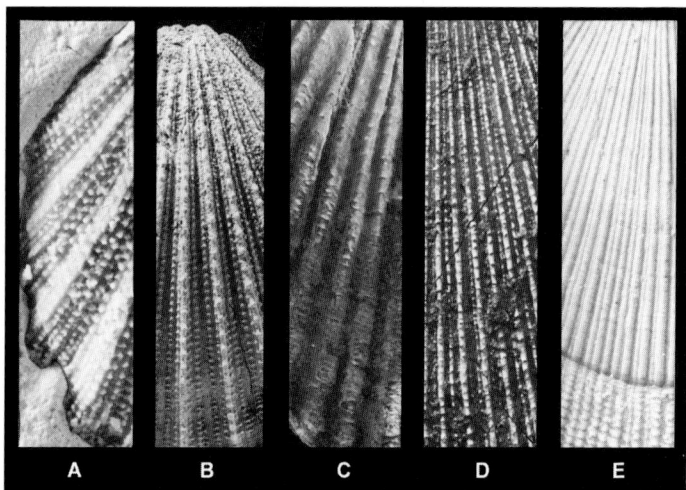

TEXT-FIG. 7.4. A, *Merklinia* sp. cf. *variabilis* (von Hagenow); ×3·5. B, *Merklinia aspera* (Lamarck); *c.* ×3. C, *Chlamys*? *subacuta* (Lamarck); *c.* ×4. D, *Chlamys*? *elongata* (Lamarck); *c.* ×1. E, *Mimachlamys cretosa* (Defrance), left valve; *c.* ×1·5.

occur on edge of shell. Intercostal intervals broader than the ribs. Auricles well developed; growth lines on the anterior auricle of right valve are parallel to the byssal sinus; on other auricles the radial ornament is most pronounced.

Remarks. This species is distinguished by its relatively low number of undivided, sharper, spine/scale-bearing ribs. British specimens are, on average, smaller than those from elsewhere; they are often preserved only as steinkerns, when fewer ribs are present, side ribs are lacking, spines and scales are lost and the areas appear smooth.

Occurrence. Cenomanian; Devon and Isle of Wight.

Genus MIMACHLAMYS

Mimachlamys cretosa (Defrance)
Plate 20, figures 7–8; Text-figures 7.4E, 7.5A

Description. Medium to large species with flattened ribs; its valves covered with a very variable number of ribs, which may bear small scales (often little more than small ridges) at their intersection with the concentric growth lines. New ribs formed by intercalation. The auricles are well developed; the anterior auricle is large with a broad deep sinus.

Remarks. Ornament varies from coarse- to fine-ribbed depending upon the nature of the sediment (Dhondt 1973). A subspecies, *M. cretosa denticulata* (von Hagenow), occurs in the Maastrichtian of Europe and is identified by its 'chlamys' shape and very spiny ornament. *M. robinaldina* (Barremian–Cenomanian) is distinguished from *M. cretosa* in being more convex, bearing spinules instead of scales on its fewer ribs, and covered by striae.

Occurrence. Turonian, S. *plana* Zone, to Campanian, *O. lunata* Zone; occurs at many localities in the counties around London and particularly in Norfolk and Suffolk.

Mimachlamys mantelliana (d'Orbigny)
Plate 20, figures 10–11; Text-figure 7.5D–E

Description. Medium-sized, very flattened species covered with 11–30 narrow, flattened, radial ribs and with a few distinctive raised, concentric growth lines.

Remarks. Distinguished from other species by being much flatter, particularly the right valve, and in having well-developed elevations formed by the few growth lines. The concentric structures on the auriculae are typically more prominent than the radial elements. *M. mantelliana* can be separated from the closely related *M. cretosa* by the lack of scales on its ribs, while the lack of elevated growth lines distinguishes the flatter examples of *M. cretosa*.

Occurrence. Santonian–Maastrichtian, *O. lunata* Zone; widespread in southern England and East Anglia; Dhondt (1973*a*, p. 98) stated that all the specimens she had studied had been found in a very fine white chalk.

Mimachlamys henrici Dhondt
Plate 20, figure 12; Text-figure 7.5C

Description. Medium-sized species distingushed from other representatives of the genus by its relatively flattened shell covered with numerous smooth ribs that are separated by narrow grooves.

Remarks. The number of ribs is very variable. It is possible that *M. henrici* may only be a subspecies of *M. robinaldina*.

Occurrence. Cenomanian; mainly Devon and Dorset, but also found in Kent, Sussex and the Isle of Wight.

Mimachlamys fissicostata (R. Etheridge)
Plate 20, figure 16; Text-figure 7.5B

PLATE 20

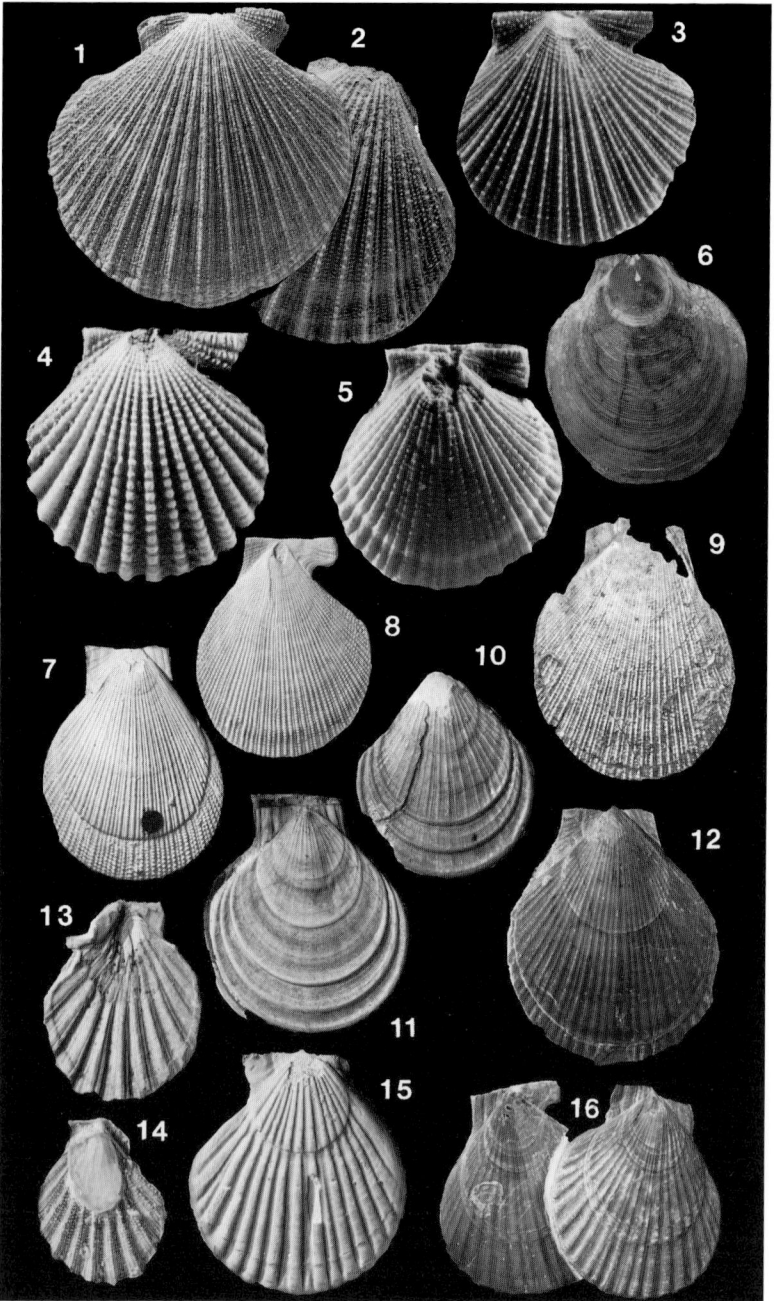

Description. Medium-sized shell covered with 11–23 broad ribs, which divide occasionally after a strongly developed growth ridge.

Remarks. Distinguished from other *Mimachlamys* species by its smooth ribs; *M. henrici* has far more ribs and virtually no intercostal intervals.

Occurrence. Lower Cenomanian; southern England and East Anglia; only known from very fine chalky sediments.

Genus CAMPTONECTES

Camptonectes (*Mclearnia*) *dubrisiensis* (H. Woods)
Plate 21, figure 1

Description. Medium- to large-sized, almost orbicular shell covered by light concentric ornament; relatively large and long anterior auricles.

Remarks. Distinguished from other species in that the greater part of its disc is smooth and only near the margins of the shell is the concentric ornament pronounced. Size serves to separate it from the much larger and

---

EXPLANATION OF PLATE 20

Figs 1–2. *Merklinia aspera* (Lamarck). Cenomanian, near Shaftesbury, Wiltshire. 1, right valve; *c.* ×0·3. 2, right valve showing tubercles on ribs; ×0·75.

Fig. 3. *Microchlamys sarumensis* (Woods). Campanian, *B. mucronata* Zone, Catton, Norwich, Norfolk; right valve; ×3.

Fig. 4. *Microchlamys arlesiensis* (Woods). Cenomanian, Dover, Kent; right valve showing distinctive scales on ribs; ×3.

Fig. 5. *Microchlamys campaniensis* (d'Orbigny). Coniacian, *M. cortestudinarium* Zone, Dover, Kent; right valve; ×4.

Fig. 6. *Syncyclonema nilssoni* (Goldfuss). Campanian, *B. mucronata* Zone, Norwich, Norfolk; left valve; ×0·5.

Figs 7–8. *Mimachlamys cretosa* (Defrance). Salisbury, Wiltshire. 7, Upper Coniacian–Lower Santonian, *M. coranguinum* Zone, Porton; left valve; ×0·75. 8, Lower Campanian, *G. quadrata* Zone, East Harnham; right valve; ×0·75.

Fig. 9. *Chlamys*? *elongata* (Lamarck). Lower Cenomanian, Folkestone, Kent; left valve; ×0·5.

Figs 10–11. *Mimachlamys mantelliana* (d'Orbigny). Upper Campanian, *B. mucronata* Zone, Norwich, Norfolk. 10, right valve; ×0·75. 11, left valve; *c.* ×1.

Fig. 12. *Mimachlamys henrici* Dhondt. Lower Cenomanian, *S. varians* Zone, Bluebell Hill, Burham, Kent; left valve; ×0·75.

Figs 13–14. *Merklinia* sp. cf. *variabilis* (von Hagenow). Upper Chalk, locality unknown. 13, interior of right valve showing byssal sinus; ×0·75. 14, right valve showing ornament; ×0·75.

Fig. 15. *Microchlamys brittanicus* (Woods). Upper Coniacian–Lower Santonian, *M. coranguinum* Zone, Croydon, Surrey; holotype, left valve; ×1·2.

Fig. 16. *Mimachlamys fissicostata* (Etheridge). Cenomanian, Dover, Kent; right and left valves; ×0·5.

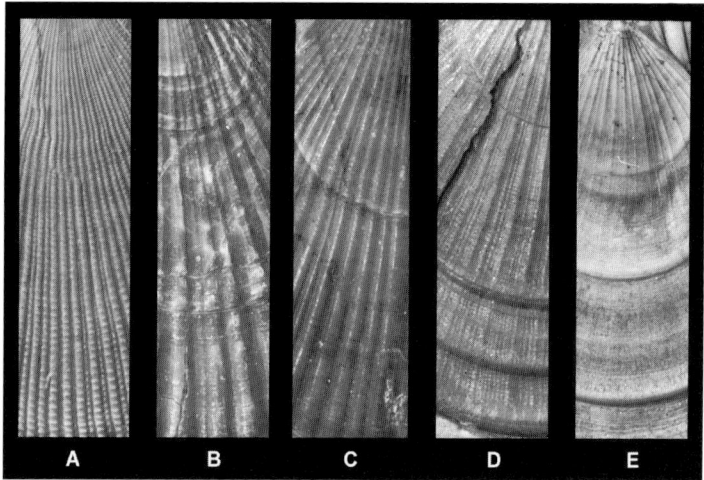

TEXT-FIG. 7.5. A, *Mimachlamys cretosa* (Defrance), right valve; ×1·5. B, *Mimachlamys fissicostata* (Etheridge); *c.* ×0·5. C, *Mimachlamys henrici* Dhondt, left valve; ×1·5. D–E, *Mimachlamys mantelliana* (d'Orbigny). D, ×1·5. E, *c.* ×2.

older Early Cretaceous *C. (M.) cinctus* (J. Sowerby); the much smaller *C. virgatus* (Nilsson) (Cenomanian–Campanian), which has a coarser microstructure of diverging striae; and the much smaller Cenomanian ?*C. milleri* (J. de C. Sowerby) from Kent in which the *Camptonectes* sculpture is less pronounced and its diverging lines are almost straight and rarely divide.

Occurrence. Lower and Middle Cenomanian, *H. subglobosus* Zone; south-east England and Cambridgeshire.

Genus MERKLINIA

Merklinia sp. cf. *variabilis* (von Hagenow)
Plate 20, figures 13–14; Text-figure 7.4A

Description. Small- to medium-sized species. Both valves with 9–12, commonly 11, ribs; these are subdivided into 3–7 almost equal riblets; their intercostal intervals are very narrow; the principal and side-riblets bear well-developed spines (long and small respectively). Valves acline to slightly prosocline with relatively large unequal auricles. Anterior auricle of right valve has a deep byssal sinus and 3–5 radial riblets parallel to hinge margin; posterior auricle of right valve smaller, triangular and almost right-angled. Anterior auricle of left valve is large and also has radial riblets.

Remarks. Complete specimens easily recognised by their slightly prosocline disc shape, the number of ribs (13) and by the extreme subdivision of the ribs into fine costellae. Distinguished from other *Merklinia* species in having longer spines on the principal ribs and a height greater than its width. *M. aspera* (Lamarck) (Albian–Cenomanian) has many more ribs (Pl. 20, figs 1–2; Text-fig. 7.4B).

Occurrence. Turonian–Maastrichtian; southern England from Devon to Kent, also East Anglia.

Genus MICROCHLAMYS

Remarks. Small symmetrical shells with radial ornament, which is often reticulate or cancellate; small, almost equal auricles and byssal notch partly reduced; scalloped, indented ventral margin. Distinguished from *Merklinia* by the nature of its rib ornament, the smaller and relatively equal auricles ('ears') and the shape of its ventral margin.

Microchlamys arlesiensis (Woods)
Plate 20, figure 4; Text-figure 7.6C

Description. Very small species with moderately convex valves of an orbicular acline shape. Left valve ornamented by elevated straight, narrow ribs. Right valve covered with straight, rounded ribs, which are mostly wider than the intercostal intervals. Thin concentric ornament produces scales where it crosses the ribs.

Remarks. Distinguished by its undivided ribs, 14–18 in number.

Occurrence. Cenomanian; south-east England; relatively rare.

Microchlamys sarumensis (Woods)
Plate 20, figure 3; Text-figure 7.6B

Description. Small species with numerous (45–65) spiny, scaled ribs and relatively large auricles.

Remarks. Distinguished from other species by the large number of dividing ribs and the concentric spiny scales on these ribs. *M. arlesiensis* (Cenomanian) is larger and its ribs do not divide; *M. campaniensis* (Coniacian–Maastrichtian) has fewer ribs and lacks scaly ornament.

Occurrence. Santonian–Upper Campanian, *B. mucronata* Zone; throughout south-east England; also known from Wiltshire and Norfolk; comparatively rare.

Microchlamys campaniensis (d'Orbigny)
Plate 20, figure 5; Text-figure 7.6A

Description. Small, strongly convex shells with a distinctive 'trellis' sculpture, which is best seen on juveniles or on the umbo. Trellis formed by well-developed and divided radial ribs and the slightly elevated concentric growth striae that cross them; various combinations and strengths of this pattern occur at most localities.

Remarks. Distinguished from all other species by its trellis pattern of ornament. *M. arlesiensis* differs in having undivided and fewer ribs (14–18), and characteristic scaly structures on its ribs; *M. pulchellus* has a more convex shell.

Occurrence. Coniacian–Lower Maastrichtian, *O. lunata* Zone; relatively rare in Kent and Norfolk.

Microchlamys brittanicus (Woods)
Plate 20, figure 15; Text-figure 7.6D–E

Description. Orbicular species ornamented by broad, simple, rounded, rod-like ribs separated by intercostal areas that are the same width or wider than the ribs. Very fine concentric lines present throughout shell, sometimes become stronger and form bars. Ribs infrequently developed intercostally. Concentric ridges present, but more characteristically, the ribs merge to form a smooth, thickened rim to the shell. Anterior auricle of left valve ornamented by radial ribs, and its growth lines are perpendicular to the hinge-line. Anterior auricle of right valve has a small byssal sinus and its growth lines follow this. Both valves have identical ornament, but the primary ribs of the right valve tend to divide near the margins.

Remarks. Distinguished by its broad, simple, rounded ribs that commonly do not divide. Other comparable species appear to be restricted to the Upper Cretaceous of Europe. In *M. septemplicatus* (Nilsson) (Campanian–Maastrichtian), which also shows a thick marginal rim, there are only a few very wide ribs.

Occurrence. Santonian, *M. coranguinum* Zone; Surrey, Kent, Hampshire and Wiltshire; scarce.

Genus SYNCYCLONEMA

Syncyclonema nilssoni (Goldfuss)
Plate 20, figure 6

Description. Large, smooth, flattened shell with concave apical margins and curved hinge margins; subequal auricles. Ornamented by concentric growth lines. Anterior auricle of right valve large and wing-like with deep byssal sinus.

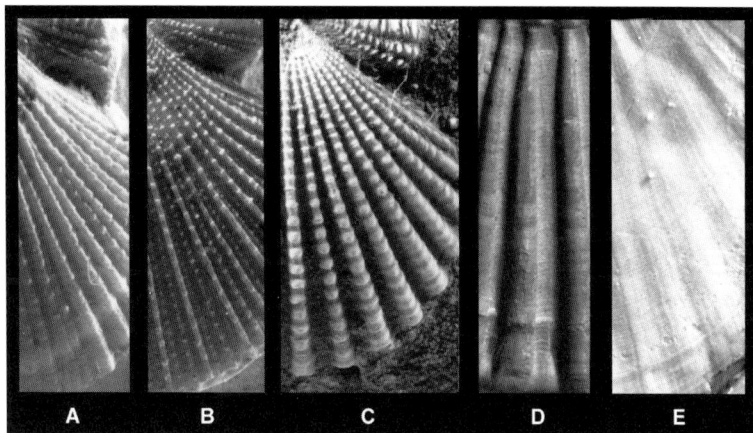

TEXT-FIG. 7.6. A, *Microchlamys campaniensis* (d'Orbigny), right valve; ×7. B, *Microchlamys sarumensis* (Woods), right valve; ×5. C, *Microchlamys arlesiensis* (Woods), right valve; ×4. D–E, *Microchlamys brittanicus* (Woods). D, left valve; ×2·5. E, right valve; *c*. ×2.

Remarks. Distinguished by its larger size, wider apical angle of 93–123 degrees (average 111 degrees), and curved apical margins. The chlamyid shape and byssal sinus serve to separate it from species of *Entolium*. In most *Syncyclonema* species the radial *Camptonectes*-like vermiculations are not visible without magnification of at least ×25.
Occurrence. Campanian, *G. quadrata* Zone; Hampshire and Norfolk.

Subfamily NEITHEINAE

Genus NEITHEA

Neithea (*Neithea*) *quinquecostata* (J. Sowerby)
Plate 21, figure 10

Description. Medium-sized to large shell with six principal ribs and 3–6 (generally four) intercalary ribs; frequently shows numerous concentric growth lines. Its equal auricles are small and do not protrude far past the umbo. There are numerous equal radial riblets (filae), which also occur on the areas. Right valve moderately convex with slightly incurved umbo; left valve flattened and never concave; number and strength of intercalary ribs variable.
Remarks. This is the commonest pectinid in the British Cretaceous. It can be distinguished from other Chalk species by its relatively finer ribbing and its small auricles, which are set well back from the umbo.

Occurrence. Barremian–Campanian; widespread throughout southern England.

Neithea (*Neithea*) *sexcostata* (S. Woodward)
Plate 21, figures 7–9

Description. Much smaller than *N. quinquecostata* with a very convex right valve and flattened left valve. Six very salient principal ribs on both valves with deep intercostal intervals occupied by irregular, often poorly delineated ribs; all of the ribs are striated horizontally by riblets. The inwardly bent areas and the small equal auricles are similarly ornamented.
Remarks. Distinguished by its smaller size and the combination of salient primary ribs with more than four irregular, poorly delimited intercalaries.
Occurrence. Cenomanian–Campanian; southern England including Norfolk, Wiltshire, Dorset and Devon; frequent in the fine White Chalk, commonly occurring when other pectinids are rare.

Neithea (*Neithea*) *regularis* (Schlotheim)
not figured

Description. *Neithea s.s.* with six well-developed principal ribs but distinguished from *N. quinquecostata* by having only three intercalary ribs (although four may occur in the outer intervals). The areas and subequal auricles are covered with numerous weak radial filae; there are fewer riblets on the lower parts. Left valve always flattened.
Remarks. The unequal auricles and the obvious concentric striations and riblets on the areas serve to distinguish *N. regularis* from other species.
Occurrence. Turonian; in Europe *N. regularis* is not found below the uppermost Cenomanian; its only British occurrence is in the Turonian of County Antrim, Northern Ireland.

Neithea (?*Neithea*) *gibbosa* (Pulteney)
Plate 21, figure 11

Description. Medium-sized to large species with large, unequal auricles; six well-developed principal ribs and generally only three intercalaries. Shell shape and ornament similar to *N. regularis,* but right valve less convex, umbo wider and rib intervals flatter.
Remarks. Distinguished in general from *N. quinquecostata* by its coarser appearance and much larger size, while all the ribs seem to be of the same type and its large auricles are closer to the hinge line. Exceptionally, specimens found in the Upper Greensand at Blackdown, Devon, have a greater number of intercalary ribs than normal.
Occurrence. Albian–Cenomanian; Devon, Dorset and Wiltshire.

Family SPONDYLIDAE

Genus SPONDYLUS

Remarks. Recent species of *Spondylus* are restricted to tropical and subtropical seas and it is assumed that their ecological needs during the Cretaceous were the same, even though fossil species occur in higher latitudes than today. A study of the type species, *Spondylus gaedorupus* Linnaeus, has established that shell shape and convexity, as well as features of ornament such as the number and strength of both ribs and spines, can be extremely variable at all growth stages (Dhondt and Dieni 1990). In common with other sessile bivalves, *Spondylus* varies in shape owing to adjustment to the living space available on substrates, and may not develop characteristic supplementary structures. Consequently the method of attachment of the right valve found in fossil species ranges from complete cementation to fixation by means of concentric lamellae or buttresses of varying size according to the extent of substrate available. Dhondt and Dieni (1990) considered *Dianchora* to be a *Spondylus* that has lost its inner shell layer during fossilization.

Spondylus spinosus (J. Sowerby)
Plate 22, figure 1

Description. Ovate, slightly inaequilateral shell, rounded ventrally, and with prominent umbonal region. Valves may be equally convex, but normally the right valve is flattened owing to its attachment, which is achieved by means of irregular concentric frills. Right valve larger than left valve; strong radial ribs with a series of strong, long, tapering spines that are grooved on their upper surface. Left valve very convex, ornamented by strong radial ribs that occasionally develop spines; the juvenile shell may also have tuberculate spines on the rib crests. The spines on the right valve do not occur to a pattern, although the two most lateral series are usually the best developed and are evenly spaced around the commissure.

Remarks. Carter (1972) discussed the ontogeny of *S. spinosus* and interpreted its possible functional morphology and palaeoecology. Two extreme forms have been recognised: (1) *aequalis*-type, in which intermediate ribs are developed only near the margin of the left valve and, together with the well-marked and closely packed growth lines, give a frilled appearance to the shell. The ribs on the right valve are not divided, and spines occur on both valves (e.g. '*S. brightonensis*' from the *M. coranguinum* Zone). (2) *duplicatus*-type, in which the left valve has many intermediate ribs and the ribs on the right valve have a narrow median furrow.

PLATE 21

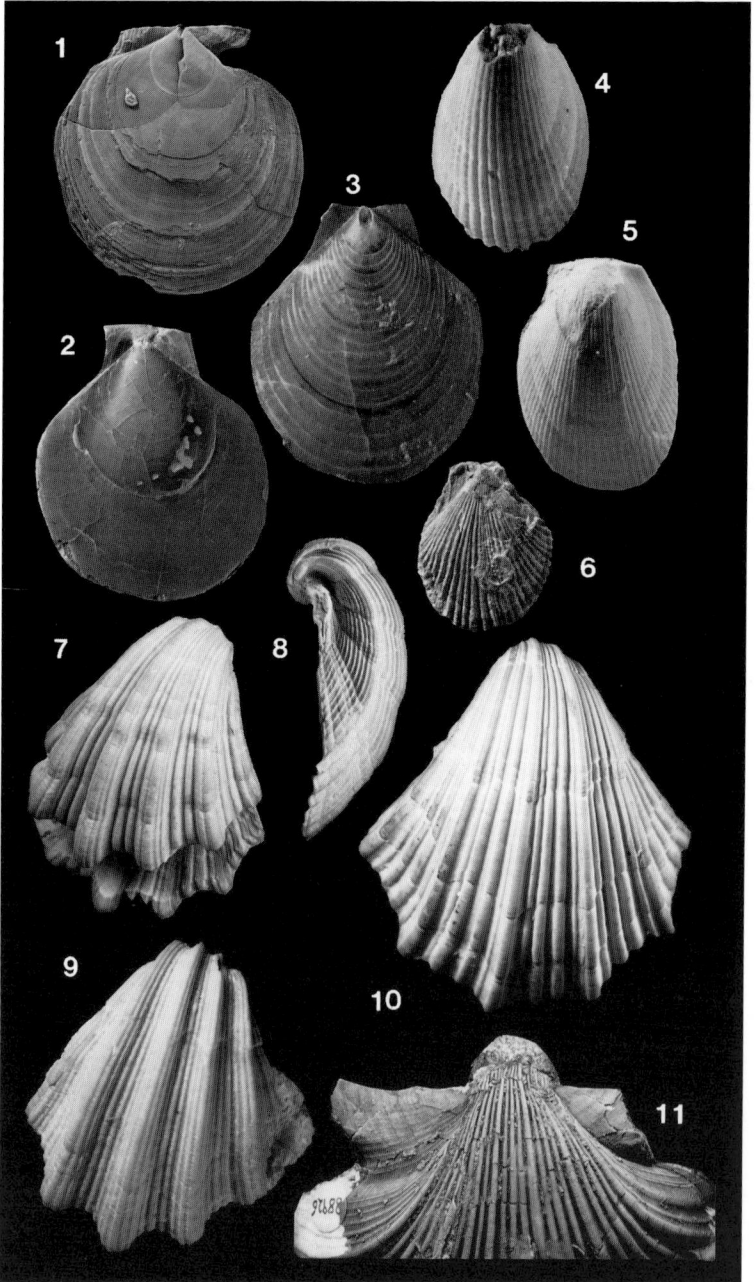

Occurrence. Cenomanian, *S. gracile* Zone, to Late Campanian, *B. mucronata* Zone; common at most horizons and localities from the Dorset coast to Norfolk.

Spondylus striatus (J. Sowerby)
Plate 22, figure 6

Description. Large, obliquely ovate shell. Right valve usually attached over most of its surface, free part ornamented by radial ribs and shallow grooves with fine concentric lamellae. Left valve moderately convex with prominent umbo and small, smooth ears; ornamented by numerous equal, smooth, flattened radial ribs separated by deep grooves and concentric growth lines.

Remarks. *S. striatus* is larger and more obliquely-shaped than *S. latus*, and its ribs are less regular.

Occurrence. Aptian–Cenomanian; Devon, Dorset and Wiltshire.

Spondylus fimbriatus Goldfuss
Plate 22, figure 3

Description. Regular oval, slightly oblique, convex shells. Right valve can be the most inflated, but is generally flatter because it was attached to a substratum; its attached area has concentric ornament, but the free shell

---

EXPLANATION OF PLATE 21

Fig. 1. *Camptonectes* (*Mclearnia*) *dubrisiensis* (Woods). Cenomanian, Bluebell Hill, Burham, Kent; right valve; ×0·5.

Fig. 2. *Entolium membranaceum* (Nilsson). Upper Cenomanian, Bluebell Hill, Burham, Kent; left valve; *c.* ×1.

Fig. 3. *Entolium orbiculare* (J. Sowerby). Middle Cenomanian, *A. rhotomagense* Zone, between Folkestone and Dover, Kent; left valve; *c.* ×1.

Fig. 4. *Limatula wintonensis* (Woods). Upper Santonian, *U. socialis* Zone, Ringwould, near Dover, Kent; right valve; ×2·5.

Fig. 5. *Limatula decussata* (Goldfuss). Lower Campanian, *G. quadrata* Zone, East Harnham, Salisbury, Wiltshire; right valve; ×2·5.

Fig. 6. *Chlamys*? *subacuta* (Lamarck). Upper Albian, Weymouth, Dorset; left valve; ×0·5.

Figs 7–9. *Neithea* (*Neithea*) *sexcostata* (Woodward). Campanian. 7–8, *G. quadrata* Zone, Salisbury, Wiltshire; both right valves; ×1. 9, Norwich, Norfolk; right valve; ×2.

Fig. 10. *Neithea* (*Neithea*) *quinquecostata* (J. Sowerby). Campanian, *G. quadrata* Zone, near Salisbury, Wiltshire; right valve; ×1.

Fig. 11. *Neithea* (?*Neithea*) *gibbosa* (Pulteney). Lower Cenomanian, *N. carcitanense* Zone; Warminster, Wiltshire; right valve; ×1.

is ornamented by nearly equal radial ribs separated by narrow grooves; short, irregularly placed spines occur on the ribs. Left valve ornamented by numerous, slightly irregular, radial ribs which give it a 'wavy' appearance; these irregularly bear numerous spines of varying sizes and angles.

Remarks. Previously recorded as *S. dutempleanus* d'Orbigny, but owing to the considerable variability that occurs in both the ribs and spines ornamenting spondylids, this species is now treated as a junior synonym of *S. fimbriatus* Goldfuss (Dhondt and Dieni 1990).

Occurrence. Cenomanian–Campanian; localities at all horizons in the counties around London, and from Devon and Dorset to Norfolk.

Spondylus latus (J. Sowerby)
Plate 22, figures 2, 4–5

Description. Distinguished from other *Spondylus* species by its numerous smooth ribs, smooth ears, and virtual lack of spines. It is also considered

---

EXPLANATION OF PLATE 22

Fig. 1. *Spondylus spinosus* (J. Sowerby). Chalk, South Downs; right valve showing irregular arrangement of spines on radial ribs; ×0·75.

Figs 2, 4. *Spondylus* cf. *latus* (J. Sowerby). Campanian, *O. pilula* Zone, Peacehaven, Sussex; ×1·2.

Fig. 3. *Spondylus fimbriatus* Goldfuss. Lower Campanian, *G. quadrata* Zone, East Harnham, Salisbury, Wiltshire; left valve; ×1·25.

Fig. 5. *Spondylus latus* (J. Sowerby). Upper Campanian, *B. mucronata* Zone, Earlham Lime Works, Norwich, Norfolk; right valve (note buttresses supporting this valve); *c.* ×1.

Fig. 6. *Spondylus striatus* (J. Sowerby). Lower Cenomanian, *N. carcitanense* Zone, Warminster, Wiltshire; left valve; ×0·5.

Figs 7–8. *Spondylus serratus* Woods. 7, Senonian, locality unknown; left valve (note fine scaly ribs); ×1. 8, Campanian, Norwich, Norfolk; right valve (note lamellose concentric ribs); *c.* ×1.

Fig. 9. *Pseudoptera caerulescens* (Nilsson). Upper Santonian, *U. socialis* Zone, Palm Bay, Margate, Kent; left valve; *c.* ×1·5.

Fig. 10. *Plicatula inflata* J. de C. Sowerby. Lower Cenomanian, Burham, Kent; right valve; ×1.

Fig. 11. *Plicatula barroisi* Peron. Upper Turonian, *S. planus* Zone, Dover, Kent; right valve; ×5.

Fig. 12. *Hypoxytoma tenuicostata* (Roemer). Lower Campanian, ?*G. quadrata* Zone, Wells, Norwich, Norfolk; left valve; ×1·2.

Figs 13–14. *Gyropleura inaequirostrata* (S. Woodward). Campanian, Norwich, Norfolk. 13, ×3. 14, view of shell encrusting an echinoderm to show shape and striations; ×1·5.

PLATE 22

to have a higher number of ribs than other species, while its left valve is often longitudinally elongate rather than suboval or circular.

Occurrence. Cenomanian–Campanian; localities from Devon and Dorset to Yorkshire; particularly along the Sussex coast in south-east England; Cenomanian forms are usually larger than later specimens.

Spondylus serratus Woods
Plate 22, figs 7–8

Description. Distinguished from other *Spondylus* species by its regular pattern of ribs, which is quite different from that seen in other species, and by its concentrically aligned fine tubercles or small spines. Distinguished from *S. latus* by its broader intercostal grooves and ribbed ears.

Remarks. A free-living species in its adult stages.

Occurrence. Santonian, *U. socialis* and *M. testudinarius* zones; Wiltshire and Kent; rare.

Family ENTOLIIDAE
Subfamily ENTOLIINAE

Genus ENTOLIUM

Entolium orbiculare (J. Sowerby)
Plate 21, figure 3; Plate 26, figure 16

Description. Very brittle, thin, almost flat shell; suborbicular in shape, and subequivalve, with relatively small, well-developed, subequal auricles; apical margins usually straight, but can be curved. Right valve is smooth and more convex than left valve, and has an obsolete byssal sinus. Left valve bears a variable number of concentric ridges separated by shallow furrows; its auricles are equal, triangular and obtusely angled, though still rounded, and project dorsally more than those of the right valve.

Remarks. Distinguished from the smooth *E. membranaceum* by the concentric ridges on its left valve, and in having more dorsally elevated auricles and a slightly narrower apical angle.

Occurrence. Berriasian–Turonian; found throughout southern England and as far north as Lincolnshire and Norfolk; one of the commonest and most widely distributed of pectinids.

Entolium membranaceum (Nilsson)
Plate 21, figure 2

Description. Completely smooth shell, both inside and out; auricles are almost exactly equal, the byssal sinus is absent but well-developed auricular crurae are present.

Remarks. Distinguished from *E. orbiculare* by the complete lack of any macrosculpture and its dorsally projecting auricles. The well-developed auricular crurae indicate that the species belongs to *Entolium*. Its equal auricles and straight apical margins distinguish this species from the similarly smooth *Syncyclonema nilssoni* (Goldfuss) (Campanian).

Occurrence. Turonian–Maastrichtian; unexpectedly rare in the English Chalk; only known from Kent, Lincolnshire and Northern Ireland.

Superfamily PLICATULOIDEA
Family PLICATULIDAE

Genus PLICATULA

Plicatula barroisi Peron
Plate 22, figure 11

Description. Small, rounded, ovate shell, slightly oblique. Right valve inflated, apical part truncated by attachment; left valve flattened or slightly concave, often with a subcircular opening near umbo. Both valves ornamented by numerous strong, rounded, radial ribs that are separated by deep grooves; these ribs may bifurcate at the valve margins or at the strongly marked growth rings.

Occurrence. Cenomanian–Campanian, particularly common in the Upper Turonian *S. plana* Zone; Hampshire and south-east England.

Plicatula inflata J. de C. Sowerby
Plate 22, figure 10

Description. Oval, somewhat triangular shell, very oblique with rounded margins. Right valve moderately convex, but increasing with age; left valve flat or concave. Both valves ornamented with a few regular, radial, slightly curving ribs that bear recumbent spines that are stronger and longer near the posterior and anterior margins.

Remarks. Distinguished from the Albian *P. gurgitus* Pictet and Roux in having far fewer and more regular ribs that do not bifurcate and carry fewer spines. The presence of spines separates it from *P. barroisi*.

Occurrence. Albian–Cenomanian; south-east England.

Family DIMYIDAE

Genus ATRETA

Atreta nilssoni (Hagenow)
Plate 26, figure 11

Description. Small, semi-oval or semicircular shell in outline, but may be a little oblique. Right valve attached by nearly the whole of its surface; the interior with slightly raised radial ribs that become more numerous near the sharp, raised margin; a broad, smooth, sloping border bounded by a raised edge occurs beyond this margin; occasionally radial ribs may develop outside the border. Left valve slightly convex, ornamented by closely packed, irregular, concentric lamellae.

Remarks. This species is commonly attached to *Echinocorys, Inoceramus* and *Belemnitella*. The fixed right valve is the one normally found, while the left valve has only been collected from the *S. plana* Zone and exposures of the Upper Chalk. The monomyarian muscle scar in *Atreta* distinguishes it from the similarly attached bivalve *Dimyodon,* which is dimyarian.

Occurrence. Cenomanian–Maastrichtian; widespread throughout England.

Order LIMOIDA
Family LIMIDAE

Genus CTENOIDES

Ctenoides divaricata (Dujardin)
Plate 17, figure 5

Description. Convex, somewhat flattened shell with variable outline ranging from ovate to oblong; antero-dorsal margin slopes steeply, posterior margin more gradually. Unequal ears relatively short and high. Cardinal area narrow; hinge with moderately strong lateral teeth. Ornament of numerous small, rounded, radial ribs that diverge from one or two centres. Ribs slightly raised, often wavy and irregular, particularly near growth ridges.

Remarks. Recognised by the divergence of its ribs and its short antero-dorsal margin; *C. tecta* (Goldfuss) (Cenomanian–Maastrichtian) has much finer ribs and a long, straight, antero-dorsal margin; *C. rapa* (Cenomanian) is larger, has relatively more convex and ovate valves, and fine radial ribs present on both ears.

Occurrence. Upper Turonian, *S. plana* Zone, to Lower Campanian, *G. quadrata* Zone; south-east England including Wiltshire and Norfolk.

Ctenoides rapa (d'Orbigny)
Plate 17, figure 6

Description. Moderately convex shell with ovate outline; much higher than long; all margins evenly rounded. Umbones small, pointed, and close together. Ears moderately large and unequal, with fine radial ribs. Ornament of numcrous fine, rounded, radial ribs that diverge and have a wavy course, which is often further deflected by the growth lamellae. Ribs become narrower and sharper near both margins and may also bear small spines; crossed by fine linear concentric ridges.

Remarks. Distinguished from both *C. divaricata* and *C. tecta* by its larger size, relatively much more convex and ovate valves and the fine radial ribs present on its ears; its ribs may also bear small spines.

Occurrence. Upper Albian–Cenomanian; Devon.

Genus LIMA

Lima aspera (Mantell)
Plate 23, figure 9

Description. Ovate shell, higher than long; rounded outline but both margins straight (antero-dorsal shortest). Small pointed umbones close together. Deep, narrow, anterior area; small, triangular anterior ears; posterior ears obtusely triangular, elongate, and with radial ribs. Ornament of numerous, slightly undulating, moderately wide, flattened radial ribs that diverge from a median line and are often bent by growth lines. Ribs separated by narrow grooves; pits may occur in these on dorsal part of valve, or transverse grooves on ventral part. Apart from concentric ridges the rib surface is generally smooth, but short indentations or short spiny projections can occur irregularly on the inner edge of each rib.

Remarks. Distinguished from the Cenomanian *L. subovalis* J. de C. Sowerby by its wider ribs and the projections on their inner side; the divergence of the ribs is also much greater than in that species.

Occurrence. Cenomanian; in Europe more typically found in the Turonian; south-east England, Cambridgeshire and Norfolk.

Genus PLAGIOSTOMA

Plagiostoma hoperi Mantell
Plate 23, figures 6–7

Description. Convex, rounded, inaequilateral shell; longer than high; antero-dorsal margin rather long, slightly convex and nearly straight. Umbones close together. Apical angle 115–117 degrees. Small unequal ears with growth lines; posterior ear larger and long. Anterior area large and deep, with sharp border; radial grooves present. Shell smooth; numerous linear grooves with pits near umbo as well as in anterior and posterior margins; variable distribution of this ornament over the shell. Growth rings at intervals.

Remarks. Distinguished from other species by having a relatively smooth shell surface. *P. cretaceum* is distinguished by its smaller apical angle and shorter, higher, less convex valves; the entire shell surface is ornamented by deep grooves. Typically, *P. hoperi* is longer than high, whereas *P. cretaceum* is higher than long.

Occurrence. Upper Turonian, *S. plana* Zone, to Maastrichtian; widespread throughout England.

Plagiostoma cretaceum Woods
Plate 23, figures 8, 10

Description. Oval, very inaequilateral, not very convex shell; higher than long; antero- and postero-dorsal margins nearly straight, rounded ventrally. Small umbones close together. Apical angle 90–100 degrees. Small unequal ears, posterior larger. Moderate-sized anterior area very deep, sharp edge, and numerous radial ribs. Ornamented by numerous well-defined, slightly wavy radial grooves that have distinct pits; the flattened, rounded interspaces have the appearance of 'ribs'. Fine concentric ridges and a few growth rings also generally visible.

Remarks. Distinguished from *P. hoperi* by its smaller apical angle, higher and shorter, less convex valves and by the entire shell surface being ornamented by deeper grooves.

EXPLANATION OF PLATE 23

Figs 1–2. *Limaria elongata* (J. de C. Sowerby). 1, Campanian, *G. quadrata* Zone, Brighton, Sussex; left valve; ×2. 2, Cenomanian, Bluebell Hill, Burham, Kent; left valve; ×0·75.

Figs 3–4. *Plagiostoma marrotianum* d'Orbigny. Upper Campanian, *B. mucronata* Zone, Cunnell's Pit, Norwich, Norfolk; right valve; ×0·5.

Fig. 5. *Limaria reichenbachi* Geinitz. Lower Cenomanian, *M. dixoni* Zone, Wilmington, Devon; left valve; ×1·25.

Figs 6–7. *Plagiostoma hoperi* Mantell. Lower Campanian, *B. quadrata* Zone, East Harnham, Salisbury, Wiltshire. 6, lunule, escutcheon and ear; ×1. 7, left valve; ×0·8.

Figs 8, 10. *Plagiostoma cretaceum* Woods. 8, ?Campanian, Norwich; right valve; ×1. 10, Santonian, *M. cortestudinarium* Zone, Cuxton, Kent; left valve; ×1.

Fig. 9. *Lima aspera* (Mantell). Lower Cenomanian, Margett's Pit, Burham, Kent; left valve; *c.* ×1.

Fig. 11. *Plagiostoma globosum* J. de C. Sowerby. Cenomanian, Dover, Kent; left valve; ×1.

Fig. 12. *Limea (Pseudolimea) granulata* (Nilsson). Upper Campanian, *B. mucronata* Zone, Mousehold, near Norwich, Norfolk; right valve; ×2.

Fig. 13. *Limea (Pseudolimea) composita* (J. de C. Sowerby). Lower Cenomanian, *N. carcitanense* Zone, Warminster, Wiltshire; left valve; ×2.

PLATE 23

Occurrence. Upper Turonian, *S. plana* Zone, to Upper Campanian, *B. mucronata* Zone; southern England including Norfolk.

Plagiostoma globosum J. de C. Sowerby
Plate 23, figure 11

Description. Ornament consists of regular radial and concentric reticulate pattern of numerous pits, which can simulate raised ribs. Near the ventral margin the pits are more elongate and their concentric arrangement becomes more irregular. The radial arrangement is most apparent at the anterior and posterior margins. The smooth valves are faintly marked by growth lines.

Remarks. Distinguished from all other Upper Cretaceous species found in Britain by its reticulate 'pseudoribbing' pattern. Its more inflated valves also serve to separate it from both *P. hoperi* and *P. cretaceum*.

Occurrence. Albian–Cenomanian; various localities in the counties around London and in Devon.

Plagiostoma marrotianum d'Orbigny
Plate 23, figures 3–4

Description. Distinguished from other Chalk species by its strong ribs and rather large ears. It is less rectangular in shape than species of *Limaria*. The rounded, raised and smooth ribs can flatten towards the ventral margin, and in older gerontic specimens this produces a smooth 'skirt'.

Occurrence. Upper Campanian, *G. quadrata* Zone; relatively rare in the UK.

Genus LIMARIA

Limaria elongata (J. de C. Sowerby)
Plate 23, figures 1–2

Description. Subquadrangular to oblong shell of moderate convexity, rounded anteriorly. Antero-dorsal margin nearly straight and parallel to postero-ventral margin. Umbones close together. Ears of moderate size. Ornament extremely variable, consisting of 19–20 strong ribs separated by asymmetric, smooth, shallow furrows, or deeply rounded grooves, which are of a similar size to the ribs. The ribs are far more complex than in most other Cretaceous species, although the presence of finer secondary ribs is limited to the dorsal part of the shell. The ribs may be (1) sharp and serrate, with secondary ribs on their flanks or at the foot of each side; (2) merely smoothly rounded; (3) spinose as in the variety *Limaria elongata*

var. *echinata*, in which there is a row of short, rounded, stumpy spines on the ridge of each rib and another row of slightly smaller spines in the furrows. Fine, concentric growth lines cross both ribs and grooves.
Occurrence. Cenomanian–Lower Turonian; Lower Campanian, ?*G. quadrata* Zone; southern England from Devon to Kent.

Limaria reichenbachi Geinitz
Plate 23, figure 5

Description. Convex, oblong, oblique shell, rounded ventrally; antero-dorsal margin long, nearly straight, and almost parallel to postero-ventral margin. Ears rather small with anterior ear a little larger than the posterior. Ornament of 7–10 very straight, rounded, radial ribs separated by rounded grooves of equal width; other narrower secondary ribs occur on these and in the grooves. Concentric growth lines cross ribs and grooves.
Remarks. Distinguished by its ornament of very strong radial ribs. *L. cantabrigiensis* (Woods) (Cenomanian) has frequent strong, concentric ridges that produce projections wherever they cross ribs. *L. brittanica* (Woods) (Coniacian–Campanian) has two categories of ribs and only thin growth lines.
Occurrence. Albian–Cenomanian; Devon and Wiltshire; rare.

Genus LIMATULA

Limatula wintonensis (Woods)
Plate 21, figure 4

Description. Small, inflated, nearly equilateral shell that is pointed dorsally. Ornamented by 10–16 rounded ribs and narrow intercostal intervals; ribs confined to the median part of the shell. Strong, regular, concentric growth ridges occur, and these may be concentrated close together near a growth ring.
Remarks. Distinguished from the European *L. semisulcata* (Nilsson) (Campanian–Maastrichtian) by having narrower ribs; from *L. decussata* by having fewer ribs, smooth areas near the shell margins, and ridges rather than tubercules on the ribs; from *L. fittoni sensu* Woods (Albian–Cenomanian) by having ridges on its ribs; and from the European *L. kunradensis* Marquet (Maastrichtian) by its semicircular shape.
Occurrence. Upper Turonian, *S. plana* Zone, to Lower Campanian, *G. quadrata* Zone; found throughout southern England.

Limatula decussata (Goldfuss)
Plate 21, figure 5

Description. Short, wide, convex shell, almost an equilateral, rounded-oblong. Umbones small; ears rather small and nearly equal. Shell covered with numerous keeled ribs separated by narrow grooves, but strong ribs not always present over entire shell. Numerous concentric ridges give rise to tubercules at the summit of the larger ribs.

Remarks. Distinguished from other species by the existence of ribs over the entire shell; its tuberculate keeled ribs also separate it.

Occurrence. Santonian–Lower Maastrichtian; Wiltshire and Norfolk.

Genus LIMEA

Subgenus LIMEA (PSEUDOLIMEA)

Remarks. *Limea* (*Pseudolimea*) can be differentiated from *Limatula* by its obliquity and relative width, and from *Limaria* by its small size and the absence of a gape (Dhondt 1989).

Limea (*Pseudolimea*) *granulata* (Nilsson)

Plate 23, figure 12

Description. Very convex, oval (subovate–subcircular), slightly oblique, medium-sized shell with rounded outline. Height only a little greater than length. Apical angle very large. Umbones small, incurved and close together. Ears of moderate size, nearly equal, with spiny ribs. Ornament of numerous (20–24) strong, trifid ribs separated by narrow furrows. Ribs bear three rows of scale-like spines, one with larger spines at the sharp summit and the others at either side.

Remarks. Preservation affects the appearance of the ribs; the scale-like spines range from erect spines to sloping scales, tubercles or rounded granules. Distinguished by characteristic tripartite rib arrangement and small knobs on the end of all ribs.

Occurrence. Upper Cenomanian, *N. juddii* Zone, to Maastrichtian; commonest in Upper Campanian, *B. mucronata* Zone; Norfolk, Suffolk and Wiltshire.

Limea (*Pseudolimea*) *composita* (J. de C. Sowerby)

Plate 23, figure 13

Description. Usually larger than *P. granulata*, and distinguished from that species by having tubercles on the ribs rather than scale-like spines; the medium-sized shell is less oblique and more equilateral (suborbicular–subovate); the umbones are more prominent; the height is greater than the width. Trifid ribs seem to be narrower and have more prominent summits.

Remarks. It is not easy to differentiate *L. (P.) composita* from *L. (P.) granulata* because they have virtually the same number of tripartite/trifid ribs, and the differences in rib structure described above are only apparent on well-preserved specimens. The nature of their intercostal intervals provides the easiest method of separating the two species: this is virtually smooth in *L. (P.) composita* but covered with scale-like granulation in *L. (P.) granulata*. *L. (P.) denticulata* (Nilsson) (Santonian–Maastrichtian) has a medium to large convex shell, which is suborbicular to subovate in outline, and is ornamented by 20–40 ribs, but it is distinguished by its undivided triangular ribs with sharp summits. *L. (P.) geinitzi* (von Hagenow) (Maastrichtian) is much smaller and has 40–65 undivided ribs that are far more rounded in cross-section.

Occurrence. Appears to be restricted to the Cenomanian; south-west England, Devon and Wiltshire; ?Norfolk.

FOSSIL OYSTERS

Remarks. The oysters are probably the most difficult single group of macrofossils to classify. However, as they are also the most successful group of cemented bivalves, and their strong calcitic shell ensures that they are usually preserved, they are the commonest molluscan fossils. Unfortunately, their gregarious life-style, together with the considerable variation that occurs in the form of their shell, results in a wide range of shape. Since there is also a strong tendency for quite distinct taxa to appear similar in external form, and the shells of individuals frequently conform to the shape of the material upon which they are cemented, the task of naming specimens becomes even more difficult. It is advisable to have several specimens available before attempting to identify species of fossil oysters. The nature of the ligament pit and the position and shape of the adductor muscle scar, or the impression made by its pad, are useful features to note.

The volume on oysters by Stenzel (1971) in the Treatise on Invertebrate Paleontology is the most comprehensive reference available; information can also be obtained from other text-books, such as those by Easton (1960) and Carter (1968). The publication of Malchus (1990) contains a detailed analysis of the morphology of oyster shells. He used shell form, shapes of ligament areas, patterns of sub-ligamental planes and chomata types and 27 varieties of five basic types of adductor to classify oysters and determine their probable relationships. His classification is adopted here.

Although oysters are notoriously variable, Cooper (1992) has described deep-water specimens that show relatively stable shell characteristics, although these are still influenced by the surface area of available hard substrates. Yamaguchi (1994) described the process and structure of oyster cementation and suggested that it had developed early in the evolution of the Pteriomorpha. Following Palmer (1989), Malchus (1995) discussed the larval shell characters of some fossil Ostreoida. This led him to question the current subfamily classification based on 'brooding' and 'non-brooding' features.

Family PALAEOLOPHIDAE
Subfamily PALAEOLOPHINAE

Genus RASTELLUM

Rastellum (Arctostrea) colubrinum (Lamarck)
Plate 24, figure 1

Description. Shell ranging from straight, long and tapering to crescentic, hooked, or semicircular; crest of right valve may be flattened; short hinge line. Very steep parallel flanks and a very serrated commissure; close regular plicae with tips of zigzags sharp, long and narrow; slender hyote spines sometimes present at crests.
Remarks. *Rastellum quercifolium*, a rare Cenomanian, ribbed, convex form, is easily recognised by its subquadrate to rounded outline.
Occurrence. Middle–Upper Cenomanian and ?Turonian; very common in marginal facies throughout England.

Family GRYPHAEIDAE
Subfamily EXOGYRINAE

Genus AMPHIDONTE

Amphidonte obliquatum (Pulteney)
Plate 24, figures 2–3

Description. Shell shape variable depending on attachment. Left valve very inequivalve; inequilateral ovate shell almost coiled in the plane of its commissure; usually higher than long; posterior margin can be extended; inflated and often with a small attachment surface. Umbo spiral and curved inwards and backwards (prosogyral); shell usually attached behind umbo. Distinct rounded carina from umbo to postero-ventral extremity. Surface smooth with growth lines only, but irregular ridges may occur; juvenile stages can have radial ribs. Band of ridges and pits on inside of valve. Right valve thin, operculiform, slightly concave; small spiral umbo.

Remarks. Distinguished from other species by its less obvious carina and more rounded valve. The posterior part of the valve is less flattened than that of the very smooth *A. suborbiculatum* (Cenomanian), and its anterior part is much narrower and steeper, while the umbo of the left valve is more spiral. Many specimens have traces of irregular ornament bands.

Occurrence. Albian–Turonian; southern England from Devon to Kent.

Genus COSTAGYRA Vialov

Remarks. *Costagyra* is characterized by having a few strong radial ribs on its left valve that are separated by large concave intervals; scales or spines may occur on the ribs, and the ribs may be prolonged as digitations at the shell margin. Malchus (1990) considered this to be a subgenus of *Exogyra*.

Costagyra digitata (J. Sowerby)
Plate 24, figure 4

Description. Thick, medium-sized, inequivalve, exogyroid shell with a rounded or sharp carina and prominent, widely-spaced radial folds, some of which originate from the carina. Spiral umbo curved backwards. Adductor muscle scar pad orbicular; chomata typically lath-like. Right valve has narrow radial ribs separated by wide concave interspaces.

Remarks. Distinguished from *Exogyra* by the 7–12 radial folds on the left valve, which may also project beyond the valve periphery forming spinose projections. Differs from *C. laciniata* (Upper Cenomanian–Campanian) in lacking an elongate attachment area and by its ornament of strong ribs.

Occurrence. Albian–Cenomanian; Devon and Dorset; scarce.

Subfamily GRYPHAEOSTREINAE

Genus GRYPHAEOSTREA

Gryphaeostrea canaliculata (J. Sowerby)
Plate 24, figures 5, 7

Description. Small, very inequivalve shell, usually higher than long. Ligamental area is a deep, narrow, spiral groove that widens and straightens with growth. Left valve highly convex; commissural shelf lacks chomata; appears to have a rounded carina owing to mid-whorl convexity; umbo/beak opisthogyrally spiral. Elongate attachment scar posterior to umbo; anterior wall spirally curved and rising obliquely from substratum. Smooth growth squamae occur in later stages and spoon-shaped shelly 'claspers' may grow out from these. Right valve flat, suborbicular to triangular, and coiled in a plane normal to the hinge; umbo flat; ornament of regularly spaced comarginal growth lamellae.

PLATE 24

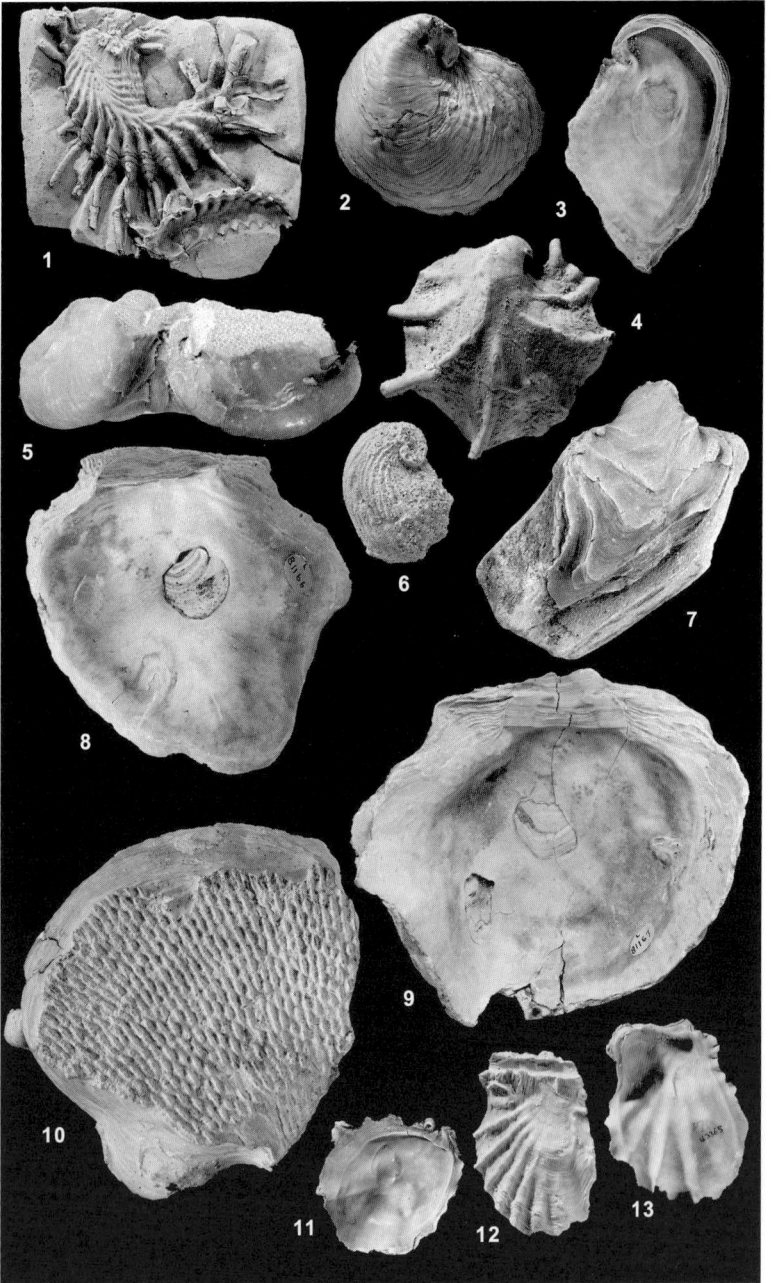

Remarks. Recognised by its groove-like ligament area, the regularly spaced squamae of the right valve and the combination of a gryphaeiform left valve and an exogyriform-coiled right valve.
Occurrence. ?Aptian–Maastrichtian; *G. canaliculata* var. *striata* Rowe, with numerous closely-packed growth lines, is abundant in the Lower Campanian, *G. quadrata* Zone; widespread throughout southern England.

Subfamily PYCNODONTEINAE

Genus PYCNODONTE
Subgenus PYCNODONTE (PHYGRAEA)

Pycnodonte (*Phygraea*) *vesiculare* (Lamarck)
Plate 24, figures 8–10

Description. Extremely variable; shell small to large; left valve generally highly convex, umbo incurved. Umbo of left valve only just above long, straight, dorsal margin; auricles present. Usually with small–medium attachment area and a posterior ear demarcated by a radiating sulcus. Shell outline subcircular to semicircular; chomata arborescent and vermiculate; radial ribs variable and not particularly strong. Right valve flat to concave,

---

EXPLANATION OF PLATE 24

Fig. 1. *Rastellum* (*Acrostrea*) *colubrinum* (Lamarck). Lower Cenomanian, Folkestone, Kent; left valve; ×1·0.
Figs 2–3. *Amphidonte obliquatum* (Pulteney). Cenomanian, Folkestone–Dover, Kent. 2, left valve; ×0·7. 3, internal view of left valve; ×0·7.
Fig. 4. *Costagyra digitata* (J. Sowerby). Cenomanian, beach below Rousdon, east of Seaton, Devon; left valve; ×0·75.
Figs 5, 7. *Gryphaeostrea canaliculata* (J. Sowerby). 5, Campanian, near Salisbury, Wiltshire; two left valves typically attached to an elongate substrate; ×1. 7, Lower Cenomanian, Dover, Kent; right valve, showing irregular shape and growth rugae; ×0·75.
Fig. 6. *Amphidonte* sp. Lower Cenomanian, *M. dixoni* Zone, Wilmington, Devon; left valve showing radial ribs that characterize juveniles; ×0·75.
Figs 8–10. *Pycnodonte* (*Phygraea*) *vesiculare* (Lamarck). Upper Campanian, *B. mucronata* Zone, Norwich, Norfolk. 8, left valve; ×0·5. 9, right valve showing ligamental area, bourrelets, vermiculate chomata and adductor muscle pad; ×0·5. 10, external view of left valve, showing evidence of attachment to the sponge *Ventriculites radiatus*; ×0·5.
Figs 11–13. ?*Hyotissa semiplana* (J. de C. Sowerby). 11–12, Cenomanian or Turonian, Dover, Kent. 11, internal right valve showing ligament area, vermiculate chomata and orbicular adductor muscle pad; ×0·5. 12, external view of left valve; ×0·5. 13, ?Campanian, Norwich, Norfolk; external view of right valve; ×0·5.

but frequently has transferred image of cementation area; surficial riblets occur.

Remarks. This is a very widely distributed species in fine-grained, outer shelf sediments. Its sedentary life and adaptations to various substrates have resulted in a wide variety of shell shapes, which often occur in deposits of the same environment. The form *P. hippopodium* (Nilsson) is larger, has a greatly reduced attachment area, but possesses an anterior 'wing'.

Occurrence. Cenomanian–Maastrichtian; relatively common throughout the UK.

Genus HYOTISSA

?Hyotissa semiplana (J. de C. Sowerby)
Plate 24, figures 11–13

Description. Medium-sized to large subequal valves; Quenstedt muscle scar immediately under ligament area; chomata present on inner valve margins; vesicular wall structure. Left valve slightly larger and more convex than right valve; variable outline dependent on large attachment area, but generally suborbicular or possibly falciform. Right valve corresponds to left valve but was also influenced by the nature of the surface to which it was attached. Characteristic ornament on unattached part of valves consists of strong, rounded, radial folds with wider interspaces and well-marked growth lines. Indication of slight projections on folds, but these are caused by growth lines and are not hyote. Plicae of commissure rounded and asymmetrical.

Remarks. Many of the features of this species are not exactly compatible with those described for true *Hyotissa*, which leads to some uncertainty over its generic assignment.

Occurrence. Cenomanian–Maastrichtian; widespread throughout the counties around London, Wiltshire and Norfolk.

Family LIOSTREIDAE

Genus ACUTOSTREA

?Acutostrea boucheroni (Woods, *non* Coquand)
Plate 25, figure 7

Description. Small shell, considerably higher than long, generally with the anterior and posterior margins gradually diverging from the umbo, but may be irregular with ventral part expanded. Left valve very convex and often semi-cylindrical; larger specimens develop a posterior wing. Long triangular resilifer and a long narrow sub-umbonal cavity; faint chomata; orbicular or reniform muscle scar; very small attachment area at umbo. Surface marked with growth rings. Right valve smooth, thin, faintly convex or flat.

Remarks. Malchus (1990) listed the English material described by Woods under *Curvostrea tevesthensis* (Coquand), but after referring to their small size, thin shell and comma-shaped muscle scar, suggested they should possibly be placed in the group of *A. incurva*.

Occurrence. Santonian–Lower Campanian, *G. quadrata* Zone; southern England, Norfolk and Yorkshire.

Acutostrea cf. *incurva* Nilsson
Plate 25, figure 4

Description. Medium-sized; usually higher than long, with shape variable depending on attachment. Both valves characterized by small hooked or beak-like tip to umbos and a triangular resilifer. Shell margins have varying chomata. Left valve slightly convex, sometimes with irregularly concentric growth lamellae. Right valve even thinner shelled; flat, with fine concentric growth lines and sometimes with a fine radial outer layer.

Occurrence. Cenomanian–Lower Campanian, *B. mucronata* Zone; southern England, Devon, Wiltshire to Norfolk.

Genus AGEROSTREA

Agerostrea lunata (*sensu* Woods, *non* Nilsson)
Plate 25, figs 1–3

Description. Regularly curved inequivalve shell, usually falciform. Left valve moderately convex, especially near the small prosogyrous umbo which protrudes slightly; very small attachment area; ligament pit curves posteriorly; wing or ear-like extension on either side of umbo. Ornament of juvenile stage consists of concentric lines or ridges; in the adult it consists of a few broad, rounded folds occurring only near the anterior and ventral margins. Right valve initially smooth and slightly concave but develops folds comparable to those of the left valve.

Remarks. This British species is more falciform than *A. lunata* Nilsson. Its few commissural folds occur only near the valve margins and are not as high or wide as those of Nilsson's Swedish material. It also has a short, straight hinge line which contributes to the appearance of anterior and posterior ears, whereas ?*Agerostrea lunata s.s.* has more pointed umbones. It is far less plicate than the very narrow *A. ungulata* (Schlotheim) (Maastrichtian) of European localities.

Occurrence. Lower Maastrichtian, '*O.*' *lunata* Zone; Norfolk.

PLATE 25

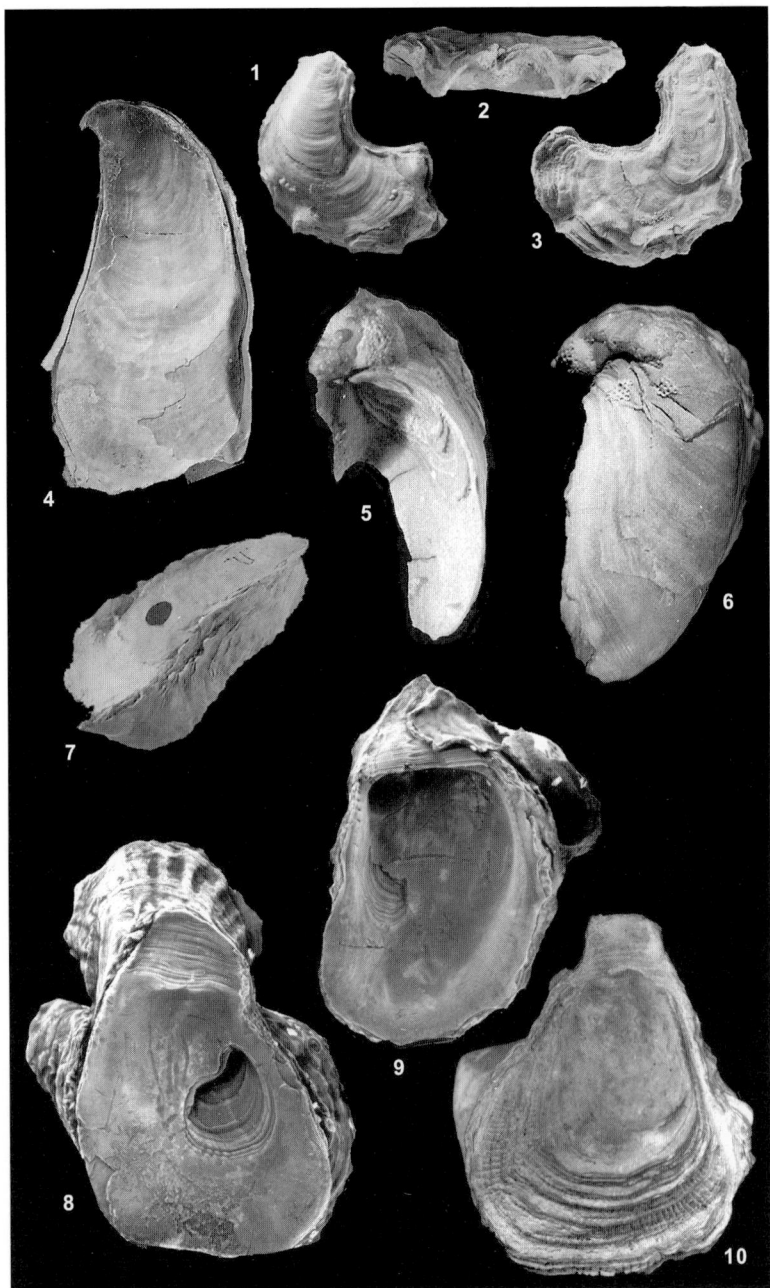

Family OSTREIDAE
Subfamily CRASSOSTREINAE

Genus CUBITOSTREA

Cubitostrea sarumensis (Woods)
Plate 25, figures 8–10

Description. Very variable in shape and convexity. Ligamental area high, hive-shaped and curved slightly backward. Left valve thick and very convex; inner margin on either side of the hinge with a line of small pits (chomata); reniform adductor muscle insertion close to posterior margin. Right valve triangular to ovate. Ornament of concentric growth lamellae and typically with numerous fine radial threads, which increase in strength near margins; these together with the multiple growth lines produce a reticulate pattern.

Remarks. This is questionably placed here because Malchus (1995) considered *Cubitostrea* to be primarily Palaeogene. The weakly convex and differently ornamented specimens attributed to this species by Freneix (1986) were placed in the genus *Striostrea* (Cenozoic) based on the similarity of microstructure on its right valve, together with the form, position and insertion of its musculature. Its quadrate outline makes it resemble *Hyotissa semiplana*, but its larger, beehive-like ligament area and its muscle position distinguish it. It has denser ornament than *Crassostrea cunabula* (Cenomanian), the latter typically lacking chomata.

Occurrence. Lower Campanian, *G. quadrata* Zone; Wiltshire.

EXPLANATION OF PLATE 25

Figs 1–3. *Agerostrea lunata* (*sensu* Woods). Lower Maastrichtian, *B. lanceolata* Zone; Trimingham, Norfolk. 1, left valve; ×1·3. 2–3, right valves; ×1·5.

Fig. 4. *Acutostrea* cf. *incurva* Nilsson. Upper Campanian, *B. mucronata* Zone, Norwich, Norfolk; right valve; ×1·0.

Figs 5–6. *Gryphaeostrea canaliculata* (J. Sowerby). Upper Campanian, *B. mucronata* Zone, Norwich, Norfolk. 5, right valves; ×1·5. 6, right valve; ×1·5.

Fig. 7. ?*Acutostrea boucheroni* (Woods). Santonian, *U. socialis* Zone, Thanet coast, Kent; both valves with flattish right valve above; ×1.

Figs 8–10. *Cubitostrea sarumensis* (Woods). Lower Campanian, *G. quadrata* Zone, East Harnham, Salisbury, Wiltshire. 8, interior of upper right valve (the lower valves of two smaller individuals are also attached to this specimen); ×1. 9, interior of lower fixed left valve showing triangular ligamental area, reniform muscle scar and the chomata on both sides of the shell margin; ×1. 10, shell with both valves conjoined, right valve uppermost and covering most of left valve; ×1.

Order VENEROIDA
Superfamily HIPPURITOIDEA
Family MONOPLEURIDAE

Genus GYROPLEURA

Gyropleura inaequirostrata (S. Woodward)
Plate 22, figures 13–14

Description. Inflated shell; typical pachydont tooth arrangement without differentiated cardinals and laterals; posterior and anterior muscles form elongate scars on raised platforms; somewhat reduced posterior myophoric septum. Large exogyriform right valve, coiled in low spiral; attached by most of its anterior; pointed umbo prominent and incurved anteriorly; anterior slightly concave. Left valve capuliform, umbo near hinge margin. Ornament of strong, lamellar, radial ribs, which are undulose wherever they cross growth lines; ribs have fine, scale-like serrate summits and are separated by broad, flat and smooth interspaces.

Remarks. Distinguished from *G. cornucopiae* (Cenomanian) by having fewer ribs with scale-like summits.

Occurrence. Upper Campanian, *B. mucronata* Zone; Norfolk.

Family RADIOLITIDAE
Subfamily SAUVAGESIINAE

Genus DURANIA

Durania mortoni (Mantell)
Plate 17, figure 14

Description. Strongly inequivalve. Attached valve elongate cylindrical-conical, slightly arched or almost straight. 'E' and 'S' siphonal band pillars with strong longitudinal angular ribs, sometimes in groups of two or three, crossed by growth lines; concave bands with fine ribs; convex interbands with strong ribs. Radiating grooves bifurcate between inner and outer margins of upper surface of outer wall; honeycomb cellular structure between inner and outer walls. Free valve operculiform.

Occurrence. Lower Cenomanian–Upper Turonian, *S. plana* Zone; southern England from Dorset and Wiltshire to Kent.

Superfamily CARDITOIDEA
Family CARDITIDAE
Subfamily CARDITESINAE

Genus LUDBROOKIA

?*Ludbrookia cottaldina* (d'Orbigny)
Plate 17, figure 10; Plate 26, figure 15

Description. Moderately convex inequilateral shell, generally quadrate but some wider forms; a little higher than long. Lunule very short, escutcheon narrow and elongate. Long, straight, posterior margin; short, curved, anterior margin; slightly rounded ventral margin; crenulate inner margin. Ornamented by numerous thin, simple, radial ribs, a few of which divide; interspaces between ribs of variable width and crossed by linear ridges.

Occurrence. Cenomanian; mainly known as internal casts from Lower Cenomanian localities on the Isle of Wight, and in Buckinghamshire and Surrey; also from the Middle Cenomanian, *H. subglobosus* Zone, in Sussex.

Subfamily CARDITINAE
Genus ?CARDITA

'Cardita' cancellata Woods
Plate 26, figure 8

Description. Ovate-trapezoidal, almost equilateral convex shell with rounded margins; length greater than height. Umbones moderately prominent and curved slightly anteriorly. Adductor muscles placed more laterally than in other Cretaceous carditids. Fine, crenulate inner margin. Carditid dentition. Ornament composed of numerous concentric and radial ribs which, with the concentric element more pronounced, produce a pattern of small, elongate tubercles.

Remarks. Often found only as internal moulds, *'Cardita' cancellata* is distinguished from ?*Ludbrookia cottaldina* (d'Orbigny) (Cenomanian) by its much finer ornament and reticulate pattern.

Occurrence. Turonian, *S. plana* Zone; Kent, Hertfordshire and Buckinghamshire; also in derived material from the Haldon Hills, Devon.

Genus ?BEGUINA

'Beguina' turoniensis (Woods)
Plate 17, figures 8–9

Description. Very small, subquadrate shell; higher than long; inequilateral and much inflated along mid-line from umbo to postero-ventral extremity. Umbones prominent, curved anteriorly. Hinge with posterior cardinals greatly elongated and anterior cardinal widened, laterals wanting. Ornament of fine radial riblets and growth lines; riblets become higher and fence-like; very fine spines on riblets; very deep crenulate margin that appears as a peripheral series of spines on internal moulds.

Remarks. Probably common, but its small size and poor preservation has resulted in very few specimens being present in museum collections. It

does not have the typical upward-pointing acicular cardinal teeth of the Cardiidae to which it has previously been assigned, but is very like the Recent genus *Beguina* in shell morphology.
Occurrence. Turonian, *S. plana* Zone; Berkshire, Oxfordshire, Hertfordshire and Kent.

Superfamily uncertain
Family MACTROMYIDAE

Genus UNICARDIUM

'*Unicardium*' *ringmeriense* (Mantell)
Plate 17, figure 7

Description. Rounded-subquadrate, tumid shell with large, prominent, broad, prosogyrous umbones; length greater than height. Anterior margin moderately convex and curving rapidly to ventral margin. Ovate anterior adductor muscle scars. Lunule poorly defined. Ornament of concentric ridges, generally narrow but can be quite strong.
Remarks. Its rounded and more globose shell is more characteristic of *Unicardium* than the elliptical, equilateral genus *Mactromya*.
Occurrence. Albian–Lower Cenomanian; Sussex and Devon.

Superfamily VENEROIDEA
Family VENERIDAE
Subfamily TAPETINAE

Genus CYCLORISMA

Cyclorisma rhotomagensis (d'Orbigny)
Plate 26, figure 9

Description. Moderately convex shell; subovate outline; longer than high. Small incurved prosogyrous beaks, moderately prominent, situated one-third of length from anterior extremity. Lacks umbonal ridge; lunule and escutcheon not well defined. Postero-dorsal margin long and slightly humped; ventral margin evenly rounded; anterior margin sharply rounded; antero-dorsal margin slightly concave. Short triangular pallial sinus. Hinge not known. Ornament of fine concentric growth lines with periodic stronger growth margins.
Occurrence. Lower–Middle Cenomanian; Devon, Dorset and Isle of Wight.

Family ARCTICIDAE

Genus PROVENIELLA

Proveniella quadrata (d'Orbigny)
Plate 26, figure 12

Description. Subrhomboidal to trapeziform, inflated, inequilateral shell with a rather strong carina on posterior slope. Length greater than height. Cyprinoid hinge, notably with strong laminar lateral teeth. Prominent umbones at anterior curved inwards; excavated lunule; entire pallial line. Rounded anterior margin merges into slightly curved ventral margin; truncated posterior extremity and postero-dorsal region flattened. Ornament of closely-packed concentric striae/growth lines.

Remarks. Usually found only as internal casts. The umbones are in a less anterior position in *Cyprina ligieriensis* (Cenomanian), and the angle between its posterior and dorsal margins is larger. *P. regularis* (d'Orbigny) (Albian) is distinguished by its more regular, less inequilateral and more globose shell; also its less posterior truncation and less curved umbones.

Occurrence. Albian–Cenomanian; Devon, Wiltshire, Bedfordshire and Sussex.

Family ?TRAPEZIIDAE

Genus MIOCARDIOPSIS

?Miocardiopsis trapezoidalis (Roemer)
Plate 26, figures 13–14

Description. Ovate-trapezoidal, moderately convex shell with forward-curving umbones; a sharp, gently curving carina extends from umbo to the posterior extremity, separating the smooth posterior area. A few curved, blade-like teeth are subparallel to the hinge margin.

Remarks. Recognised by its strong ridge, which sets off a small posterior area, and the rather humped shell shape with rounded umbos.

Occurrence. Cenomanian–Turonian, and Upper Campanian, *B. mucronata* Zone; Devon (Cenomanian); Buckinghamshire, Hertfordshire, Oxfordshire, Bedfordshire, Kent, Surrey and Yorkshire (Turonian); Norfolk and Northern Ireland (Campanian).

Superfamily SOLENOIDEA
Family ?SOLENIDAE

Genus 'LEPTOSOLEN?'

'*Leptosolen*?' *rectangularis* (Woods)
Plate 26, figure 7

Description. Oblong, quadrate, inequilateral, moderately convex shell; anterior not as high as posterior. Postero-dorsal margin nearly straight and parallel to ventral margin; posterior margin truncated, also nearly straight or slightly concave; the anterior margin rounded. Inconspicuous umbos at anterior end. Rounded, curved carina extends from umbo to postero-ventral extremity. Relatively smooth shell with growth lines and faint radial ribs.

Remarks. This species has a prominent vertical inner rib or buttress behind the anterior adductor muscle and below the umbo. Such a feature is typical of a number of the Solenacea, particularly *Leptosolen,* but is not restricted to this superfamily. It is separated from *Miocardiopsis trapezoidalis* by its more rectangular shape, the more diagonal carina and larger posterior area.

Occurrence. Turonian, *S. plana* Zone; Berkshire.

EXPLANATION OF PLATE 26

Figs 1–2. *Pholadomya* (*Procardia*) *decussata* (Mantell). Cenomanian, locality unknown. 1, anterior view; 2, right valve; both ×0·75.

Figs 3, 5. '*Thracia*' *carinifera* (J. de C. Sowerby). 3, Upper Albian, Chardstock, Devon; ×1·5. 5, Lower Cenomanian, Ventnor, Isle of Wight; left valve; ×1·75.

Figs 4, 6. *Cuspidaria* cf. *caudata* (Nilsson). Campanian, *B. mucronata* Zone, Norwich, Norfolk. 4, *c*. ×3. 6, right valve; ×3.

Fig. 7. '*Leptosolen*?' *rectangularis* (Woods). Turonian, *S. plana* Zone, Hitchin, Hertfordshire; silicone rubber cast of external mould of right valve; ×2·5.

Fig. 8. '*Cardita*' *cancellata* Woods. Turonian, *S. plana* Zone, Hitchin, Hertfordshire; silicone rubber cast of incomplete external mould of left valve; ×3.

Fig. 9. *Cyclorisma rhotomagensis* (d'Orbigny). Cenomanian, Haven Cliff, east of Seaton, Devon; left valve; *c*. ×1·5.

Fig. 10. *Pholadomya* (*Pholadomya*) *stewarti* Tate. Chalk, Kilcorig, near Broadmount, north-west of Lisburn, County Antrim, Northern Ireland; right valve; ×1·5.

Fig. 11. *Atreta nilssoni* (Hagenow). Coniacian, *M. cortestudinarium* Zone, Chatham, Kent; ×1·5.

Fig. 12. *Proveniella quadrata* (d'Orbigny). Cenomanian, *N. carcitanense* Zone; Warminster, Wiltshire; internal mould of right valve and umbo of left valve; ×0·75.

Figs 13–14. ?*Miocardiopsis trapezoidalis* (Roemer). Turonian, *S. plana* Zone, Hitchin, Hertfordshire. 13, postero-dorsal view; 14, left valve; both *c*. ×1·5.

Fig. 15. ?*Ludbrookia cottaldina* (d'Orbigny). Chalk, horizon unknown, Lewes, Sussex; left valve; ×3.

Fig. 16. *Entolium orbiculare* (J. Sowerby). Cenomanian, Folkestone, Kent; right valve, showing hinge and auricular crurae; ×0·75.

PLATE 26

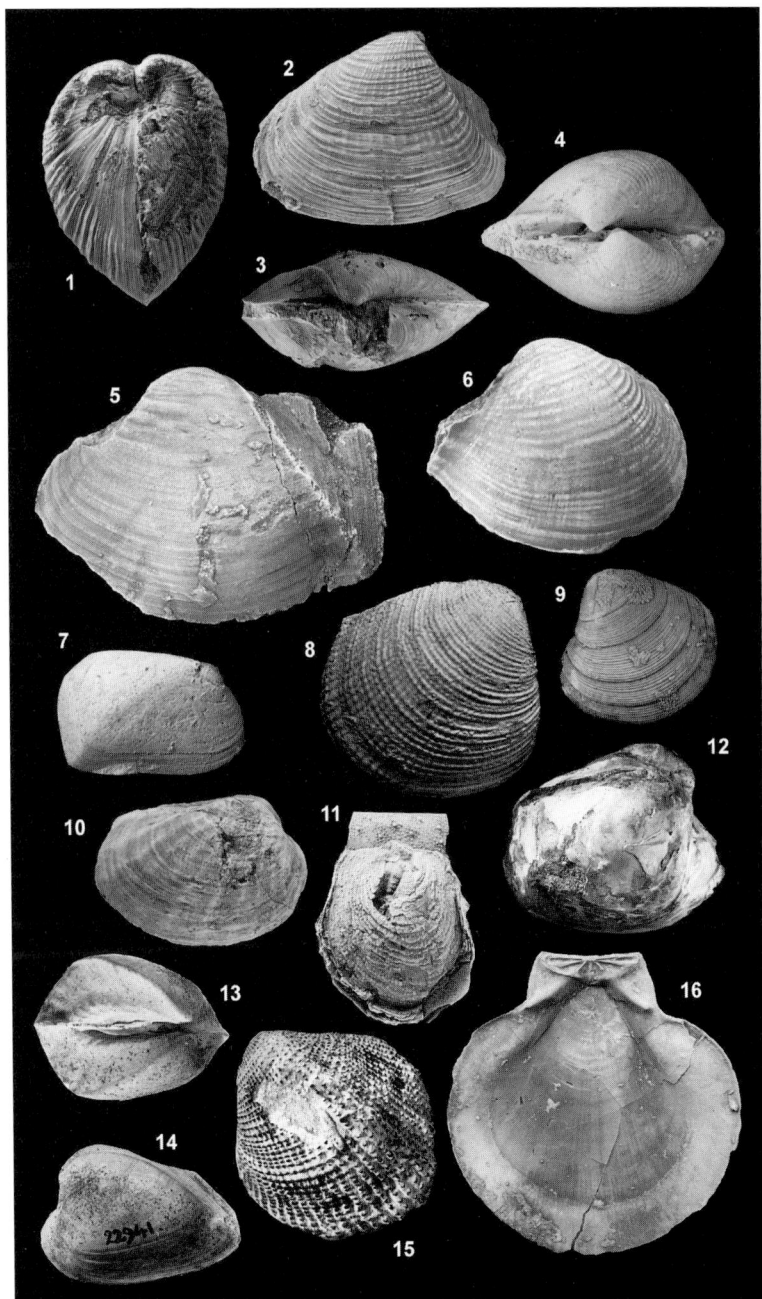

Order PHOLADOMYOIDA
?Superfamily MODIOMORPHOIDEA
Family PERMOPHORIDAE
Subfamily MYOCONCHINAE

Genus MYOCONCHA

Myoconcha (*Modiolina*) *cretacea* d'Orbigny
Plate 17, figure 2

Description. Medium-sized to large, thick, mussel-shaped, equivalve shell. Escutcheon carinate. Hinge with simple interlocking cardinal teeth. Ligament elongate and external, posterior to umbones. Anterior adductor set behind subvertical myophoric buttress, smaller than posterior adductor; pallial line entire. Ornament of fine radial ribs with fine concentric striae occurring in their interspaces; posterior radiating cords bear small scale-like spicules.
Occurrence. Cenomanian; common in some phosphatic 'basement' beds as internal moulds; Devon and Dorset.

Superfamily PHOLADOMYOIDEA
Family PHOLADOMYIDAE

Genus PHOLADOMYA

Pholadomya (*Pholadomya*) *stewarti* Tate
Plate 26, figure 10

Description. Small to medium-sized, thin, subrectangular to ovate shell. Posterior and anterior margins well rounded with slight gapes. Radiating ribs below umbones. Margins of escutcheon well rounded. Hinge without teeth. External ligament on short nymphs. Adductor muscle scars subequal. Pallial line with a moderately deep sinus.
Occurrence. Campanian; Northern Ireland, rare.

Subgenus PROCARDIA

Pholadomya (*Procardia*) *decussata* (Mantell)
Plate 26, figures 1–2

Description. Medium-sized to large, thin, convex, cockle-shaped shell. Posterior margin rounded with a slight gape; anterior margin much truncated and separated from flank by a distinct but rounded angulation. Escutcheon subcarinate; ligament external. Hinge without teeth. Adductor muscle scars subequal; pallial line with a moderately shallow sinus. Ornamented by low, narrow, radial ribs that are separated by broad,

shallow interspaces and comarginal rugae; early shell has a granular appearance; ribs absent from posterior and flattened anterior parts of shell.
Remarks. *P. cordata* (Tate) (Campanian) is similar to *P. decussata* but has a more rounded shell section and posterior margin, and a deeper escutcheon. Often crushed.
Occurence. Cenomanian; common in south-east England.

Superfamily PHOLADOIDEA
Family PHOLADIDAE
Subfamily JOUANNETIINAE

Genus JOUANNETIA

Jouannetia rotundata (J. de C. Sowerby)
Plate 17, figure 13

Description. Small, rotund shell with large anterior gape covered by callum; posterior gape small. Shell very short, acutely triangular with ventral condyle and posterior internal septum sloping to the posterior from within the umbo. Anterior slope with reticulate sculpture formed by sharp, radiating ribs and flanged growth laminae. Disc with two carinate ribs behind sulcus and sharp lamellose growth laminae. No other accessory plates or calcareous tube present.
Remarks. This species bored into shale and weakly lithified rock. It is generally preserved as internal moulds but external moulds are ocasionally found.
Occurrence. Turonian, *S. plana* Zone; not uncommon in south-east England, Berkshire and Surrey.

Superfamily LATERNULOIDEA
Family ?THRACIIDAE

Genus 'THRACIA'

'Thracia' carinifera (J. de C. Sowerby)
Plate 26, figures 3, 5

Description. Thin, ovate shell with slightly attenuated, sharply truncated posterior siphonal area set off by a distinct carina radiating from the umbo; well-rounded anterior margin. Umbones broad, incurving closely together. Almost equivalve. Adductor scars subequal; pallial line with shallow sinus. Escutcheon carinate; hinge plate narrow, uninterrupted below umbones except for a pair of small, backward-sloping chondrophores for an internal ligament; lithodesma apparently lacking. Relatively smooth;

ornament of broad, slightly raised, concentric rugae, most apparent on anterior part; numerous comarginal fine growth lines on postero-dorsal area.

Occurrence. Cenomanian; south-east England and Isle of Wight in marginal chalk and sandy chalk facies.

Superfamily CUSPIDARIOIDEA
Family CUSPIDARIIDAE

Genus CUSPIDARIA

Cuspidaria cf. *caudata* (Nilsson)
Plate 26, figures 4, 6

Description. Fairly small, rounded, oviform shell with elongate siphonal posterior extension. External ligament elongated; resilium in a small, spoon-shaped fossette. Ornament of low comarginal growth rugae and a very slight carina setting off the courcelet.

Remarks. The presence of several species has been demonstrated in the Chalk of Denmark, but the distinctive posterior shell features are not preserved in British material.

Occurrence. ?Turonian; Campanian; Northern Ireland and Norfolk.

8. GASTROPODS

by R. J. CLEEVELY *and* N. J. MORRIS

A general description of the morphological features of gastropods and accounts of their classification can be found in the appropriate sections in Moore (1960), Fretter and Graham (1962), Wenz (1969) and Murray (1985). For a glossary of terms, see Arnold (1965). More detailed accounts of a few families or species may be found in Woods (1896–97: Mollusca of the Chalk Rock); Cox (1960: Pleurotomariidae); Abbass (1962: Turritellidae and Mathildidae; and 1973: Procerithiidae); and Cleevely (1980: Epitoniidae).

DESCRIPTIONS

Subclass PROSOBRANCHIA
Order VETIGASTROPODA
Suborder TROCHINA
Superfamily TROCHOIDEA
Family TROCHIDAE

Genus TURCICA

?*Turcica schlueteri* (Barrois and de Guerne)
Plate 27, figure 5

Description. Conical shell without an umbilicus. Whorls flat-sided with deep depression at the suture, producing a slight, concave shelf. Ornamented on whorl side by up to seven closely spaced, fine, spiral, tuberculate cords, the two marginal cords generally being the strongest; the tubercles are connected by prosocline collabral ribs with fine growth lines in between. Convex base has numerous spiral ribs covered with finer, closely spaced tubercles; strongly prosocline growth lines. Aperture entire. **Occurrence**. Turonian, *S. plana* Zone; Kent, Bedfordshire, Berkshire and Hertfordshire.

Genus PLANOLATERALUS

Planolateralus berocscirense (Barrois and de Guerne)
Plate 27, figures 7, 10

Description. Trochoid shell composed of flat-sided whorls with relatively deep sutures. Slightly convex base with noded umbilical margin. Ornament consists of three primary and two secondary tuberculate spirals with a peripheral row of small tubercles; only closely packed growth lines occur on the base. Typical subrounded calliomphallid aperture.

Occurrence. Turonian, *S. plana* Zone; Berkshire, Oxfordshire, Hertfordshire and Surrey.

Planolateralus dievarum (Cossmann)
Plate 27, figure 6

Description. Conical, slightly elongate shell with a distinct canaliculate suture. Ornamented by five unequal and granulose spiral cords; the imbricate abapical cord is the most prominent.

Occurrence. Turonian, *S. plana* Zone; Hertfordshire.

Subfamily MARGARITINAE

Genus PERIAULAX

Periaulax heberti (Barrois and de Guerne)
Plate 27, figs 2–3

Description. Conical shell with height slightly greater than width. Whorls slightly convex and ornamented by ten granulose spiral cords that continue

---

EXPLANATION OF PLATE 27

Figs 1, 4. *Helicacanthus guerangeri* (d'Orbigny). Cenomanian, Wilmington, Devon; ×1·5.

Figs 2–3. *Periaulax heberti* (Barrois and de Guerne). Turonian, *S. plana* Zone, Hitchin, Hertfordshire. 2, ×2; 3, ×2·5.

Fig. 5. ?*Turcica schlueteri* (Barrois and de Guerne). Turonian, *S. plana* Zone, Dover, Kent; ×1·3.

Fig. 6. *Planolateralus dievarum* (Cossmann). Turonian, *S. plana* Zone, Hitchin, Hertfordshire; *c.* ×2.

Fig. 7. *Planolateralus berocscirense* (Barrois and de Guerne). Turonian, *S. plana* Zone, Aston Hill, Oxfordshire; ×2.

Fig. 8. *Calliomphalus steinlai* (Geinitz). Turonian, *S. plana* Zone, Hitchin, Hertfordshire; ×1·2.

Fig. 9. '*Cirsocerithium*' *reussi* (Geinitz). Turonian, *S. plana* Zone, Hitchin, Hertfordshire; ×5.

Fig. 10. *Planolateralus* sp. cf. *berocscirense* (Barrois and de Guerne). Turonian, *S. plana* Zone; Sherborne Farm Pit, Surrey; ×1·5.

Figs 11–12. '*Discohelix*' *binghami* (Baily). Cenomanian, *A. rhotomagense* Zone, Lewes, Sussex; *c.* ×10.

All except 11 and 12 are of silicone rubber casts taken from external moulds.

PLATE 27

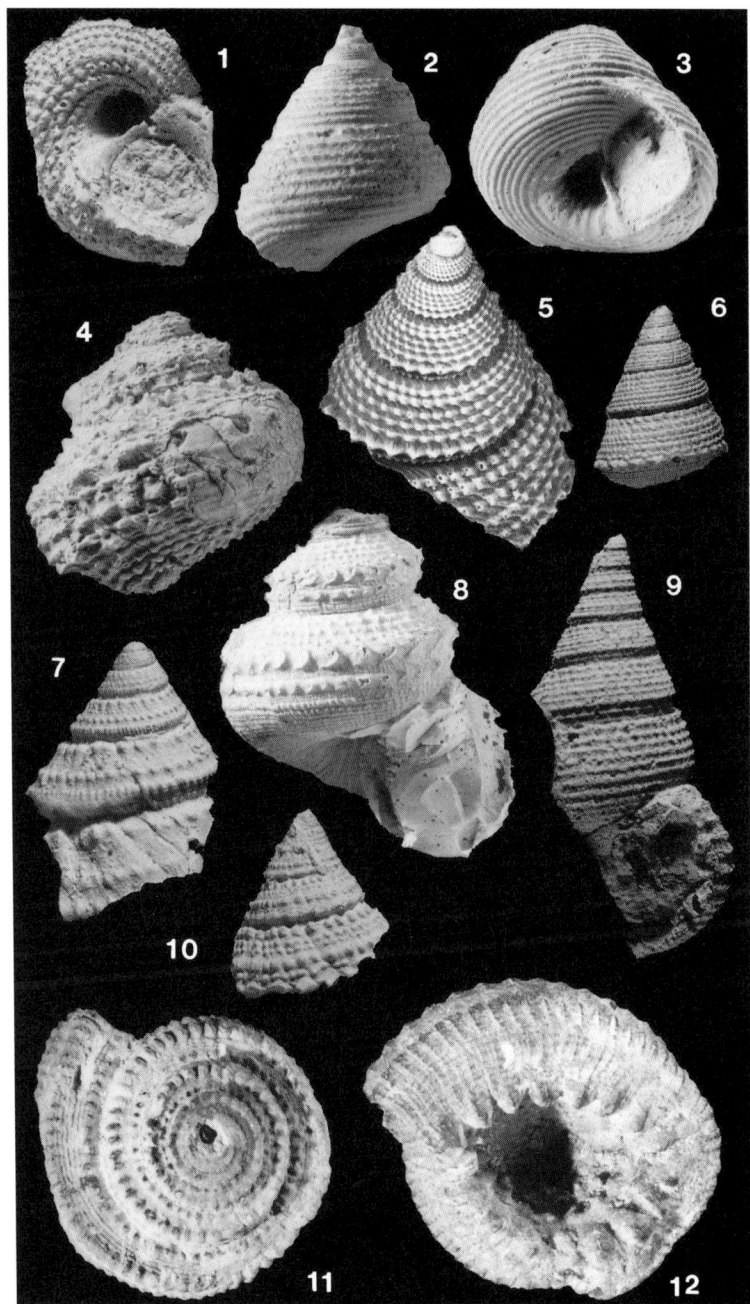

onto the slightly convex base. Large umbilicus with axially-ridged margin. Subquadrangular to rounded aperture.

Occurrence. Cenomanian and Turonian, *S. plana* Zone; possibly also Campanian; Kent, Sussex, Hampshire, Berkshire, Hertfordshire, and ?Northern Ireland (Campanian).

Periaulax striatus (Woodward)

Text-figure 8.1E–F

Description. Species with beaded and rounded spiral cords that are alternately slightly stronger and slightly weaker. The whorl profile is not so rounded as in *P. heberti*, and there is a slight angle separating a ramp from the remaining shell surface. Growth lines do not coalesce into small blocks on the umbilical edge as they do in *P. heberti*.

Remarks. May be synonymous with *P. rimosus* (Binkhorst) var. *granulosus* from the Maastrichtian of the Netherlands.

Occurrence. Upper Campanian; Norfolk.

Subfamily MONODONTINAE

'*Monodonta*' sp.

Text-figure 8.1C

Description. A slightly elongate shell with fine spiral chords. The whorl outline is rounded, conical and with a change in curvature at the base marked by a more prominent spiral rib. There is a small umbilicus and the columellar lip has a low tubercle.

Remarks. This belongs to a group of species including '*Trochus*' *cordieri* d'Archiac, which may require a new generic name.

Occurrence. Campanian, *B. mucronata* Zone; County Derry, Northern Ireland.

?Superfamily AMBERLEYOIDEA

?Family AMBERLEYIDAE

Genus CALLIOMPHALUS

Calliomphalus steinlai (Geinitz)

Plate 27, figure 8; Text-figure 8.1D

Description. Turbinate, turreted shell with deep sutures. Convex, rather angular whorls; posterior part flattened and ornamented by four finely tuberculate spiral cords; a row of large, prominent, interconnecting tubercles occurs at the keel, below which is a relatively flat, smooth area that is succeeded by three or four further tuberculate spiral cords. Rounded base with numerous fine spiral tuberculate cords; moderate sized umbilicus. Subcircular aperture.

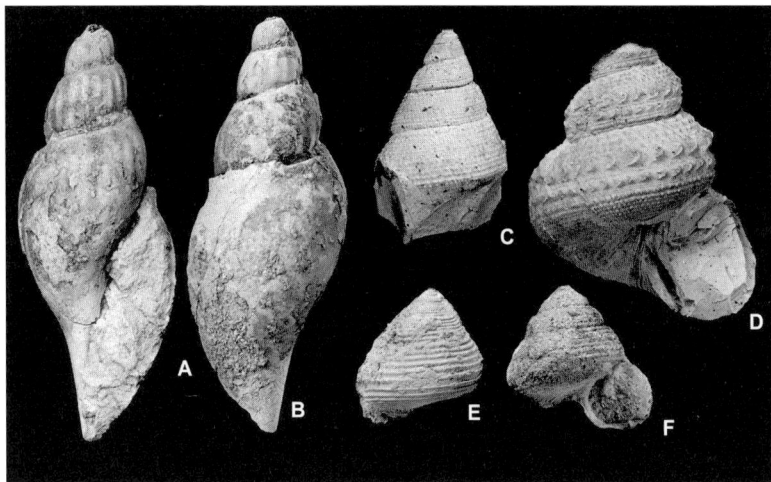

TEXT-FIG. 8.1. A–B, *Anomalofusus* sp., Upper Campanian, *B. mucronata* Zone, Glovers Quarry, Carmean, Moneymore, County Derry, Northern Ireland; ×1. C, '*Monodonta*' sp., Upper Campanian, *B. mucronata* Zone, Benbradagh Hill, Slopes of Curraghlane, north-east of Dungiven, County Derry; ×1·5. D, *Calliomphalus steinlai* (Geinitz), Upper Turonian, *S. plana* Zone, Hitchin, Hertfordshire; ×1. E–F, *Periaulax striatus* (Woodward), Campanian, Norwich, Norfolk; ×2.

Remarks. Found as external and internal moulds. Previously described as '*Turbo*' *geinitzi* Woods.

Occurrence. Turonian, *S. plana* Zone; Kent, Berkshire, Hertfordshire and Bedfordshire.

Family NODODELPHINULIDAE

Genus HELICACANTHUS

Helicacanthus guerangeri (d'Orbigny)
Plate 27, figures 1, 4

Description. Small turbinate shell with flat-sided angular whorls. Orna-mented by several series of tuberculate spiral cords, which are particularly strong at the adapical suture and at the mid-whorl angles. Rounded convex base has similar finer tuberculate ornament. Wide, smooth, open umbilicus and rounded aperture.

Remarks. Distinguished from *H. mailleana* d'Orbigny (Cenomanian), which also has an adapical row of spines, by the more delicate, neater and regular ornament on the whorl side. In *H. mailleana* the spines are fewer as well as being much coarser and stronger; only the growth lines ornament the whorl side. *H. cenomanensis* (Cenomanian) differs in lacking spiral ornament on the adapical shelf.

Occurrence. Cenomanian; Devon.

Family uncertain

Genus ?DISCOHELIX

'*Discohelix*' *binghami* (Baily)
Plate 27, figures 11–12

Description. Depressed conical shell, much wider than high; pentagonal aperture; orthocline–opisthocline growth lines. Umbilicus as high as wide. Distinctive ornament consisting of three strong, spiral ridges with regular collabral ribs forming noded tubercles on the ridges; the ridge at the adapical suture is the weakest with those occurring at the angulation and mid-whorl being of similar strength. Finer spiral threads occur between the primary ridges. Margin of umbilicus bounded by strong, rounded tubercles, which span several collabral ribs and growth stages.

Remarks. The lack of the typical discoidal shell and opisthocyrt growth lines found in *Discohelix s.s.* raises doubt over the generic determination of this species and its inclusion in this family.

Occurrence. Cenomanian; Kent and Sussex.

Superfamily PLEUROTOMARIOIDEA
Family PLEUROTOMARIIDAE

Genus BATHROTOMARIA Cox

Bathrotomaria dixoni Cox
Plate 28, figures 1, 3; Text-figure 8.2A

Description. Medium-sized, depressed turbiniform shell. Whorls obtusely angular with gently sloping ramp and slightly conical outer face. Base feebly convex with peripheral bulge and moderately broad umbilicus. Broad, cord-like selenizone at ramp angle. Growth lines opisthocyrt. Ornament of strong collabral ribs on the ramp crossed by five or more spiral threads producing tubercles at their intersection. On the lower part of the whorl the collabral element is weaker and the spiral threads are equal in strength, producing a cancellate pattern with small granules on the spirals.

Remarks. The sharply rounded periphery and strong ribs on the ramp distinguish *B. dixoni* from *B. perspectiva* (Cenomanian–Lower Santonian); while the wide selenizone and axial ribs serve to separate it from *B. linearis* (Cenomanian). Specimens described as *Leptomaria seatonensis* Cox appear to be identical to *B. dixoni*.

Occurrence. Cenomanian; southern England from Devon to Kent.

Bathrotomaria linearis (Mantell)
Text-figure 8.2E

Description. Medium-sized to large, depressed, conical shell; diameter is generally twice height, but very variable. Whorls obtusely angular, with sloping outer face and inclined ramp. Selenizone forming a smooth convex cord of moderate width at the mid-whorl angulation. Last whorl sub-angular with a slight peripheral bulge. Flattened base, broad umbilicus. Ornament of spiral threads and cords with very fine collabral threads that may cause cords to become obscurely cancellate or nodose.

Remarks. This species is distinguished from *B. perspectiva* by the subangular and peripheral bulge of its last whorl, and from *B. dixoni* by the wider ramp, narrow selenizone and lack of axial ribs.

Occurrence. Cenomanian; widespread in southern England.

Bathrotomaria perspectiva (Mantell)
Plate 28, figures 2, 4–5; Text-figure 8.2B–C

Description. Medium-sized to large shell varying from trochiform to almost subdiscoidal in shape, depending on the nature of its early whorls; always much wider than high. Whorls range from obtusely angular to strongly convex; gently inclined, slightly concave to convex, broad ramp. Selenizone moderately wide, strongly convex and cord-like, situated well above mid-whorl. Convex base ornamented by concentric cords of unequal size; umbilicus wide. Ornament consists of spiral and collabral grooves that produce a reticulate pattern of lozenges, granules or beads; 'granulate'-type (Text-fig. 8.2B) and 'depressa'-type (Text-fig. 8.2C).

Remarks. Although shell height is variable, ranging from elevated trochiform to a very depressed subdiscoidal shape, all forms have the same basic pattern of ornament. This species differs from *B. dixoni* in its broader sutural ramp and the lack of transverse riblets on this ramp, and from *B. linearis* by its relatively wider selenizone and the presence of axial ribs.

Occurrence. Cenomanian–Lower Santonian, *M. coranguinum* Zone; widespread throughout southern England.

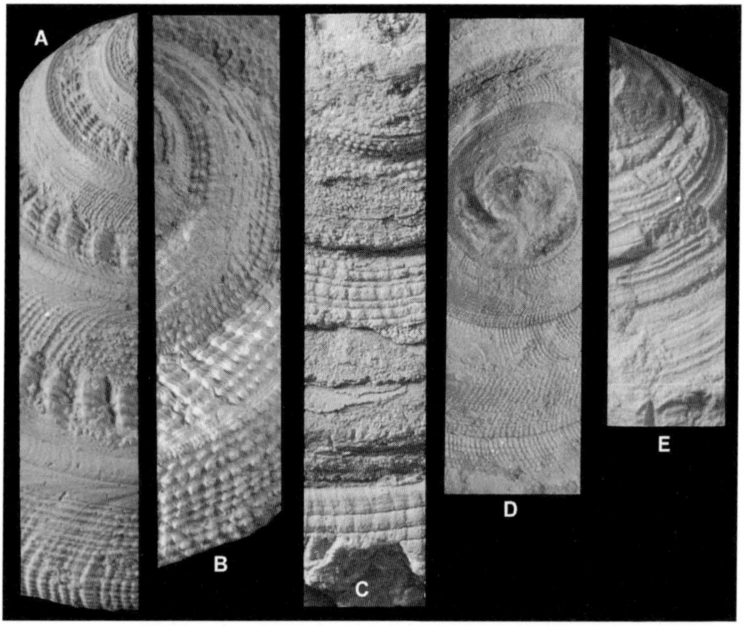

TEXT-FIG. 8.2. Ornament of species of *Bathrotomaria*. A, *B. dixoni* Cox, Middle
Cenomanian, near Culverhole, east of Seaton, Devon; ×3. B–C, *B. perspectiva*
(Mantell). B, 'granulata'-type, showing fine granose tubercles, Lower Cenomanian,
S. varians Zone, Bridport, Dorset; ×3. C, 'depressa'-type, showing lozenges
formed in spirals, Upper Turonian, *S. plana* Zone, Oldbury Farm, Wiltshire; ×2·5.
D, *B. velata* (Goldfuss), Cenomanian, Glynde, Sussex; ×1·1. E, *B. linearis*
(Mantell), Lower Cenomanian, *S. varians* Zone, Hamsey, near Lewes, Sussex;
×2·3.

Bathrotomaria velata (Goldfuss)
Text-figure 8.2D

Description. Fairly large, depressed turbiniform shell in which the diameter is twice the height; spire angle about 120 degrees. Gently sloping,
almost horizontal ramp. Base convex with subangular peripheral bulge and
wide umbilicus. Very narrow, convex, transversely ridged, cord-like
selenizone situated between two linear grooves at the ramp angle; labral slit
may extend back for some distance from the aperture. Delicate ornament
composed of very small granules occurring at the intersections of very
narrow, closely spaced spiral cords and collabral threads. Base ornamented
by numerous narrow spiral cords that may be granulose near periphery.
Occurrence. Cenomanian; Sussex, Kent and Devon.

Genus CONOTOMARIA

Conotomaria percevali Cox
Plate 28, figure 6

Description. Moderately large, conical, top-like shell in which height almost equals diameter; spire angle exceeds 60 degrees. Whorls flat-sided, but last whorl may develop an ill-defined shoulder and become subangular at basal periphery. Base feebly convex with narrow umbilicus. Selenizone narrow, flush between spiral threads and situated in upper half of whorl side, separated from suture by a smooth band. Ornament beneath selenizone of broad, flat, spiral bands separated by narrow intervals containing a spiral thread. Base ornamented by unequal spiral bands separated by relatively deep, narrow intervals.

Remarks. The high position of the selenizone on a flat whorl side distinguishes this from other species. In *C. chardstockensis* Cox (Cenomanian) the selenizone is almost at mid-whorl; in *C. laticarinata* Cox (Cenomanian) it occurs at a sutural ledge; and in *C. mailleana* (d'Orbigny) (Cenomanian) it occurs high on a whorl where the shape changes from convex to concave. The broad, flat, spiral cords also help to separate *C. percevali* from other species, which generally have much finer or narrower spirals. Its regular conical shape distinguishes it from species of the more angular genus *Bathrotomaria* Cox.

Occurrence. Cenomanian; Devon; ?Wiltshire.

Order CAENOGASTROPODA
Superfamily TURRITELLOIDEA
Family TURRITELLIDAE

Genus TURRITELLA

Turritella (*Turritella*) *dibleyi* Newton
Plate 29, figure 1

Description. Moderately large, slender turritellid; whorl outline strongly convex. Orthocline to feebly prosocline growth lines with a broad, shallow sinus. Ornamented by numerous and very regular spiral and collabral threads, with the latter becoming strong on later whorls and almost forming varices at the intersections.

Remarks. *T.* (*T.*) *unicarinata* (Campanian) from County Antrim, Northern Ireland, and Norwich, Norfolk has a similar ornament but with fewer spiral cords, most of which are of the same order of size, and no collabral rugae. It also has simple arcuate growth lines.

PLATE 28

Occurrence. Lower and Middle Cenomanian; Kent and Cambridgeshire.

Superfamily CERITHIOIDEA
Family PROCERITHIIDAE

Genus ?CIRSOCERITHIUM

'*Cirsocerithium*' *reussi* (Geinitz)
Plate 27, figure 9

Description. Small, elongate, conical shell with well-rounded convex whorls and impressed sutures with slight adapical shelf. Aperture entire with no anterior canal; rounded D-shape with apertural varix; narrow umbilicus; opisthocyrt growth lines. Ornament consists of numerous granulose or finely tuberculate spiral cords (the first being the strongest) and axial ribs. The latter do not proceed on to the base.
Remarks. Distinguished from true *Cirsocerithium* species by the lack of a partially formed anterior canal. Distinguished from *C. subspinosum* (Deshayes) (Albian) which also has an apertural varix and a similar straight and thickened parietal lip, by its granulose ornament and the presence of an umbilicus. The Cenomanian ?*C. nooryi* Abbass is littoriniform and has coarser granulose spirals.
Occurrence. Turonian, *S. plana* Zone; Hertfordshire and Surrey.

Family CERITHIIDAE

Genus EXECHOCIRSUS

Exechocirsus saundersi (Woods)
Plate 29, figure 5

Description. Medium-sized, slightly elongate cerithiid with almost flat-sided whorls. Distinct, impressed, linear suture. Flat to slightly convex

---

EXPLANATION OF PLATE 28

Figs 1, 3. *Bathrotomaria dixoni* Cox. 1, Lower Cenomanian, Culverhole, Devon; holotype; ×1. 3, Cenomanian, Dover, Kent; apical view of slightly flattened internal mould of paratype; ×0·75.
Figs 2, 4–5. *Bathrotomaria perspectiva* (Mantell). 2, Turonian, *S. plana* Zone, Aston Hill, Oxfordshire; angular form; ×1. 4, Coniacian, *M. cortestudinarium* Zone, Dover, Kent; internal mould of a large shell; ×0·75. 5, ?*B. perspectiva*; Turonian, *S. plana* Zone, Hitchin, Hertfordshire; *c.* ×2.
Fig. 6. *Conotomaria percevali* Cox. Cenomanian Limestone near Axmouth, Devon; holotype; ×1.

base with a granular subcarinate cord near the periphery and numerous spiral threads. Obovate–quadrangular aperture with a short, slightly curved, anterior canal and a callused inner lip. There is a strong varix diametrically opposite the aperture. Ornament of four equal spiral cords with interspaces of varying width containing fine spiral threads, and 16 orthocline axial ribs; rounded tubercles or beads occur at the intersections. Other more variably ornamented forms occur.

Occurrence. Turonian, *S. plana* Zone; common throughout the counties around London.

Superfamily STROMBOIDEA
Family APORRHAIDAE

Genus PERISSOPTERA

Perissoptera mantelli Gardner
Plate 29, figure 4

Description. High-spired, turreted, typical aporrhaid shell; last whorl expanded to produce a 'wing' with posterior spike. Whorls ornamented by opisthocline axial ribs and fine spiral threads; last whorl angular with relatively smooth side, the ribs forming large nodes.

Remarks. Usually found as poorly preserved internal moulds; possibly synonymous with *Perissoptera parkinsoni* (Mantell), the only difference being in the form of the last whorl and the relatively straighter axial ribs in that Albian species.

Occurrence. Lower Cenomanian, Wiltshire and Hampshire.

Genus ?ROSTELLARIA

'Rostellaria' pricei Woodward
Plate 29, figure 2

Description. Large, elongate spire of relatively flat-sided whorls; the outer lip of last whorl expanded to form smooth-edged semicircle. There is an elongate anterior canal. Ornament (when preserved) consists of a narrow band of fine spiral threads at the adapical suture of each whorl.

Remarks. The poor preservation of various large, elongate, gastropod moulds occurring in Upper Cretaceous deposits leads to considerable uncertainty over their identification. This species, however, can be distinguished from the similarly shaped *Chemnitzia woodwardii* Seeley (=*Turritella turbinata* J. de C. Sowerby) in that the latter has slightly more convex whorls which are ornamented by many fine spiral threads.

Occurrence. Cenomanian; Kent, scarce.

Superfamily JANTHINOIDEA
Family EPITONIIDAE

Genus CROSSOTREMA

Crossotrema crebricostatum (Gardner)
Plate 29, figure 3

Description. Moderately large, elongate, turreted shell with strongly convex whorls; a deep, impressed suture is obscured by the axial lamellae, which alternate with those of adjacent whorls. Aperture not known, but shell grows from within aperture and is reflected backwards and outwards to produce a narrow flange which forms the ornamental lamellae. Intricate ornament consists of delicate, axial, collabral, opisthocyrt lamellae and spiral cords, which occur in pairs and produce crenulations in the lamellae. Fine collabral threads produce slight beads where they cross the cords and a lattice where they cross the separating grooves.

Remarks. The elaborate frilled ornament distinguishes this unique epitoniid. The lack of varices separates it from all other crossotremids, while the lack of a basal disc also serves to separate it from other Cretaceous members of the family. A further distinction is the equal development of both axial and spiral ornament.

Occurrence. Campanian, *B. mucronata* Zone; the unique holotype was found in Norfolk.

Superfamily MURICOIDEA
Family FASCIOLARIIDAE

Genus ANOMALOFUSUS

Anomalofusus sp.
Text-figure 8.1A–B

Description. Elongated, high-spired species with a lanceolate aperture and a moderately long anterior canal. There are about 12 sinuous, vertical ribs per whorl, and these diminish and almost disappear on the last whorl.

Remarks. Usually found preserved as internal moulds.

Occurrence. Campanian, *B. mucronata* Zone; County Derry, Northern Ireland.

Subclass OPISTHOBRANCHIA
Order TECTIBRANCHIA
Superfamily ACTEONOIDEA
Family RINGICULIDAE

PLATE 29

Genus AVELLANA

Avellana incrassata (J. Sowerby)
Plate 29, figures 6–8

Description. Small, globose, low-spired shell with oblique, lunate, entire aperture that has a thick reflected lip and denticulate inner margin. Three strong folds occur; a columellar fold, which is virtually horizontal, and two parietal folds that extend to the aperture. Ornament consists of numerous, equally spaced, incised, spiral furrows covered with many axial striae. **Occurrence**. Cenomanian–Santonian/Campanian; widespread throughout southern and eastern England from Devon to Norfolk; ?Campanian in Northern Ireland.

Fig. 1. *Turritella* (*Turritella*) *dibleyi* Newton. Lower Cenomanian, Margett's Pit, Burham, Kent; holotype; *c.* ×2.

Fig. 2. '*Rostellaria*' *pricei* Woodward. Cenomanian, locality unknown; ×0·5.

Fig. 3. *Crossotrema crebricostatum* (Gardner). Campanian, *B. mucronata* Zone, Norwich, Norfolk; holotype; ×2.

Fig. 4. *Perissoptera mantelli* Gardner. Lower Cenomanian, Devizes, Wiltshire; ×1·5.

Fig. 5. *Exechocirsus saundersi* (Woods). Turonian, *S. plana* Zone, Latimer, Buckinghamshire; ×2.

Figs 6–8. *Avellana incrassata* (J. Sowerby). Turonian, *S. plana* Zone, Guildford–Godalming bypass, Surrey. 6, showing ornament of spiral cords with axial elements in the furrows between these; also traces of the thickened outer lip of the aperture at bottom left. 7, internal mould. 8, showing columella and parietal folds. 6, 8, ×2; 7, ×4.

9. AMMONITES

by C. W. WRIGHT and W. J. KENNEDY

In the Chalk ammonites are abundant only at a few horizons. Most levels
in the Cenomanian produce large faunas in one area or another but in the
chalky facies there is no continuous series of faunas at any one locality.
The Lower and Middle Turonian tend to have sparse faunas throughout,
but ammonites generally are rare. The Upper Turonian in the Midlands
and south-east has a fairly abundant fauna in the *S. plana* Zone, but
elsewhere there are few ammonites. One species is moderately common in
the Santonian and a small fauna is found in the upper *I. lingua* Zone in
Yorkshire. Slightly higher faunas occur in the *B. mucronata* chalk of
Norfolk and in the White Limestone of Northern Ireland.

Ammonites occur mainly in three different kinds of preservation in the
Chalk. The most widespread is as composite internal moulds. After death
the shell has been filled with sediment which has partially consolidated;
the shell has then disappeared entirely or been replaced by a film of
pyrites, commonly oxidised to limonite, and the external mould has been
pressed on to the internal mould. Such specimens are often entire but the
inner whorls are poorly preserved; they frequently show distortion, with
symmetrical or asymmetrical secondary compression. In hardgrounds,
such as the Chalk Rock or those in the Norwich Chalk, the filling of the
shell has been firmly cemented at an early stage of burial and the moulds
retain their original proportions.

In the Glauconitic Marls at the base of the Chalk and at certain other
horizons there occur, often abundantly, phosphatised internal moulds,
normally of fragments or nuclei, well preserved or abraded but uncrushed.
In south-west England the Basement Beds above the diachronous base of
the Chalk yield more or less well-preserved entire or fragmentary
specimens either as phosphatic moulds or with phosphatised shell on a fill
of cemented sediment.

British Chalk ammonites were first monographed by Sharpe (1853–57).
This work was later updated by Wright and Wright (1951), who produced
a supplement to Sharpe's monograph. A considerable amount of work on
Chalk ammonite faunas has been done recently and is still in progress. For
more detailed treatment of some groups the following should be consulted:
Wright (1979) and Wright and Kennedy (1981, 1984–96); for terminology,
see Arkell *et al.* (1957) and Wright (1996).

DESCRIPTIONS

Class CEPHALOPODA
Order AMMONOIDEA
Suborder AMMONITINA
Superfamily DESMOCERATACEAE
Family DESMOCERATIDAE
Subfamily PUZOSIINAE

Remarks. Normally with round or oval whorl section, constrictions and weak or no ornament. The suture in Late Cretaceous forms is finely divided and has a strongly recurved part near the umbilical seam.

Genus PUZOSIA

Remarks. Rather evolute; whorl section subcircular to compressed oval or with flat sides, venter rounded; with sinuous to falcoid constrictions, between which are few to many weak ribs, on the outer part of sides in *P.* (*Puzosia*) or arising and branching near the edge of the umbilicus in *P.* (*Anapuzosia*).
Occurrence. Lower Albian–Upper Turonian.

Puzosia (Puzosia) mayoriana (d'Orbigny)
Plate 30, figure 4

Description. Whorl section varies from subcircular to slightly compressed, with inner part of sides flat or slightly convex. Constrictions 4–8 to a whorl, straight on the inner flank, then sinuous or sharply bent forward. Between constrictions are 15–25 weak ribs.
Occurrence. Upper Albian–Upper Cenomanian, *M. geslinianum* Zone; generally distributed in southern England but uncommon.

Genus PARAPUZOSIA

Remarks. Large, compressed, high-whorled, with sharply rounded umbilical rim, flat or slightly convex sides and rounded venter. Constrictions are confined to early whorls but major ribs associated with them persist and sooner or later multiply.
Occurrence. Lower Cenomanian–Upper Campanian.

Subgenus AUSTINICERAS

Remarks. Major ribs fewer than about ten to a whorl until the last, where they multiply.
Occurrence. Lower Cenomanian–Middle Turonian.

Parapuzosia (*Austiniceras*) *austeni* (Sharpe)
Text-figure 9.1

Description. Up to 1500 mm diameter. Sides more or less flat, converging to narrowly arched venter. Early whorls with five constrictions; at about 100 mm diameter these are replaced by major ribs.

Occurrence. Lower Cenomanian, *M. mantelli* Zone, to Turonian, *C. woollgari* Zone; rare in the Lower Cenomanian, common above; generally distributed but rare in the north.

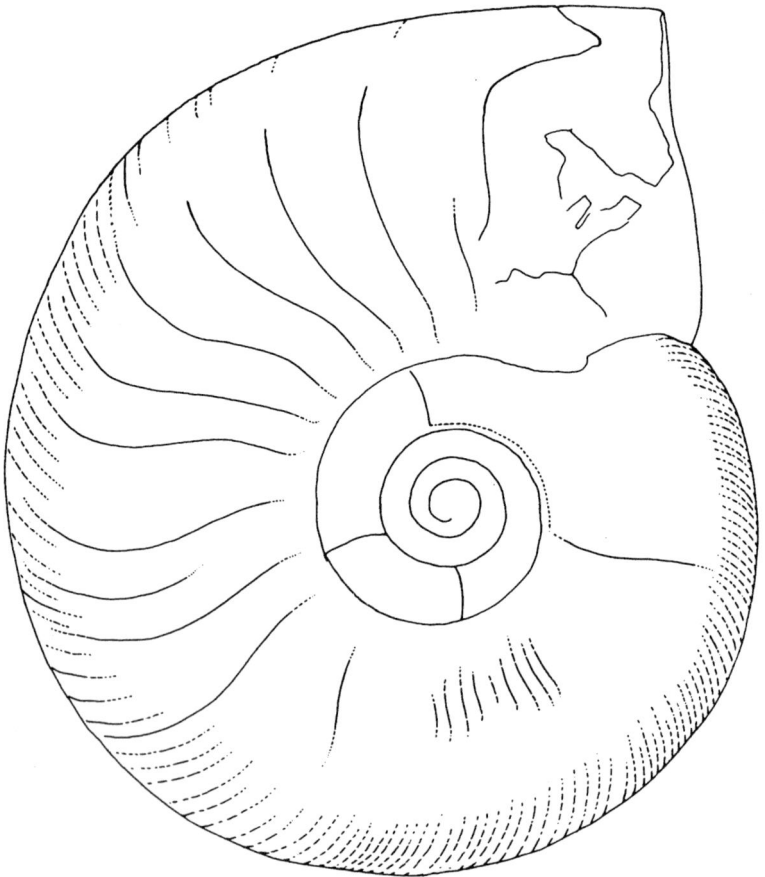

TEXT-FIG. 9.1. *Parapuzosia* (*Austiniceras*) *austeni* (Sharpe). Middle Cenomanian, *A. rhotomagense* Zone, Guildford, Surrey; diagrammatic; ×0·4.

Subgenus PARAPUZOSIA

Remarks. Major ribs numerous after initial nearly smooth stage with constrictions.

Occurrence. ?Coniacian, Santonian–Campanian.

Parapuzosia (*Parapuzosia*) *leptophylla* (Sharpe)
Text-figure 9.2

Description. Very large. Generally badly preserved, and compressed and smooth in appearance. Occasional well-preserved specimens show strong primary ribs branching on the outer part of the sides.

Occurrence. Santonian, *M. coranguinum–M. testudinarius* zones; particularly common in Kent but occurs generally.

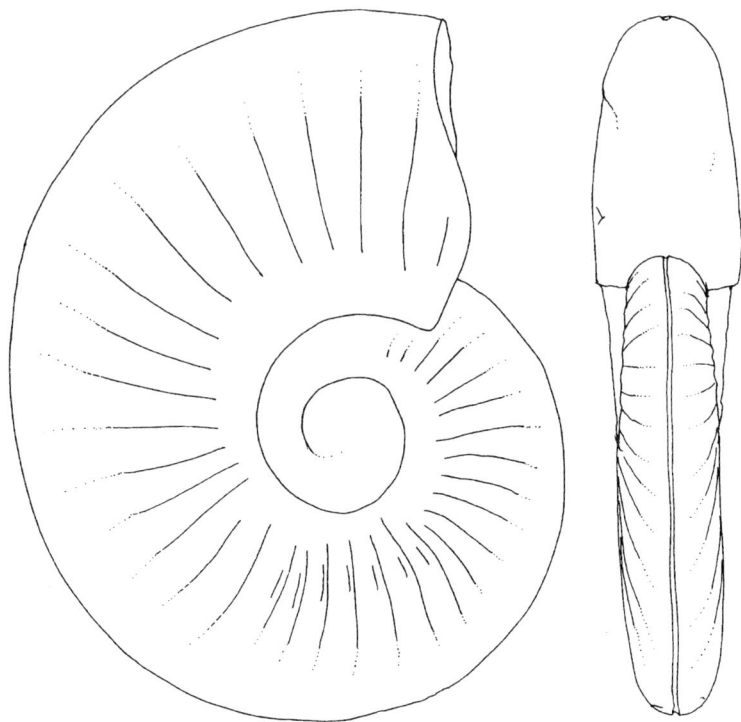

TEXT-FIG. 9.2. *Parapuzosia* (*Parapuzosia*) *leptophylla* (Sharpe). Santonian, *M. coranguinum* Zone, Greenhithe, Kent; diagrammatic; ×0·35.

Subfamily HAUERICERATINAE

Genus HAUERICERAS

Remarks. Compressed, high-whorled with sharp keel; smooth except for constrictions.

Occurrence. Coniacian–Maastrichtian.

Hauericeras pseudogardeni (Schlüter)
Text-figure 9.3

Description. Rather involute, with constrictions curved strongly forward on outer part of sides. Only very poorly preserved internal moulds are known in England.

Occurrence. Lower Campanian, *I. lingua* Zone; rare at Sewerby, East Yorkshire (North Humberside).

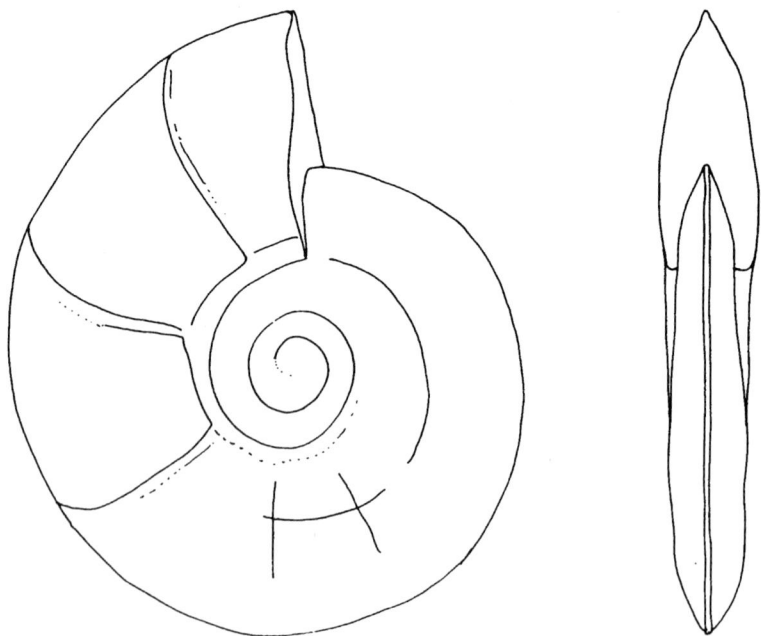

TEXT-FIG. 9.3. *Hauericeras* (*Hauericeras*) *pseudogardeni* (Schlüter). Campanian, *I. lingua* Zone; diagrammatic; ×0·3.

Family PACHYDISCIDAE

Genus LEWESICERAS

Remarks. Whorl section slightly compressed to subcircular; early whorls with strong, rather irregular ribs springing from umbilical tubercles, and constrictions; later whorls with weak or no ornament. Suture with wide-open elements, shallowly incised.

Occurrence. Lower Cenomanian–Upper Turonian.

Lewesiceras peramplum (Mantell)
Text-figure 9.4

Description. Up to 900 mm diameter. Early whorls higher than wide with flat and parallel or slightly converging sides and evenly rounded venter. Later whorls more inflated and finally subcircular to subquadrate in section. Inner whorls with five or six umbilical tubercles, narrow curved primary ribs, and 2–4 short secondaries between. Later ribs become lower and weaker with more secondaries. After diameter of 120 mm, smooth except for 11–15 low bar-like ribs on inner part of sides.

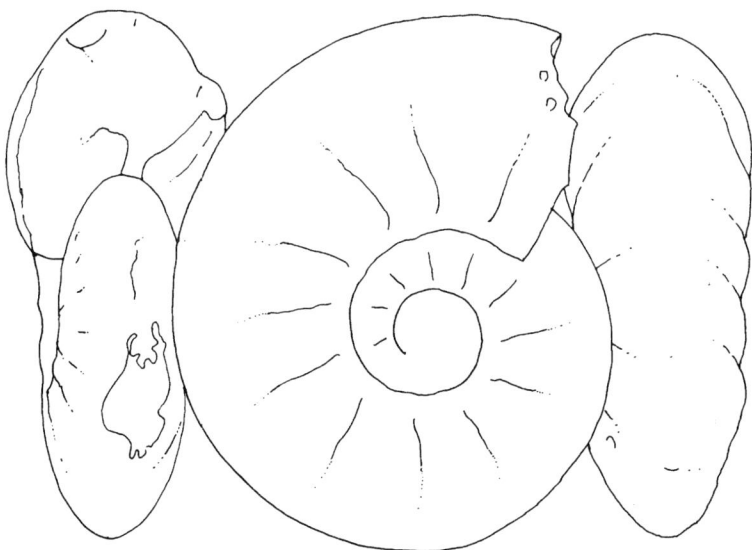

TEXT-FIG. 9.4. *Lewesiceras peramplum* (Mantell). Lower Turonian, *M. nodosoides* Zone (*I. labiatus* Zone); diagrammatic; ×0·4.

Occurrence. Turonian, *M. nodosoides* and *C. woollgari* zones (*I. labiatus* and *T. lata* zones); common and widespread.

Lewesiceras mantelli Wright and Wright
Plate 38, figures 1–2

Description. Smaller than *L. peramplum*; whorl section inflated in mid-growth, umbilical tubercles very prominent.
Occurrence. Turonian, *S. neptuni* Zone (*S. plana* Zone); the commonest ammonite in the Chalk Rock, but also widespread in normal chalk facies.

Superfamily HOPLITACEAE
Family HOPLITIDAE
Subfamily HOPLITINAE

Remarks. Small to moderate-sized, compressed to very inflated; venter flat, concave or with a central groove, normally bordered by alternate, oblique or parallel, clavate tubercles.
Occurrence. Lower Albian–Lower Cenomanian.

Genus HYPHOPLITES

Remarks. Rather small, compressed to inflated, with grooved venter; ribs sickle-shaped.
Occurrence. Upper Albian–Lower Cenomanian.

Hyphoplites falcatus (Mantell)
Plate 30, figures 1–2, 9

Description. Compressed, involute, flat-sided, with flat sickle-shaped ribs in branching pairs (*H. f. faurora* Wright and Wright), single (*H. f. falcatus*,

---

EXPLANATION OF PLATE 30

Figs 1–2. *Hyphoplites falcatus falcatus* (Mantell). Lower Cenomanian, *M. mantelli* Zone, Stoneham, near Lewes, Sussex; ×1.

Fig. 3. *Forbesiceras bicarinatum* Szász. Upper Cenomanian, *C. guerangeri* Zone, Shapwick Grange, Devon; ×1.

Fig. 4. *Puzosia* (*Puzosia*) *mayoriana* (d'Orbigny). Probably Middle Cenomanian, *A. rhotomagense* Zone, Ventnor, Isle of Wight; ×1.

Figs 5–6. *Hyphoplites curvatus curvatus* (Mantell). Lower Cenomanian, *M. mantelli* Zone, Warminster area, Wiltshire; ×1.

Figs 7–8. *Forbesiceras largilliertianum* (d'Orbigny). Lower Cenomanian, *M. mantelli* Zone, *M. couloni* horizon, Hutchin's Pit, Wilmington, Devon; ×1.

Fig. 9. *Hyphoplites falcatus interpolatus* Wright and Wright. Lower Cenomanian, *M. mantelli* Zone, Search Farm, near Mere, Wiltshire; ×1.

PLATE 30

1

2

4

3

5

6

7

8

9

Pl. 27, figs 1–2*)* or split so that the 'sickle' has more than one 'handle' (*H. f. interpolatus* Wright and Wright, Pl. 27, fig. 9); umbilical tubercles absent or weak.

Occurrence. Upper Albian, *S. dispar* Zone, to Lower Cenomanian, *M. dixoni* Zone; widespread in southern England.

Hyphoplites curvatus (Mantell)
Plate 30, figures 5–6

Description. Involute, compressed and high-whorled to evolute and depressed. Ribs branch from weak to strong umbilical tubercles and others may be intercalated. Sharp or blunt inner and outer ventrolateral tubercles are present on the phragmocone but disappear on the body chamber. *H. c. curvatus* is slightly compressed and has strong tubercles and many irregularly intercalated and branched fine ribs.

Occurrence. Lower Cenomanian, *M. mantelli* and *M. dixoni* zones; widely distributed in southern England.

Family SCHLOENBACHIIDAE

Remarks. Keeled derivatives of Hoplitinae.
Occurrence. Cenomanian.

Genus SCHLOENBACHIA

Remarks. Very variable in whorl section and strength and density of ribs and tubercles.
Occurrence. Cenomanian, *M. mantelli* – *C. guerangeri* zones.

Schloenbachia varians (J. Sowerby)
Plate 31, figures 1–2, 5–9

Description. Varies, probably continuously, from compressed, high-whorled and fine-ribbed with no lateral tubercles and only minute umbilical and outer ventrolateral ones (forma *subplana*), to inflated and depressed with a strong mid-lateral tubercle (forma *ventriosa*); intermediates may have virtually no ribs (forma *glabra)* or strong ribs with two lateral tubercles (forma *subtuberculata)*. Many variants have been described as species.

Occurrence. Lower Cenomanian, *M. mantelli* Zone, to Middle Cenomanian, *A. jukesbrownei* Zone; ?Upper Cenomanian, *C. guerangeri* Zone; throughout England and Northern Ireland, generally common, rare in Yorkshire.

Schloenbachia lymense Spath
Plate 31, figures 3–4

Description. Small, varies little; whorl-section rectangular, with a lateral bulla near the umbilical tubercle.

Occurrence. Upper Cenomanian, *C. guerangeri* Zone (Bed C of Cenomanian Limestone); Devon.

Superfamily ACANTHOCERATACEAE
Family FORBESICERATIDAE

Remarks. Very involute and compressed with sides parallel or converging to a narrow venter. Suture with long narrow elements, including an extra lobe dividing the first lateral saddle.

Occurrence. Upper Albian–Upper Cenomanian.

Genus FORBESICERAS

Occurrence. Throughout the Cenomanian except the uppermost part.

Forbesiceras largilliertianum (d'Orbigny)
Plate 30, figures 7–8

Description. Ribs very fine, flexuous, crossing the venter transversely for most of growth; later the ribs disappear and the venter becomes bicarinate and finally rounded.

Occurrence. Lower Cenomanian, *M. mantelli* Zone to Middle Cenomanian, *A. jukesbrownei* Zone; rare, throughout southern England.

Forbesiceras bicarinatum Szász
Plate 30, figure 3

Description. Virtually smooth at all stages; venter bicarinate until body chamber.

Occurrence. Upper Cenomanian, *C. guerangeri* Zone; rather common in the Cenomanian Limestone of Devon.

Family ACANTHOCERATIDAE
Subfamily MANTELLICERATINAE

Remarks. Venter flat or concave, bordered by opposite ventrolateral tubercles.

Genus MANTELLICERAS

PLATE 31

1

2

3

4

5

6

7

8

9

Remarks. Of moderate size, macroconchs up to twice as big as micro-conchs; compressed and high-whorled to depressed and inflated; inner whorls finely ribbed, body chamber generally coarsely ribbed; always with an umbilical and a pair of ventrolateral tubercles at some stage of growth, commonly also one or two laterals.
Occurrence. Lower Cenomanian.

Mantelliceras mantelli (J. Sowerby)
Plate 33, figures 1–2

Description. Ribs in early and middle growth rather sharp with more or less equally spaced umbilical, lateral and inner and outer ventrolateral tubercles, giving a polygonal whorl section; on outer whorl all but the umbilical and outer ventrolateral disappear and the ribs become broadly rounded. Ribs number 30–45 to a whorl and tubercles vary from weak to strong.
Occurrence. Lower Cenomanian, *M. mantelli* and *M. dixoni* zones; common and generally distributed in southern England in the *M. mantelli* Zone, rare in Kent and Devon in the *M. dixoni* Zone.

Mantelliceras saxbii (Sharpe)
Plate 32, figures 1–3

Description. Rather involute at first, decreasing so with age; com-pressed with more or less flat sides; ribs 27–40, generally flexuous, directed forwards on early whorls; inner and outer ventrolateral tubercles present on early whorls, the former commonly absent later.
Occurrence. Lower Cenomanian, *M. mantelli* and *M. dixoni* zones; generally distributed and common in southern England in the *M. mantelli* Zone, especially in the *M. saxbii* Subzone, rare in the *M. dixoni* Zone.

EXPLANATION OF PLATE 31

Figs 1–2. *Schloenbachia varians* (J. Sowerby) forma *subplana*. Lower Cenomanian, *M. mantelli* Zone, Sussex; ×1.
Figs 3–4. *Schloenbachia lymense* Spath. Upper Cenomanian, *C. guerangeri* Zone, Devon; ×1.
Fig. 5. *Schloenbachia varians* (J. Sowerby) forma *subtuberculata*. Lower Cenomanian, *M. mantelli* Zone, Warminster area, Wiltshire; ×1.
Fig. 6. *Schloenbachia varians* (J. Sowerby) forma *subvarians*. Lower Cenomanian, *M. mantelli* Zone, Warminster area, Wiltshire; ×1.
Figs 7–9. *Schloenbachia varians* (J. Sowerby) type form. Lower Cenomanian, *M. mantelli* Zone, Sussex; ×1.

Mantelliceras dixoni Spath
Plate 32, figures 4–5

Description. Whorl section higher than wide, oval; ribs rather distant, the inner part of the primaries very prominent, with strong umbilical tubercles and mid-lateral angulation; on the body chamber the ribs form a straight bar across the venter.

Occurrence. Lower Cenomanian, *M. dixoni* Zone; rather rare in Kent, Sussex and Cambridgeshire, common in Devon.

Subfamily ACANTHOCERATINAE

Remarks. Venter with siphonal as well as opposite ventrolateral tubercles, at least in early stages.

Occurrence. Lower Cenomanian–Turonian.

Genus ACANTHOCERAS

Remarks. Rather large, whorl section square or rectangular, compressed to depressed; inner whorls with umbilical, inner and outer ventrolateral and siphonal tubercles; on outer whorls the two ventrolaterals may fuse into a very large tubercle and the siphonal tends to disappear.

Occurrence. Middle Cenomanian.

Acanthoceras rhotomagense (Brongniart)
Text-figure 9.5

Description. Continuously variable in whorl-section, density of ribs and relative strength of ribs and tubercles. In the slightly later *A. jukesbrownei* (Spath) ribs remain long and short to the end (Text-fig. 9.6).

Occurrence. Middle Cenomanian, *A. rhotomagense* Zone; generally distributed and common in the south, increasingly rare northwards.

Genus CALYCOCERAS

Remarks. Small to moderately large, markedly dimorphic in size. Ribs generally dense and dominant over the tubercles, continuous over the venter.

Occurrence. Middle–Upper Cenomanian.

EXPLANATION OF PLATE 32

Figs 1–3. *Mantelliceras saxbii* (Sharpe). Lower Cenomanian, *M. mantelli* Zone, Isle of Wight; ×1.

Figs 4–5. *Mantelliceras dixoni* Spath. Lower Cenomanian, *M. dixoni* Zone, Dover, Kent; ×1.

PLATE 32

1   2   3

4   5

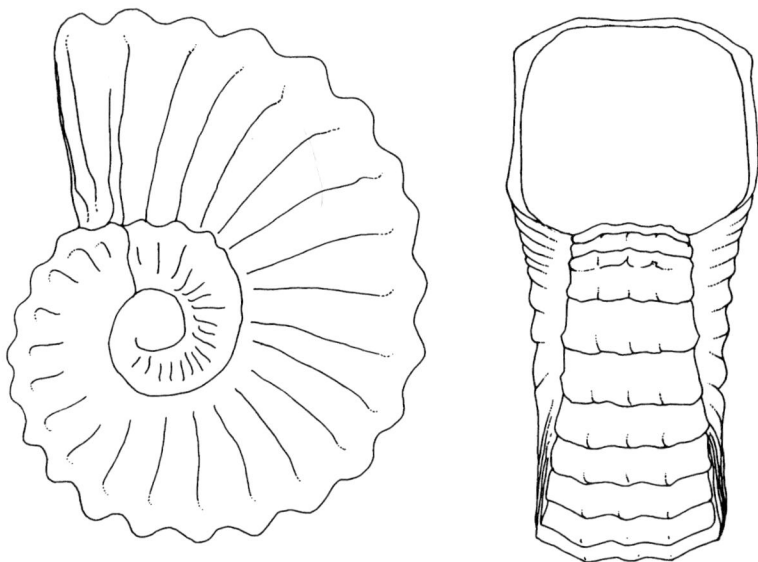

TEXT-FIG. 9.5. *Acanthoceras rhotomagense* (Brongniart). Middle Cenomanian, *A. rhotomagense* Zone; diagrammatic; ×0·4.

Subgenus GENTONICERAS

Remarks. Small; inner whorls with slight to sharp and prominent umbilical, inner and outer ventrolateral and siphonal tubercles on dense ribs; body chamber normally with sharp and high, rather distant ribs and only umbilical tubercles.

Occurrence. Middle–Upper Cenomanian.

Calycoceras (*Gentoniceras*) *gentoni* (Brongniart)
Plate 33, figures 8–10

Description. Whorl section higher than wide, sides tending to be flat, increasing very slowly with growth.

Occurrence. Middle Cenomanian, *A. rhotomagense* Zone; widespread and often common in southern and especially south-west England.

Subgenus NEWBOLDICERAS

Remarks. Rather large; whorl-section angular and tubercles generally persist to body chamber.

Occurrence. Middle–Upper Cenomanian.

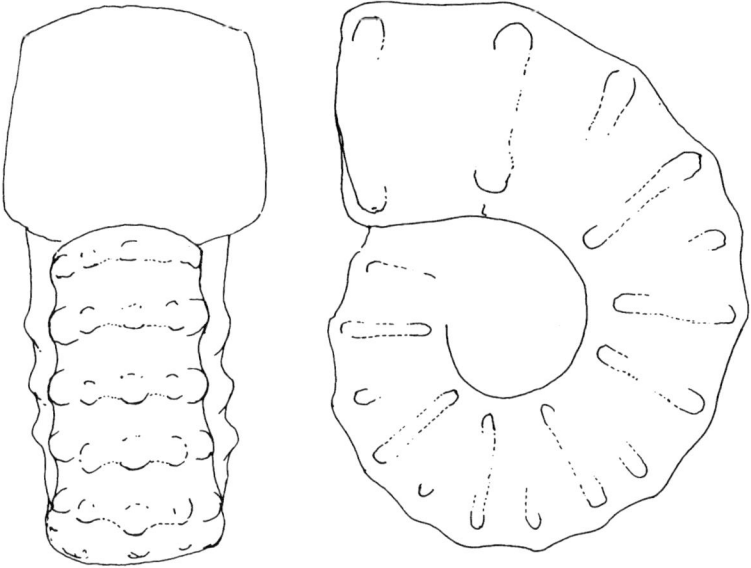

TEXT-FIG. 9.6. *Acanthoceras jukesbrownei* (Spath). Middle Cenomanian, *A. jukesbrownei* Zone; diagrammatic; ×0·4.

Calycoceras (*Newboldiceras*) *asiaticum* (Jimbo)
Plate 33, figures 5–7

Description. Variable; middle whorls polygonal in section with a wide facet between inner and outer ventrolateral tubercles. Inner whorls, particularly those of *C.* (*N.*) *a. spinosum* (Kossmat), closely resemble those of *Acanthoceras*.

Remarks. This species has commonly been referred to as *C. newboldi* (Kossmat).

Occurrence. Middle Cenomanian, *A. rhotomagense* and *A. jukesbrownei* zones; ?Upper Cenomanian, *C. guerangeri* Zone; widely distributed in southern England, particularly common in the south-west.

Subgenus PROEUCALYCOCERAS

Remarks. Inner whorls with dense ribs, subdued tubercles and rather broad flat venter, outer with coarse ribs.

Occurrence. Middle–Upper Cenomanian.

PLATE 33

1 2 3 4

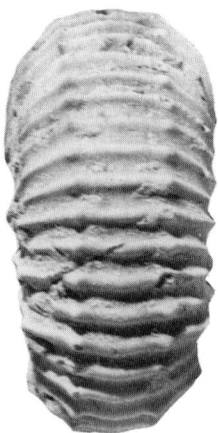

5 6 7

8 9 10

Calycoceras (*Proeucalycoceras*) *guerangeri* (Spath)
Plate 34, figures 1–2

Description. Moderate-sized; early whorls with fine ribs and convex sides converging to almost flat venter; body chamber with coarse blunt ribs and almost no tuberculation.
Occurrence. Upper Cenomanian, *C. guerangeri* Zone; widespread in southern England, rare in the south-east, common in Devon.

Subgenus CALYCOCERAS

Remarks. Inflated and depressed with very broadly rounded venter; ventral tuberculation subdued after an early stage.
Occurrence. Top of Middle Cenomanian–Upper Cenomanian.

Calycoceras (*Calycoceras*) *naviculare* (Mantell)
Plate 35, figures 9–10

Description. Rather large; innermost whorls with depressed coronate section with short flanks marked by sharp umbilical and inner ventrolateral tubercles; venter broad, almost flat, with or without weak outer ventro-lateral and siphonal tubercles. Middle and later whorls less depressed. The figured specimen is a microconch with fine, backwardly curved ribs; macroconchs are much larger and have coarse, rounded ribs.
Occurrence. Upper Cenomanian, *C. guerangeri* and *M. geslinianum* zones; southern England from Kent to Devon.

Genus PROTACANTHOCERAS

Remarks. Dwarf forms derived from *Acanthoceras*, characterised by body chamber ornament of non-tuberculate ribs in chevrons on the venter.
Occurrence. Middle–Upper Cenomanian.

EXPLANATION OF PLATE 33

Figs 1–2. *Mantelliceras mantelli* (J. Sowerby). Lower Cenomanian, *M. mantelli* Zone, Isle of Wight; × 1.

Figs 3–4. *Metoicoceras geslinianum* (d'Orbigny). Upper Cenomanian, *M. geslinianum* Zone, Devon (see also Pl. 35, figs 7–8); × 1.

Figs 5–7. *Calycoceras* (*Newboldiceras*) *asiaticum* (Jimbo). Middle Cenomanian, *A. rhotomagense* Zone, Ringstead Quarry, Dorset; × 1.

Figs 8–10. *Calycoceras* (*Gentoniceras*) *gentoni* (Brongniart). Middle Cenomanian, *A. rhotomagense* Zone, *T. acutus* Subzone, Snowden Hill Quarry, Chard, Somerset; × 1.

PLATE 34

Protacanthoceras tuberculatum tuberculatum (Thomel)
Plate 35, figures 4–5

Description. Diameter 20–37 mm; early and middle whorls similar to
nuclei of *Acanthoceras* but more involute; body chamber with strong ribs
on the venter. Ribs strong, 17–19 to a whorl, with prominent tubercles. *P.
t. mite* Wright and Kennedy has more delicate ornament and the later *P. t.
devonense* is smaller and has fewer ribs.
Occurrence. Middle–Upper Cenomanian, *A. rhotomagense–C. guerangeri*
zones; rare in south-east England, common in Dorset and Devon.

Protacanthoceras bunburianum (Sharpe)
Plate 35, figures 1–3

Description. Variably compressed, with narrow fastigiate venter bearing
fine, close, outer ventrolateral and siphonal tubercles that turn into chevron
ribs on the body chamber; ornament tends to weaken on the sides.
Occurrence. Upper Cenomanian, *C. guerangeri* Zone; common in Devon.

Genus THOMELITES

Remarks. More or less evolute, inflated to compressed, sides parallel out
to the inner ventrolateral tubercle, then converging to narrow, flat venter
bordered by clavate outer ventrolateral tubercles; siphonal tubercles are
present on inner whorls but may disappear on outer.
Occurrence. Upper Cenomanian.

Thomelites sornayi (Thomel)
Plate 35, figure 6

Description. Variable from inflated, with very strong umbilical tubercles
and a marked facet between inner and outer ventrolaterals, to compressed,
with nearly flat and parallel sides and subdued tuberculation.
Occurrence. Upper Cenomanian, *C. guerangeri* Zone; very rare in Sussex
and Dorset; common in Devon.

EXPLANATION OF PLATE 34

Figs 1–2. *Calycoceras (Proeucalycoceras) guerangeri* (Spath). Upper
  Cenomanian, *C. guerangeri* Zone, Shapwick Grange, Devon; ×1.
Figs 3–5. *Neocardioceras juddii* (Barrois and de Guerne). Upper Cenomanian, *N.
  juddii* Zone, Haven Cliff, Seaton, Devon; ×1.

PLATE 35

Ammonites 197

Genus EUCALYCOCERAS

Remarks. Rather small, high-whorled and flat-sided with dense, high, narrow ribs, that may become flat-topped with narrow interspaces on the body chamber. Umbilical tubercles and outer ventrolateral persist but inner ventrolaterals and siphonals tend to disappear in later growth.
Occurrence. Middle–Upper Cenomanian.

Eucalycoceras pentagonum (Jukes-Browne)
Plate 36, figures 6–7

Description. Early whorls with nearly square section, later ones higher with arched venter. Ribs 55–65 to a whorl.
Occurrence. Upper Cenomanian, *C. guerangeri* Zone; rather rare in Devon.

Genus NEOCARDIOCERAS

Remarks. Small, compressed, finely ribbed, with umbilical, inner and outer ventrolateral and siphonal tubercles, the latter forming a nodate keel.
Occurrence. Uppermost Cenomanian.

Neocardioceras juddii juddii (Barrois and Guerne)
Plate 34, figures 3–5

Description. Evolute, sides flat and parallel on inner two-thirds, then converging to more or less flat, narrow venter; ribs bear a thin, high, umbilical bulla, run straight for the inner two-thirds, then turn sharply forwards at a slight inner ventrolateral tubercle. The contemporary *N. j. barroisi* Wright and Kennedy is more inflated and has more rounded ribs and blunter tubercles.
Occurrence. Upper Cenomanian, *N. juddii* Zone; common in Devon.

EXPLANATION OF PLATE 35

Figs 1–3. *Protacanthoceras bunburianum* (Sharpe). Upper Cenomanian, *C. guerangeri* Zone, Whitlands Beach, Devon; ×1.
Figs 4–5. *Protacanthoceras tuberculatum tuberculatum* (Thomel). Middle Cenomanian, *A. jukesbrownei* Zone, Osmington, Dorset; ×1.
Fig. 6. *Thomelites sornayi* (Thomel). Upper Cenomanian, *C. guerangeri* Zone, Shapwick Grange, Lyme Regis, Devon; ×1.
Figs 7–8. *Metoicoceras geslinianum* (d'Orbigny). Upper Cenomanian, *M. geslinianum* Zone, Burham, Kent (see also Pl. 33, figs 3–4); ×1.
Figs 9–10. *Calycoceras (Calycoceras) naviculare* (Mantell). Upper Cenomanian, *M. geslinianum* Zone, Betchworth Limeworks, Surrey; ×1.

Genus WATINOCERAS

Remarks. Generally small; compressed, quadrate or rounded in section, with flat venter bordered by fine or coarse outer ventrolateral clavi, without distinct siphonal tubercles.
Occurrence. Lower Turonian.

Watinoceras praecursor Wright and Kennedy
Plate 36, figures 3–5

Description. Inner whorls fine-ribbed, outer with alternate ribs strong and bearing umbilical and inner and outer ventrolateral tubercles or spines. Venter flat, with little or no trace of ribs. *W. praecursor* has more evolute coiling, the fine-ribbed stage persisting longer, higher whorl section and wider venter than *W. coloradoense* (Henderson), which has not yet been found in England.
Occurrence. Lower Turonian, *W. coloradoense* Zone (basal *I. labiatus* Zone); common in Devon.

Subfamily EUOMPHALOCERATINAE

Remarks. Multituberculate offshoots of *Calycoceras* with constrictions on at least the inner whorls.
Occurrence. Middle Cenomanian–Coniacian.

Genus EUOMPHALOCERAS

Remarks. Very evolute, with depressed section, long, sloping umbilical wall and broad venter on which are numerous ribs, transverse or in chevrons, with or without outer ventrolateral and siphonal tubercles; generally with constrictions. Suture with widely splayed lateral lobe and first lateral saddle with oblique outer side.
Occurrence. Upper Cenomanian.

EXPLANATION OF PLATE 36

Figs 1–2. *Euomphaloceras euomphalum* (Sharpe). Upper Cenomanian, *C. guerangeri* Zone, beach below Whitlands, Devon, near Lyme Regis, Dorset; ×1.

Figs 3–5. *Watinoceras coloradoense praecursor* Wright and Kennedy. Lower Turonian, *W. coloradoense* Zone (basal *I. labiatus* Zone). 3–4, Haven Cliff, Seaton, Devon; ×1. 5, beach below Whitlands, Devon, near Lyme Regis, Dorset, ×1.

Figs 6–7. *Eucalycoceras pentagonum* (Jukes-Browne). Upper Cenomanian, *C. guerangeri* Zone, beach below Whitlands, Devon, near Lyme Regis, Dorset; ×1.

PLATE 36

1

2

3

4

5

6

7

Euomphaloceras euomphalum (Sharpe)
Plate 36, figures 1–2

Description. Early whorls with sharp inner ventrolateral tubercles bordering a broad, flat venter with ribs and constriction. This may persist with weak outer ventrolateral and siphonal tubercles or the outer ventrolaterals may become stronger and form obtuse chevrons on the venter.
Occurrence. Upper Cenomanian, *C. guerangeri* Zone; rather common in Devon.

Genus ROMANICERAS

Remarks. Large, up to 500 mm diameter, evolute, with round to subquadrate whorl section. Ribs either single or branching at umbilical or lateral tubercles or intercalated; long ribs have nine, 11 or 13 tubercles in all, short ones at least three. Earliest whorls have collared constrictions.
Occurrence. Middle–Upper Turonian.

Subgenus ROMANICERAS

Remarks. With nine rows of tubercles in all.
Occurrence. As for genus.

Romaniceras (*Romaniceras*) *deverianum* (d'Orbigny)
Plate 37, figures 1–2

Description. Ribs fine at first, coarser later, 32–42 to a whorl, most arising singly or in pairs from the umbilical tubercles, a few intercalated.
Occurrence. Turonian, *C. woollgari* Zone (*T. lata* Zone); rare in Kent, Sussex, Hampshire, Bedfordshire and Buckinghamshire.

Subgenus YUBARICERAS

Remarks. With 11 rows of tubercles in all.
Occurrence. As for genus.

Romaniceras (*Yubariceras*) *ornatissimum* (Stoliczka)
Plate 37, figures 3–4

Description. Variable in degree of inflation and strength of the tubercles, which disappear in the adult stage.
Occurrence. Turonian, *C. woollgari* Zone (*T. lata* Zone); rare in Sussex and Berkshire.

Subfamily MAMMITINAE

Remarks. Normally with sparse low ribs, bearing blunt umbilical and inner and outer ventrolateral tubercles, and a flat or concave venter.
Occurrence. Upper Cenomanian–Lower Coniacian.

Genus METOICOCERAS

Remarks. Compressed and involute to square-whorled and evolute, with ribs commonly flat.
Occurrence. Upper Cenomanian.

Metoicoceras geslinianum (d'Orbigny)
Plate 33, figures 3–4; Plate 35, figures 7–8

Description. Variable, either evolute with body chamber markedly uncoiling and more inflated than the spire or involute with the body chamber only slightly uncoiling and more compressed than the spire.
Occurrence. Upper Cenomanian, *M. geslinianum* Zone; widespread in central and southern England, common in Devon.

Genus MAMMITES

Remarks. Rather large with trapezoidal to rectangular whorl-section; distant blunt ribs bear distinct inner and outer ventrolateral tubercles on the inner whorls but on the outer these may fuse into large upward- or outward-directed horns.
Occurrence. Turonian.

Mammites nodosoides (Schlüter)
Text-figure 9.7

Description. On the last whorl there are very strong, blunt, umbilical tubercles and ventrolateral horns joined by feeble ribs.
Occurrence. Turonian, *M. nodosoides* Zone (*I. labiatus* Zone); generally distributed, rare in the north, more common in the south.

Family VASCOCERATIDAE

Remarks. Derivatives of Acanthoceratidae that tend to lose ribs and tubercles and to have a simplified, only slightly incised suture.
Occurrence. Upper Cenomanian–Upper Turonian.

Subfamily VASCOCERATINAE
Genus FAGESIA

Remarks. Globose, cadicone or coronate; inner whorls with few, coarse ribs springing from blunt umbilical tubercles; outer whorls with similar

PLATE 37

ornament or becoming smooth. Suture variable but commonly with deeper and more incised elements than most of the family.

Occurrence. Lower Turonian.

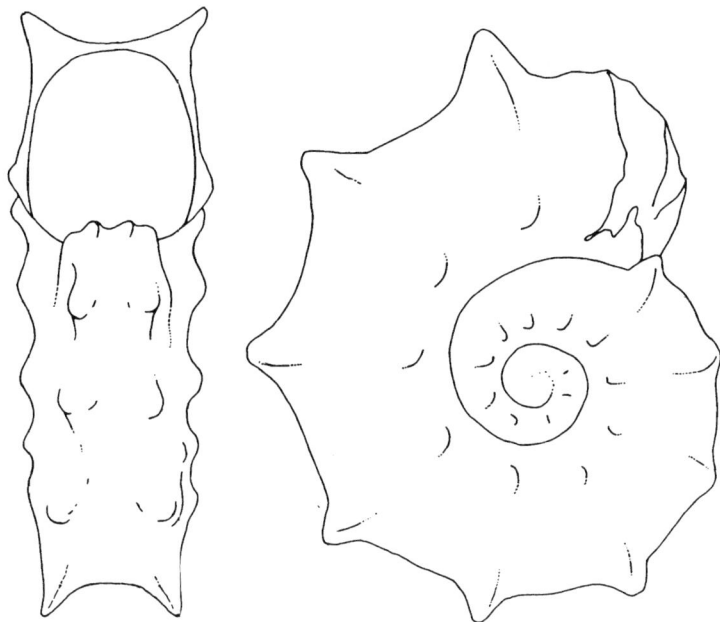

TEXT-FIG. 9.7. *Mammites nodosoides* (Schlüter). Lower Turonian, *M. nodosoides* Zone (*I. labiatus* Zone); diagrammatic; ×0·3.

EXPLANATION OF PLATE 37

Figs 1–2. *Romaniceras* (*Romaniceras*) *deverianum* (d'Orbigny). Middle Turonian, *C. woollgari* Zone (*T. lata* Zone), Amberley, Sussex; ×1.

Figs 3–4. *Romaniceras* (*Yubariceras*) *ornatissimum* (Stoliczka). Middle Turonian, *C. woollgari* Zone (*T. lata* Zone), southern England; ×1.

Figs 5–6. *Collignoniceras woollgari* (Mantell). Middle Turonian, *C. woollgari* Zone (*T. lata* Zone), Lewes, Sussex; ×1.

Fig. 7. *Subprionocyclus neptuni* (Geinitz). Upper Turonian, *S. neptuni* Zone (*S. plana* Zone), Hitch Wood pit, Stevenage, Hertfordshire; ×1.

Figs 8–9. *Subprionocyclus branneri* (Anderson). Upper Turonian, *S. neptuni* Zone (*S. plana* Zone), Hitch Wood pit, Stevenage, Hertfordshire; ×1.

Figs 10–11. *Subprionocyclus normalis* (Anderson). Upper Turonian, *S. neptuni* Zone (*S. plana* Zone), no locality data; ×1.

Fagesia catinus (Mantell)
Text-figure 9.8

Description. Early whorls more or less hexagonal in section, later broader and more depressed with wide venter bearing feeble ribs springing from very large, blunt, lateral tubercles.

Occurrence. Turonian, *M. nodosoides* Zone (*I. labiatus* Zone); Kent, Sussex, Dorset, Devon and Cambridgeshire, most common in the south-west.

Family COLLIGNONICERATIDAE

Remarks. Involute to evolute with compressed oval to square whorl section and one or three serrate or entire keels; mostly with strong ribs bearing from one to five tubercles.

Occurrence. Upper Cenomanian–Middle Campanian.

Subfamily COLLIGNONICERATINAE

Remarks. With a single serrate keel. Inner and outer ventrolateral tubercles may persist or may fuse into large horns on the outer whorl.

Occurrence. Upper Cenomanian–Middle Coniacian.

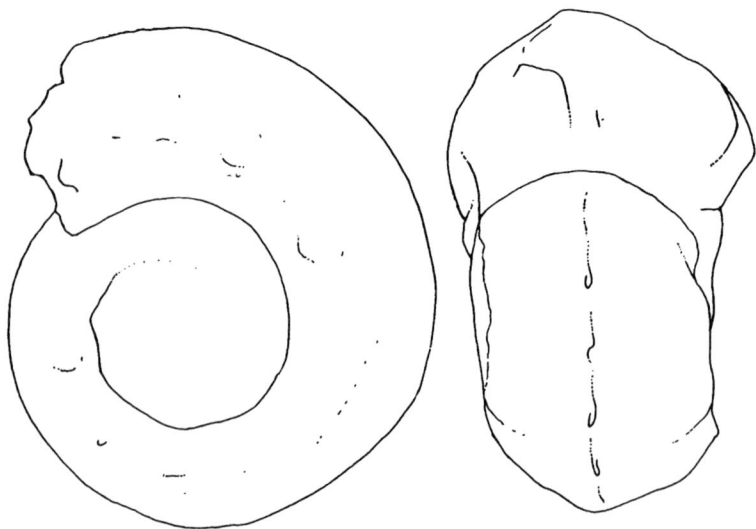

TEXT-FIG. 9.8. *Fagesia catinus* (Mantell). Lower Turonian, *M. nodosoides* Zone (*I. labiatus* Zone); diagrammatic; ×0·3.

Genus COLLIGNONICERAS

Remarks. Medium-sized; early whorls compressed, parallel-sided, with fine or coarse, mostly long ribs with umbilical bullae and outer ventro-lateral and high siphonal clavi. Later whorls similar or with strengthening tubercles and tendency to form large ventrolateral horns between which paired ribs may cross the venter.
Occurrence. Turonian.

Collignoniceras woollgari (Mantell)
Plate 37, figs 5–6

Description. Ribs on inner whorls generally fairly distant, high and bar-like. Later whorls variable; macroconchs with large ventrolateral horns.
Occurrence. Turonian, *C. woollgari* Zone (*T. lata* Zone); rare in Surrey, Sussex and Berkshire.

Genus SUBPRIONOCYCLUS

Remarks. Small, similar to *Collignoniceras* but ornament more regular and with inner as well as outer ventrolateral tubercles at least at some growth stage.
Occurrence. Upper Turonian.

Subprionocyclus neptuni (Geinitz)
Plate 37, figure 7

Description. Moderately involute with dense ribbing and fairly strong tuberculation. Intermediate between the strongly tuberculate *S. branneri* (Anderson) on one side and the compressed and finely ornamented *S. hitchinensis* (Billinghurst) and *S. normalis* (Anderson) on the other.
Occurrence. Turonian, *S. neptuni* Zone (*S. plana* Zone); common in Chalk Rock and nodular facies throughout southern and central England.

Subprionocyclus normalis (Anderson)
Plate 37, figures 10–11

Description. Compressed, moderately to very involute, with flat to fasti-giate venter. Ribs sinuous to sickle-shaped, tending to be sparse and flat.
Occurrence. Upper Turonian, *S. neptuni* Zone (*S. plana* Zone); rather rare in Surrey and Bedfordshire.

Subprionocyclus branneri (Anderson)
Plate 37, figures 8–9

Description. Ribs rather distant, tuberculation coarser than in *S. neptuni;* the siphonal tubercles tend to be very high.

Occurrence. Upper Turonian, *S. neptuni* Zone (*S. plana* Zone); moderately common in the Midlands.

Suborder ANCYLOCERATINA
Superfamily TURRILITACEAE
Family ANISOCERATIDAE

Remarks. Coiling irregular, generally in very open helix followed by straight shafts. Ventrolateral and commonly also lateral tubercles present.

Occurrence. ?Upper Aptian, Albian–Coniacian.

Genus ANISOCERAS

Remarks. Main ribs are looped in pairs or threes between the lateral and ventrolateral tubercles and across the venter.

Occurrence. Upper Albian–Upper Turonian.

Anisoceras plicatile (J. Sowerby)
Plate 41, figure 6

Description. Whorl section more or less circular. Ribs are fine, dense and rounded; looped and tuberculate ribs are separated by two or three nontuberculate ones.

Occurrence. Middle Cenomanian, *A. rhotomagense* Zone; rather common in southern England.

Genus IDIOHAMITES

Remarks. Coiling generally in one plane. The ribs joining the ventrolateral tubercles across the venter are single, not looped. Lateral tubercles rarely present.

Occurrence. Upper Albian–Upper Cenomanian.

Idiohamites alternatus (Mantell)
Plate 40, figures 1–2

Description. Whorl section compressed, ribs high, narrow and distant (only 4–6 in a length equal to the whorl height), alternately with and without ventrolateral tubercles.

Occurrence. Lower Cenomanian, *M. mantelli* Zone; rather uncommon in the south from Sussex to Devon.

Genus ALLOCRIOCERAS

Remarks. Similar to *Idiohamites* but early whorls at least helical and twisted and with sharp ribs; later whorls may have blunter ribbing.

Occurrence. Upper Cenomanian–Lower Coniacian.

Allocrioceras angustum (J. de C. Sowerby)
Plate 40, figures 11–12

Description. Whorl section oval with flat venter emphasised by outwardly directed sharp ventrolateral spines on all or alternate ribs; even if tubercles are present on all ribs the latter are alternately weak and strong.

Occurrence. Upper Turonian, *S. neptuni* Zone (*S. plana* Zone); widely distributed in south and as far north as Yorkshire.

Family HAMITIDAE

Remarks. Coiling generally in an irregular open planispire followed by two or three subparallel shafts. Ribs non-tuberculate, annular or inter-rupted on the lower side or obsolescent.

Occurrence. Lower Albian–Upper Cenomanian.

Genus HAMITES

Remarks. Ribs generally rather strong. Microconchs have constricted and collared apertures.

Occurrence. Lower Albian–Upper Cenomanian.

Hamites simplex (d'Orbigny)
Plate 41, figure 5

Description. Whorl section oval; ribs rather strong, normally 4·5 in a distance equal to the whorl height.

Occurrence. Cenomanian, *A. rhotomagense–M. geslinianum* zones; rather uncommon, throughout southern England.

Family TURRILITIDAE

Remarks. Helically coiled, loosely or tightly, typically regular but later whorls may be unstable. Generally with strong ribs, rarely smooth.

Occurrence. Middle Albian–Upper Cenomanian.

Genus MARIELLA

Remarks. Ribs single, looped or absent, with four rows of tubercles, an equal number in each row.

Occurrence. Upper Albian–Lower Cenomanian.

Subgenus MARIELLA

Remarks. Closely coiled throughout.

Occurrence. As for genus.

Mariella lewesiensis (Spath)
Plate 38, figure 6

Description. Tubercles in the four rows are more or less equal; the two lower rows are closer together than the others. Specific differences between described species in this genus are slight.
Occurrence. Lower Cenomanian, *M. mantelli* Zone; fairly common, especially in the south-east.

Genus NEOSTLINGOCERAS

Remarks. Coiled in an acute spiral; whorls have flat sides with two or three rows of small tubercles at the basal angle and a row of sparser and larger tubercles at the middle of the side.
Occurrence. Lower Cenomanian.

Neostlingoceras carcitanense (Matheron)
Plate 38, figure 3

Description. As for genus.
Occurrence. Lower Cenomanian, *M. mantelli* Zone, *N. carcitanense* Subzone; common in Sussex, the Isle of Wight, Dorset and Wiltshire.

Genus TURRILITES

Remarks. Apical angle acute; whorl section flattened or angular; ribs weak to strong, non-tuberculate or with 2–4 rows of tubercles.
Occurrence. Cenomanian.

Turrilites costatus Lamarck
Plate 39, figure 4

Description. Ribs on the upper part of the outer face and two rows of bullate tubercles below.
Occurrence. Middle Cenomanian, *A. rhotomagense* Zone, *T. costatus* Subzone, and *A. jukesbrownei* Zone; widespread in southern England but common only in its own subzone.

Turrilites acutus Passy
Plate 39, figure 6

Description. Like *T. costatus* but with an upper row of bullate tubercles rather than ribs and a more angular whorl section.

Occurrence. Cenomanian, *A. rhotomagense–C. guerangeri* zones; widespread in the south, especially common in the south-west.

Turrilites scheuchzerianus Bosc
Plate 39, figure 5

Description. With uniform non-tuberculate ribs, interrupted at mid-flank on early whorls.
Occurrence. Middle Cenomanian, *A. rhotomagense* Zone; widespread in southern England but rather uncommon.

Genus HYPOTURRILITES

Remarks. Commonly with rather large spiral angle and rounded whorls; ribs weak or absent; tubercles in three or four rows, the upper row largest and most sparse.
Occurrence. Cenomanian.

Hypoturrilites tuberculatus (Bosc)
Plate 38, figure 5

Description. Each whorl has 20 large tubercles in the upper row and about 30 in each of the two lower rows. Suture with first lateral saddle symmetrical (asymmetrical in *H. gravesianus* d'Orbigny).
Occurrence. Lower Cenomanian, *M. mantelli* Zone; rather common in Sussex and the Isle of Wight, uncommon elsewhere.

Family NOSTOCERATIDAE

Remarks. Helicoid forms in which the coiling is irregular in early or late stages or both or throughout. Ribs prominent, with or without tubercles, constrictions common. Suture normally florid.
Occurrence. Cenomanian–Maastrichtian.

Genus EUBOSTRYCHOCERAS

Remarks. Middle whorls typically in contact; ribs mostly simple; no tubercles; collared constrictions throughout.
Occurrence. Cenomanian–Santonian.

Eubostrychoceras saxonicum (Schlüter)
Plate 38, figure 4; Plate 39, figures 2–3

Description. Whorls higher than wide or circular; with deep constrictions parallel with the oblique, generally single ribs. Body chamber closely hooked round the base of the spire.

PLATE 38

1

2

3

4

5

6

Remarks. Commonly recorded as *Hyphantoceras woodsi.*
Occurrence. Upper Turonian, *S. neptuni* Zone (*S. plana* Zone); common in the Midlands and East Anglia, rare elsewhere in the south.

Genus BOSTRYCHOCERAS

Remarks. Whorls more or less circular in section; body chamber U-shaped and hanging free; small tubercles may occur on later whorls or throughout.
Occurrence. Campanian–Lower Maastrichtian.

Bostrychoceras polyplocum (Roemer)
Plate 40, figure 6

Description. Apical angle acute to slightly obtuse; whorls in contact or not.
Occurrence. Upper Campanian, *B. mucronata* Zone; rather rare in quarries around Norwich and in the White Limestone of Northern Ireland.

Genus HYPHANTOCERAS

Remarks. Regular or irregular open helical coiling, generally with pendent body chamber; thin, high primary ribs are raised into two or four tubercles and intermediate ribs are non-tuberculate.
Occurrence. Turonian–Santonian.

Hyphantoceras reussianum (d'Orbigny)
Plate 39, figure 1; Plate 40, figure 5

Description. Primary ribs in mid growth with four tubercles.
Occurrence. Upper Turonian, *S. neptuni* Zone (*S. plana* Zone); widespread in south, rare in the north.

EXPLANATION OF PLATE 38

Figs 1–2. *Lewesiceras mantelli* (Wright and Wright). Upper Turonian, *S. neptuni* Zone (*S. plana* Zone), Oldbury Castle Hill, Calne, Wiltshire; ×1.

Fig. 3. *Neostlingoceras carcitanense* (Matheron). Lower Cenomanian, *M. mantelli* Zone, *N. carcitanense* Subzone, Warminster Area, Wiltshire; ×1.

Fig. 4. *Eubostrychoceras saxonicum* (Schlüter). Upper Turonian, *S. neptuni* Zone (*S. plana* Zone), Chilterns; ×1.

Fig. 5. *Hypoturrilites tuberculatus* (Bosc). Lower Cenomanian, *M. mantelli* Zone, Kent; ×1.

Fig. 6. *Mariella lewesiensis* (Spath). Lower Cenomanian, *M. mantelli* Zone, *M. saxbii* Subzone; Isle of Wight; ×1.

PLATE 39

Family DIPLOMOCERATIDAE

Remarks. Uncoiling derivatives of Nostoceratidae, with initial whorls generally consisting of straight shafts, smooth and with constrictions. Later whorls typically with annular non-tuberculate ribs but ventrolateral spines may occur.
Occurrence. Turonian–Maastrichtian.

Subfamily POLYPTYCHOCERATINAE

Remarks. Coiling mostly in a planispire followed by one or more straight shafts; suture tends to become simple.
Occurrence. Upper Turonian–Upper Campanian.

Genus PSEUDOXYBELOCERAS

Remarks. Early part variable, followed by up to five subparallel shafts. Ribs with inner and outer or outer ventrolateral spines only, on all or some ribs.
Occurrence. Upper Turonian–Upper Campanian.

Subgenus PARASOLENOCERAS

Remarks. All ribs with outer ventrolateral spines only.
Occurrence. Coniacian–Upper Campanian.

Pseudoxybeloceras (*Parasolenoceras*) *interruptum* (Schlüter)
Plate 40, figures 9–10

Description. Whorl section depressed; ventrolateral tubercles slight.
Occurrence. Upper Campanian, *B. mucronata* Zone; rare in Norfolk and in the White Limestone, Northern Ireland.

EXPLANATION OF PLATE 39

Fig. 1. *Hyphantoceras reussianum* (d'Orbigny). Upper Turonian, *S. neptuni* Zone (*S. plana* Zone), Hitch Wood pit, Stevenage, Hertfordshire; ×1.

Figs 2–3. *Eubostrychoceras saxonicum* (Schlüter). Upper Turonian, *S. neptuni* Zone (*S. plana* Zone), Hitch Wood pit, Stevenage, Hertfordshire; ×1.

Fig. 4. *Turrilites costatus* Lamarck. Middle Cenomanian, *A. rhotomagense* Zone, *T. costatus* Subzone, Isle of Wight; ×1.

Fig. 5. *Turrilites scheuchzerianus* Bosc. Middle Cenomanian, *A. rhotomagense* Zone, Isle of Wight; ×1.

Fig. 6. *Turrilites acutus* Passy. Middle Cenomanian, *A. rhotomagense* Zone, *T. acutus* Subzone, Isle of Wight; ×1.

Family BACULITIDAE

Remarks. Small to very large, straight or slightly curved. Suture with bifid lobes.
Occurrence. Upper Albian–Maastrichtian.

Genus SCIPONOCERAS

Remarks. Oblique constrictions and normally oblique ribs present.
Occurrence. Upper Albian–Upper Turonian.

Sciponoceras baculoides (Mantell)
Plate 40, figure 13

Description. Whorl section compressed oval, narrowing towards the venter. Phragmocone virtually smooth except for strong oblique constrictions, which fade dorsally, or with short, scale-like ribs. Body chamber has strong ventral ribs.
Occurrence. Middle Cenomanian, *A. rhotomagense* Zone, to Late Cenomanian, *C. guerangeri* Zone; generally distributed.

Sciponoceras gracile (Shumard)
Plate 40, figures 3–4

Description. Compressed oval in section. Constrictions run backwards dorsolaterally then swing forwards; strong, rounded ribs are present between constrictions.

EXPLANATION OF PLATE 40

Figs 1–2. *Idiohamites alternatus* (Mantell). Lower Cenomanian, *M. mantelli* Zone, *N. carcitanense* Subzone, Warminster area, Wiltshire; ×1.

Figs 3–4. *Sciponoceras gracile* (Shumard). Upper Cenomanian, *M. geslinianum* Zone, beach below Whitlands, Devon, near Lyme Regis, Dorset; ×1.

Fig. 5. *Hyphantoceras reussianum* (d'Orbigny). Upper Turonian, *S. neptuni* Zone (*S. plana* Zone), Hitch Wood pit, Stevenage, Hertfordshire; ×1.

Fig. 6. *Bostrychoceras polyplocum* (Roemer). Campanian, *B. mucronata* Zone, Northern Ireland; ×1.

Figs 7–8. *Sciponoceras bohemicum bohemicum* (Fritsch). Upper Turonian, *S. neptuni* Zone (*S. plana* Zone), Chilterns; ×1.

Figs 9–10. *Pseudoxybeloceras* (*Parasolenoceras*) *interruptum* (Schlüter). Campanian, *B. mucronata* Zone, Northern Ireland; ×1.

Figs 11–12. *Allocrioceras angustum* (J. de C. Sowerby). Upper Turonian, *S. neptuni* Zone (*S. plana* Zone), Chilterns; ×1.

Fig. 13. *Sciponoceras baculoides* (Mantell). Middle Cenomanian, *A. rhotomagense* Zone, Ventnor, Isle of Wight; ×1.

PLATE 40

Occurrence. Upper Cenomanian, *M. geslinianum* Zone; general in southern England, especially common in upper Bed C of Cenomanian Limestone in Devon.

Sciponoceras bohemicum (Fritsch)
Plate 40, figures 7–8

Description. Mature parts of shell with flattened sides. There are close, broad, shallow constrictions with two or three indistinctly branching ribs between. An early subspecies, *S. b. anterius* Wright and Kennedy from the Upper Cenomanian *N. juddii* Zone, has the ribs and constrictions crossing the venter transversely, whereas *S. b. bohemicum* from the Upper Turonian has them crossing in a narrow curve.

Occurrence. Upper Cenomanian, *N. juddii* Zone, to Upper Turonian, *S. neptuni* Zone (*S. plana* Zone); *S. b. anterius* is widespread in southern England and the Midlands; *S. b. bohemicum* occurs in the top of the Middle Turonian in Kent, Surrey and Berkshire, and is generally distributed in the *S. plana* Zone in southern England and the Midlands.

Superfamily SCAPHITACEAE
Family SCAPHITIDAE
Subfamily SCAPHITINAE

Remarks. Microconchs have a simple aperture without lappets.
Occurrence. Upper Albian–Maastrichtian.

Genus SCAPHITES

Remarks. Small, boat-shaped, with more or less tightly coiled spire, a short straight shaft and a final recurved hook or with almost normal coiling.

EXPLANATION OF PLATE 41

Figs 1–2. *Scaphites obliquus* J. Sowerby. Lower Cenomanian, *M. mantelli* Zone, Warminster area, Wiltshire; ×1.
Figs 3–4. *Scaphites equalis* J. Sowerby. Middle Cenomanian, *A. rhotomagense* Zone, Dorset; ×1.
Fig. 5. *Hamites simplex* (d'Orbigny). Middle Cenomanian, *A. rhotomagense* Zone, Ventnor, Isle of Wight; ×1.
Fig. 6. *Anisoceras plicatile* (J. Sowerby). Middle Cenomanian, *A. rhotomagense* Zone, Ventnor, Isle of Wight; ×1.
Fig. 7. *Scaphites binodosus* Roemer (microconch). Campanian, *I. lingua* Zone, East Leys, near Bridlington, East Yorkshire (North Humberside); ×1.
Figs 8–10. *Scaphites geinitzii* d'Orbigny. 8–9, microconch; 10, macroconch. Upper Turonian, *S. neptuni* Zone (*S. plana* Zone); Chilterns; ×1.
Fig. 11. *Scaphites binodosus* Roemer (macroconch). Campanian, *I. lingua* Zone, Bessingby, near Bridlington, East Yorkshire (North Humberside); ×1.

PLATE 41

Occurrence. Upper Albian–Campanian.

Scaphites equalis J. Sowerby
Plate 41, figures 3–4

Description. Small, shaft with well-separated rectiradiate primary ribs; ventrolateral tubercles tend to develop on the hook.
Occurrence. Cenomanian, *M. mantelli* Zone, *M. saxbii* Subzone, to *M. geslinianum* Zone; generally distributed in southern England.

Scaphites obliquus J. Sowerby
Plate 41, figures 1–2

Description. Shaft with fine, close, oblique primary ribs that split into 2–5 secondaries.
Occurrence. Lower Cenomanian, *M. mantelli* Zone, to Upper Cenomanian, *C. guerangeri* Zone; generally distributed throughout southern England.

Scaphites geinitzii d'Orbigny
Plate 41, figures 8–10

Description. On the shaft primary ribs are sinuous, oblique and raised into bullate ventrolateral tubercles from which branch 3–5 secondaries.
Occurrence. Turonian, *C. woollgari* Zone (*T. lata* Zone), to Coniacian, *M. cortestudinarium* Zone; widespread in the *S. plana* Zone in southern England and the Midlands, very rare below and above.

Scaphites binodosus Roemer
Plate 41, figures 7, 11

Description. Rather large. Bullate umbilical and distinctly clavate ventrolateral tubercles on shaft and hook. Macroconchs inflated with weak tubercles; microconchs compressed with strong tubercles.
Occurrence. ?Upper Santonian, *M. testudinarius* Zone, Lower Campanian, *I. lingua* Zone; rather common in the upper part of the *I. lingua* Zone in East Yorkshire and occurs very rarely below.

10. NAUTILOIDS

by W. J. KENNEDY

Nautiloids are frequent at only a few horizons in the Chalk: their distribution broadly corresponds to that of ammonites. They are thus present in the Glauconitic Marl, Lower and Middle Cenomanian parts of the Chalk in southern England, the Cenomanian Basement Beds of south-west England and the correlative sands and limestones of Devon. They reappear in the Upper Turonian Chalk Rock, in the Upper Campanian chalk in Norfolk, and at limited horizons in the correlative White Limestone of Northern Ireland. Nautiloids had a similar aragonitic mineralogy to that of ammonites, and their preservation is identical in the range of facies in which the two co-occur. They are always a minor part of the cephalopod fauna, and in any assemblage, are rarer than ammonites.

British Chalk nautiloids were monographed by Sharpe (1853, *in* Sharpe 1853–57), and Foord (1891); Wright and Wright (1951) provided the only subsequent review. Apart from these sources, Schlüter (1871–76) provided illustrations of other Upper Cretaceous nautiloids that may be discovered in the sequence (a more accessible facsimile is Riegraf and Scheer, 1991). Generic discussions and lists of species are given by Kummel (1956; see Wiedmann 1960 for a rather different approach); an accessible source for general features of the group is Teichert *et al.* (1964).

DESCRIPTIONS

Class CEPHALOPODA
Subclass NAUTILOIDEA
Order NAUTILIDA
Family NAUTILIDAE

Genus EUTREPHOCERAS

Remarks. Very involute, nautilicone, subglobose, whorl section depressed reniform (kidney-shaped), with broadly rounded venter and flanks; umbilicus tiny, sometimes occluded. Adult aperture with broad shallow sinus. Shell surface with fine growth lines, and delicate spiral ornament in some. Suture only slightly sinuous.
Occurrence. Upper Jurassic-Miocene.

220 *Fossils of the Chalk*

Eutrephoceras sublaevigatum (d'Orbigny)
Text-figure 10.1A

Description. Whorl section depressed (wider than high) reniform, flanks and venter broadly rounded; umbilicus occluded. Shell surface ornamented by delicate growth lines only. Adult aperture preceded by broad shallow constriction. There is a shallow ventral sinus in the growth lines, and at the adult aperture. Siphuncle subcentral. Septa widely spaced, sutures feebly sinuous.
Occurrence. Upper Aptian–Upper Turonian, and perhaps higher; widespread in the *M. mantelli–A. jukesbrownei* zones in the Chalk facies of southern England; more frequent in the *M. mantelli–C. guerangeri* zones, the Basement Beds and their correlatives, in Dorset, Somerset and Devon.

Eutrephoceras expansum (J. de C. Sowerby)
Plate 42, figures 1–3; Text-figure 10.1B

Description. Whorl section depressed, venter broadly rounded. Flanks flattened, strongly divergent, producing a trapezoidal whorl section that is depressed with the greatest width at the sharp rim to the well-defined conical, pit-like umbilicus. Shell surface with prominent growth lines, projected forwards on inner flank, convex across mid-flank, directed strongly backwards on outer flank to cross venter in deep sinus that persists to adult aperture. Delicate spiral striae that intersect with growth lines to produce a distinctive reticulate pattern are occasionally preserved in juveniles. Siphuncle subcentral; septa widely separated, sutures feebly sinuous.
Remarks. Flattened flanks, sharp umbilical rim and stronger ornament distinguish *E. expansum* from *E. sublaevigatum*.
Occurrence. Throughout the Cenomanian; widespread in the Chalk facies of the *M. mantelli–A. jukesbrownei* zones of southern England; more frequent in the Basement Beds and their correlatives, *M. mantelli–M. geslinianum* zones, in Dorset, Somerset and Devon.

TEXT-FIG. 10.1. Sketch whorl sections and siphuncle positions of Chalk nautiloids (diagrammatic only, not to scale). A, *Eutrephoceras sublaevigatum* (d'Orbigny). B, *Eutrephoceras expansum* (J. de C. Sowerby). C, *Eutrephoceras darupense* (Schlüter). D, *Pseudocenoceras largilliertianum* (d'Orbigny). E, *Deltocymatoceras* sp. F, *Pseudocenoceras dorsoplicatus* (Wiedmann). G, *Pseudocenoceras?* *schroederi* (Wiedmann). H, *Pseudocenoceras fittoni* (Sharpe). I, *Anglonautilus undulatus* (J. Sowerby). J, *Angulithes triangularis* de Montfort.

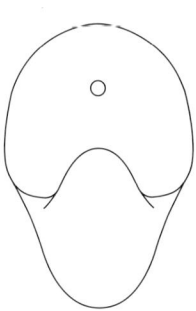

A

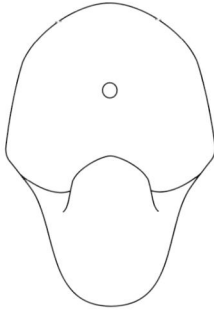

B

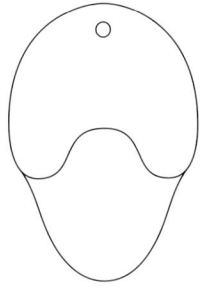

C

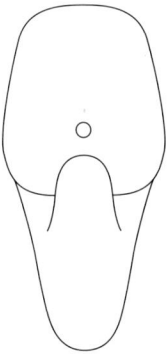

D

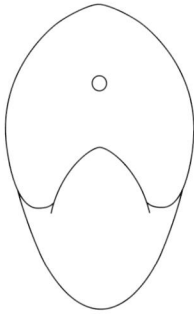

E

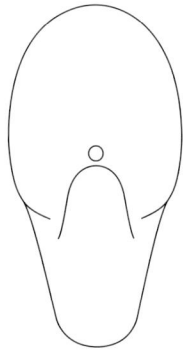

F

G

H

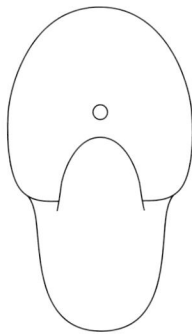

I

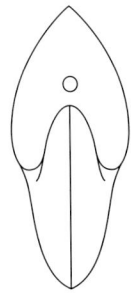

J

Eutrephoceras darupense (Schlüter)
Plate 43, figures 1–2, 4; Text-figure 10.1c

Description. Larger than *E. expansum*, whorl section broadly rounded when undeformed by post-mortem compaction, with occluded umbilicus. Siphuncle ventral of centre of whorl. Septa widely spaced; sutures feebly sinuous. **Occurrence**. Campanian; Northern Ireland; rare records of *Eutrephoceras* from the Campanian Chalk of northern and southern England may belong to this species.

Genus PSEUDOCENOCERAS

Remarks. Involute, nautilicone, whorl section compressed, subrectangular, with flattened subparallel to convergent flanks, ventrolateral shoulders broadly rounded, venter flattened to feebly convex. Umbilical wall commonly subvertical. Shell surface smooth or with growth lines only. Suture line with broad, shallow lateral lobe, transverse on venter.

Pseudocenoceras largilliertianum (d'Orbigny)
Text-figure 10.1d

Description. Whorl section compressed, with greatest breadth low on the flanks. Inner to mid-flanks feebly convex, outer flanks convergent. Ventrolateral shoulders broadly rounded, venter feebly convex. Umbilicus open, with subvertical wall and quite narrowly rounded umbilical shoulder. Adult body chamber broadens, with wide, flattened venter. Surface of shell almost smooth, with only the faintest trace of growth lines in most specimens, or none. Sutures with broad, shallow lateral lobe; transverse on venter. Siphuncle close to dorsal margin.
Remarks. Wiedmann (1960) referred the original of Sharpe (1853, pl. 6, fig. 2) to his new species *Pseudocenoceras dorsoplicatus*. It differs from *P. largilliertianum* in having low, broad undulations on the outer flank and venter of the late phragmocone. *P. fittoni* (Sharpe, 1853) has a more compressed subtriangular whorl section, relatively narrowly rounded venter, and suture line with a distinctive small saddle on the umbilical shoulder, and a broad lateral lobe that projects strongly forward over the ventrolateral shoulder.
Occurrence. Upper Albian–Upper Cenomanian, *M. geslinianum* Zone; rare in the Chalk facies of the *M. mantelli–A. rhotomagense* zones in southern England. More frequent in the Basement Beds and correlatives in Dorset, Somerset, and Devon.

Pseudocenoceras fittoni (Sharpe)
Plate 42, figures 7–9; Text-figure 10.1H

Description. Whorl section very compressed, with greatest breadth just outside umbilical shoulder, inner flanks feebly convex, outer flanks convergent, venter arched, narrowly rounded, to give distinctive triangular whorl section. Umbilicus open, with outward inclined wall and well-defined shoulder. Shell surface with delicate growth lines only. Suture with small saddle on umbilical shoulder, deep lateral lobe that is strongly projected on outer flank, transverse on venter. Siphuncle dorsal, much closer to dorsum than mid-point of whorl section.
Occurrence. Lower Cenomanian, *M. mantelli* Zone; southern England.

Pseudocenoceras dorsplicatus (Wiedmann)
Text-figure 10.1F

Description. Umbilicus large for genus, with convex wall and broadly rounded shoulder. Whorl section of phragmocone slightly depressed, with feebly convex inner and convergent outer flanks; ventrolateral shoulders broadly rounded; broad venter very feebly convex. Surface ornamented by growth lines; concave on inner flank, markedly prorsiradiate on outer, flexed forwards and strongly convex on ventrolateral shoulder, forming deep convexity over venter. Late phragmocone characterised by a 90–120 degree sector with few broad, strong dorsal folds on ventrolateral shoulders and venter, where they are deeply concave. Whorl section of body chamber broadens markedly, with very broad venter; ventrolateral folds are lost, to be replaced by much narrower, widely separated, ventrolateral and ventral ribs. Suture with broad lateral lobe; feebly convex across venter. Siphuncle dorsal, close to dorsal margin of shell.
Remarks. *Pseudoceroceras dorsplicatus* is superficially similar to *Anglonautilus undulatus*, but differs in the short ontogenetic duration of the phase of ventrolateral plications (which extend throughout the late phragmocone and all of the body chamber of *A. undulatus*), the larger umbilicus, the broad, flattened venter of the phragmocone (versus broadly rounded in *undulatus*), the quadrate (as opposed to ovoid) whorl section, and the more dorsal position of the siphuncle.
Occurrence. Albian–Cenomanian; Basement Bed and correlative facies of Dorset, Somerset and Devon.

Pseudocenoceras? *schroederi* (Wiedmann)
Text-figure. 10.1G

Description. Coiling very involute, umbilicus tiny. Whorl section compressed, with greatest breadth low on flanks; inner-mid-flank region

PLATE 42

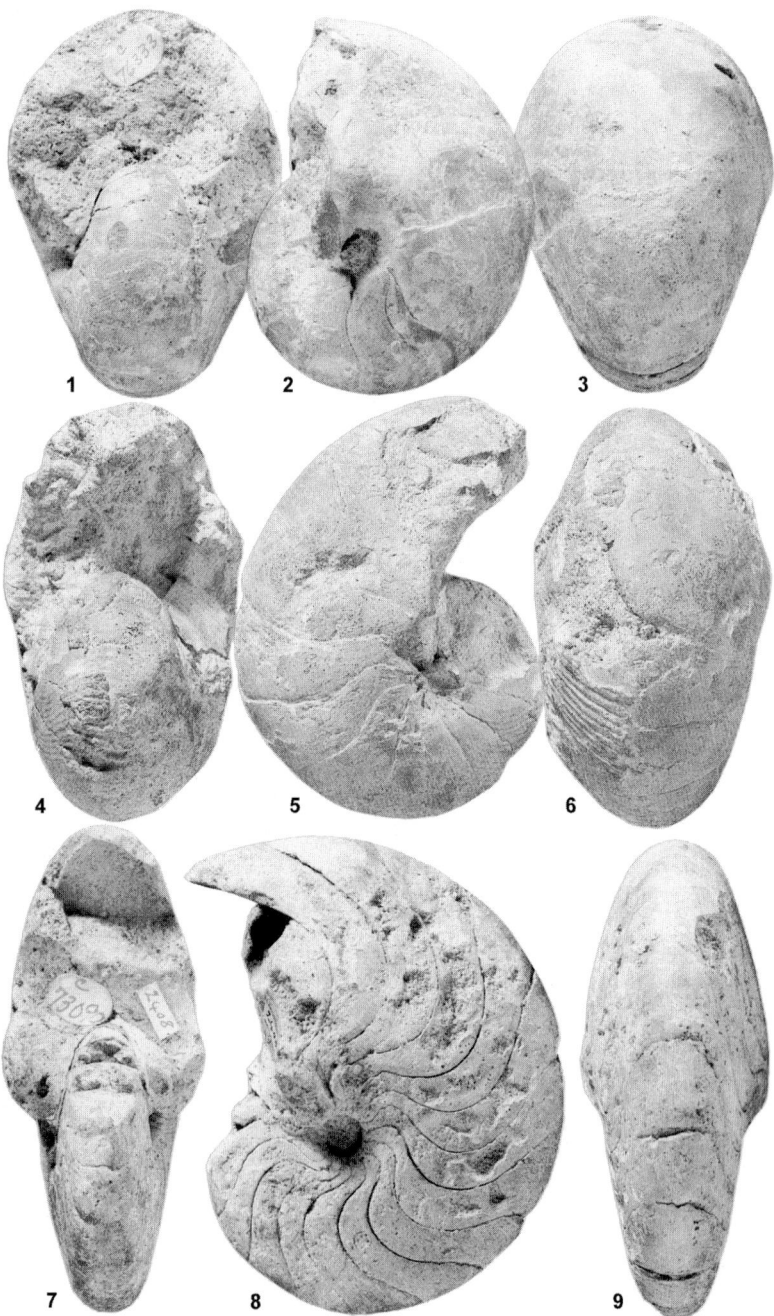

broadly convex, outer flanks convergent, ventrolateral shoulders broadly rounded. Venter broadly rounded on phragmocone, flattening on adult body chamber. Ornament of shell surface not seen in British material. Suture line with broad lateral lobe; siphuncle close to dorsal margin.

Remarks. The whorl section and form of umbilicus distinguish this species from *Pseudocenoceras* discussed above. Wiedmann (1960) referred the species to the genus *Cimomia*.

Occurrence. Lower Cenomanian, *M. mantelli* Zone, to Middle Cenomanian, *A. rhotomagense* Zone; Devon, and possibly south-east England.

Family CYMATOCERATIDAE

Remarks. The Cymatoceratidae are the strongly ribbed nautiloids of the Chalk. The family ranges from the Middle Jurassic to Oligocene according to Kummel (1964).

Genus CYMATOCERAS

Remarks. Involute, subglobose, with rounded whorl section. Flanks and venter with conspicuous ribs, generally prorsiradiate on the inner flanks, convex on outer flanks, sweeping backwards on the ventrolateral shoulder into a broad ventral concavity.

Cymatoceras deslongchampsianum (d'Orbigny)
Text-figure 10.2A

Description. Shell subglobose. Whorl section depressed, greatest breadth at umbilical shoulder. Flanks flattened, convergent, ventrolateral shoulders broadly rounded, broad venter flattened to feebly convex. Umbilicus prominent, large, with flattened, outward-inclined umbilical wall and sharp umbilical shoulder. Flanks may be feebly concave in the immediate proximity of the shoulder. Strong, sharp ribs prominent on shell surface, and on internal mould, where blunter. They are strongly prorsiradiate on the inner flank, sweep back and are strongly convex on outer flank and ventrolateral shoulder, where they may increase by branching and intercalation. They flex back on the ventrolateral shoulders into a prominent

---

EXPLANATION OF PLATE 42

Figs 1–3. *Eutrephoceras expansum* (J. de C. Sowerby). Middle Cenomanian, Snowden Hill, Chard, Somerset; ×1.

Figs 4–6. *Deltocymatoceras* sp. Upper Turonian, Aston Rowant, Oxfordshire; ×0·3.

Figs 7–9. *Pseudocenoceras fittoni* (Sharpe). Lower Cenomanian, Chardstock, Devon; ×1.

ventral concavity. Where ornament is well preserved, the radial ribs are seen to be accompanied by delicate spiral ridges that produce a distinctive reticulate ornament. Suture with small saddle or umbilical shoulder, broad, shallow lateral lobe and very shallow ventral lobe. Siphuncle dorsal of mid-point of whorl section.

Remarks. The large, prominent, conical umbilical pit with sharp rim, and the reticulate ornament are immediately diagnostic.

Occurrence. Cenomanian, *M. mantelli–C. guerangeri* zones; Chalk facies, *M. mantelli–A. rhotomagense* zones, in southern England; widespread in *M. mantelli–C. guerangeri* zones, Chalk Basement Beds and correlative facies, Dorset, Somerset and Devon.

Cymatoceras elegans (J. Sowerby)
Plate 43, figures 3, 7; Text-figure 10.2c

Description. Large. Inner flanks broadly convex, mid-outer flank region flattened, convergent, venter quite narrowly rounded. Umbilicus small. Shell surface and internal mould ornamented by strong ribs that are concave on the innermost flank, flexed back and convex on the middle to outer flank, where they increase by branching and intercalation, sweeping back and markedly convex on the ventrolateral shoulder, crossing the venter in a deep ventral concavity. Septa widely separated; sutures with broad, shallow saddle on inner flank, broad, shallow lobe on middle flank, transverse to feebly concave over venter. Siphuncle dorsal of mid-point of whorl section.

Remarks. *Cymatoceras elegans* can be separated from *C. deslongchampsianum* on the basis of the large umbilicus, sharp umbilical rim, and reticulate ornament of the latter. *Cymatoceras atlas* (Whiteaves) has a very depressed whorl section, a closed umbilicus, coarser ornament, and a siphuncle close to the ventral margin. *Cymatoceras tourtiae* (Schlüter) has a larger umbilicus, and a trapezoidal whorl section with a broad, very feebly convex venter. The siphuncle is close to the dorsal margin. *Cymatoceras bayfieldi* (Foord and Crick) has a trapezoidal whorl section, and a broad venter.

EXPLANATION OF PLATE 43

Figs 1–2, 4. *Eutrephoceras darupense* (Schlüter). Campanian, Cave Hill, County Antrim; ×0·5.

Figs 3, 7. *Cymatoceras elegans* (J. Sowerby). Lower or Middle Cenomanian, England; ×0·3.

Figs 5–6. *Angulithes triangularis* de Montfort. Lower or Middle Cenomanian, Sidmouth, Devon; ×0·3.

PLATE 43

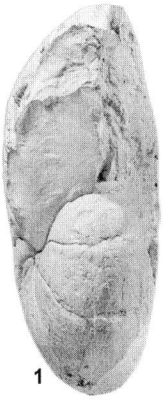

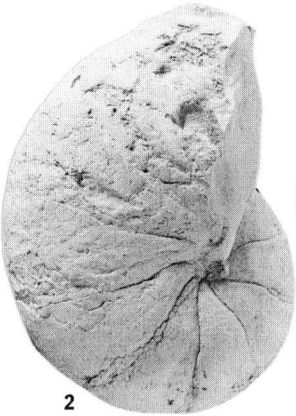

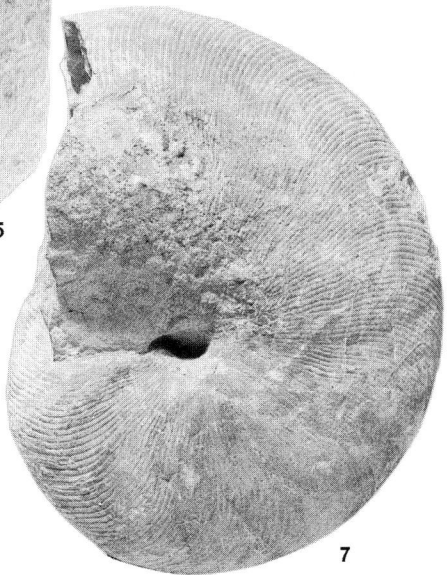

1

2

3

4

5

6

7

Occurrence. Cenomanian, *M. mantelli–A. jukesbrownei* zones; common in the chalk facies of southern England; less common in the Basement Beds and correlative facies of Dorset, Somerset and Devon.

Cymatoceras atlas (Whiteaves)
Text-figure 10.2B

Description. Large. Shell inflated, whorl section very depressed, with greatest breadth low on flanks. Flanks, ventrolateral shoulders and venter broadly rounded, to give a semilunate section. Surface of shell and mould ornamented by coarse ribs, convex across the flanks, where they may increase by branching and intercalation, sweeping backwards and convex on the ventrolateral shoulder and crossing the venter in a broad concavity. Septa widely separated. Suture with broad, shallow, lateral lobe. Siphuncle close to ventral margin.

Remarks. Whorl section, coarse ornament and position of siphuncle separate this species from other British Chalk *Cymatoceras*.

Occurrence. Cenomanian, *M. mantelli–A. jukesbrownei* zones; Chalk facies of southern England; Basement Beds of Dorset, Somerset and Devon and correlatives.

Cymatoceras tourtiae (Schlüter)
Text-figure 10.2D

Description. Whorl section depressed trapezoidal, flanks feebly convex, convergent. Ventrolateral shoulders broadly rounded, broad venter very feebly convex. Umbilicus large, with feebly convex, outward-inclined wall; umbilical shoulder blunt, quite narrowly rounded. Surface of shell and mould ornamented by coarse ribs, convex across the flanks, where they may increase by branching and intercalation, sweeping backwards and very convex across the ventrolateral shoulder, crossing venter in broad concavity. Suture with small, shallow, saddle across umbilical wall, and innermost flank broad, shallow lateral lobe, near-transverse and straight on venter. Siphuncle close to dorsal margin.

Remarks. The combination of trapezoidal whorl section and siphuncle close to the dorsal margin distinguish *C. tourtiae* from all other Cenomanian *Cymatoceras* found in the British Chalk. It most closely resembles the Campanian *C. bayfieldi*, which differs in having a more broadly and evenly rounded whorl section, widely spaced septa, a shallow lateral lobe in the suture line, and a siphuncle that is not as close to the dorsal margin of the shell.

Occurrence. Lower Cenomanian, *M. mantelli* Zone; Isle of Wight.

Cymatoceras cenomananese (Schlüter)
Text-figure 10.2E

Description. Whorl section depressed trapezoidal, flanks flattened, convergent, ventrolateral shoulders broadly rounded, venter broad, very feebly convex. Umbilicus small, with feebly convex, outward-inclined wall and broadly rounded umbilical shoulder. Internal moulds bear weak, even ribs that are strongly convex across the flanks, sweeping back across the ventrolateral shoulders to form a deep ventral concavity. Septa widely

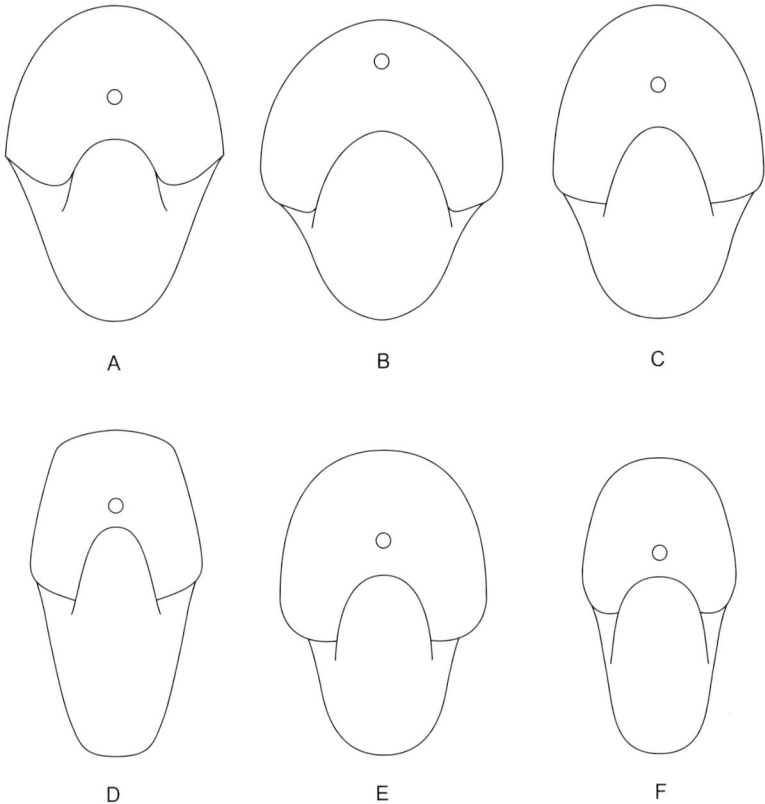

TEXT-FIG. 10.2. Sketch whorl sections and siphuncle positions of Chalk nautiloids (diagrammatic only, not to scale). A, *Cymatoceras deslongchampsianum* (d'Orbigny). B, *Cymatoceras atlas* (Whiteaves). C, *Cymatoceras elegans* (J. Sowerby). D, *Cymatoceras tourtiae* (Schlüter). E, *Cymatoceras cenomanense* (Schlüter). F, *Cymatoceras bayfieldi* (Foord and Crick).

Fossils of the Chalk

spaced, suture with lateral lobe that projects strongly forwards on ventro-lateral shoulder; near-transverse on venter. Siphuncle ventral of mid-point of septal face.

Remarks. *Cymatoceras cenomanense* most closely resembles *C. elegans*, from which it differs in having a broad, flattened venter.

Occurrence. Lower Cenomanian, *M. mantelli* Zone; Wiltshire, Isle of Wight.

Cymatoceras bayfieldi (Foord and Crick)
Text-figure 10.2F

Description. Whorl section depressed, rounded-trapezoidal, flanks feebly convex, convergent, ventrolateral shoulders broadly rounded, broad venter flattened, and only very feebly convex. Umbilicus large for genus, umbilical wall feebly convex, umbilical shoulder more narrowly rounded. Strong, even ribs are straight and prorsiradiate on the innermost flank, flexing back and broadly convex on the remainder of the flank, sweeping strongly backwards over the ventrolateral shoulder to form a deep concavity over the venter. Suture line with shallow lateral lobe, and near-transverse on venter. Siphuncle dorsal of mid-point of septal face.

Occurrence. Campanian; Norfolk and Northern Ireland.

Genus DELTOCYMATOCERAS

Remarks. Involute, flanks broadly convex, converging to bluntly acute venter; no ventrolateral shoulders. Ornament comprises prominent ribs, convex on flanks (where they may branch), sweeping back on ventro-lateral shoulders into broad ventral concavity, effaced at mid-venter. Suture with prominent saddle on umbilical shoulder, and broad shallow lateral lobe.

Deltocymatoceras sp.
Plate 42, figures 4–6; Text-figure 10.1E

Description. As for genus.

Remarks. Large specimens of *Deltocymatoceras* from the English Upper Chalk are closest to *D. galea* (Fritsch, 1872).

Occurrence. Upper Turonian, *S. neptuni* Zone; Hertfordshire, Oxfordshire, Bedfordshire, and Kent.

Genus ANGLONAUTILUS

Remarks. Coiling involute, shell globose, with broadly rounded flanks and venter. Early whorls smooth; late phragmocone and body chamber with coarse, fold-like ribs that are convex across the outer flanks and ventrolateral shoulders, and cross the venter in a pronounced concavity. They efface and disappear on the inner flanks. Suture with shallow lateral and ventral lobes. Siphuncle subcentral.

Anglonautilus undulatus (J. Sowerby)
Text-figure 10.1ı

Description. Whorl section about as wide as high, inner flanks broadly convex, outer flanks convergent, ventrolateral shoulders and venter broadly rounded. Umbilicus small. Early phragmocone whorls are smooth but for growth lines. Late phragmocone whorls and body chamber with coarse fold-like ribs (hence the specific name *undulatus*), that arise on outer flank, are coarse and convex on the ventrolateral shoulders, and cross the venter in a broad concavity. Suture with small saddle on umbilical wall, and a broad lateral lobe. Siphuncle dorsal of mid-point of whorl cross section.
Occurrence. Lower Aptian–Upper Cenomanian, *C. guerangeri* Zone; Dorset, Somerset and Devon.

Family HERCOGLOSSIDAE

Genus ANGULITHES

Remarks. Coiling involute, whorl section depressed to compressed, umbilicus small, flanks strongly convergent, with narrowly rounded to bluntly acute venter giving characteristic triangular whorl section. Ornament of growth lines only.

Angulithes triangularis de Montfort
Plate 43, figures 5–6; Text-figure 10.1ᴊ

Description. Whorl section compressed. Inner flanks broadly convex, outer flanks convergent, venter angular, giving a triangulate whorl section. Umbilicus near-closed with broadly rounded wall and shoulder. Surface of internal mould smooth. Ornament of shell surface not seen in British material. Suture with small saddle on umbilical shoulder. Lateral lobe markedly convex across flanks, strongly projected on ventrolateral shoulder into narrowly rounded ventral saddle. Siphuncle large, subcentral.
Occurrence. Lower Cenomanian, *M. mantelli* Zone; Kent and Devon.

11. BELEMNITES

by PETER DOYLE

Despite long being recognised as valuable zonal indices in the Upper Cretaceous, Chalk belemnites have received scant attention in Britain. The works of Miller (1826*a*, *b*) and Sharpe (1853) are of historical interest, and valuable contributions to our knowledge were made by some early twentieth century authors (e.g. Rowe 1904; Crick 1906, 1907, 1910). Recent work on British Chalk belemnites includes that of Jarvis (1980) and Christensen (1974, 1991, 1993, 1995). A single family, the Belemnitellidae, dominates the Chalk from the Cenomanian upwards, replacing the Belemnopseidae which is common in Jurassic and older Cretaceous deposits, and which has its highest occurrence in the Cenomanian. The Belemnitellidae has been reviewed most recently by Christensen (1997*a*). The family was relatively common and widespread in the Northern Hemisphere during the Late Cretaceous, although most diverse in Europe. Its distribution has been reviewed by Christensen (1976, 1997*b*) and Doyle (1992), with its value in biostratigraphy most prominently discussed by Christensen (1996), who has also made several recent revisions of British belemnitellid faunas: *Actinocamax, Gonioteuthis, Belemnellocamax* and *Belemnitella* from Coniacian–Maastrichtian chalks (Christensen 1991, 1995), and *Belemnocamax* from the Cenomanian (Christensen 1993).

The Belemnitellidae is distinct from all other belemnite families in possessing a ventral alveolar slit or fissure rather than a mere alveolar groove, as found in the Belemnopseidae (e.g. as in *Neohibolites*). The distance between the posterior end of the fissure, where it joins the alveolus, and the tip of the phragmocone is known as the 'Schatsky distance', and is a valuable taxonomic feature for distinguishing the genera *Belemnitella* and *Belemnella*. It is revealed by splitting the rostrum carefully in its dorso-ventral plane with a small chisel. The Belemnitellidae also commonly displays a series of dendritic vascular markings which spread out over the ventral surface of some forms (e.g. *Belemnitella* and *Belemnella*). Other belemnitellid genera possess an imperfectly calcified alveolar region, resulting in either the production of an anterior conical fracture (as in *Actinocamax, Belemnellocamax* and *Belemnocamax*), or in a secondary conical cavity known as a 'pseudoalveolus' (as in *Gonioteuthis*).

DESCRIPTIONS
Class CEPHALOPODA
Subclass COLEOIDEA
Order BELEMNITIDA
Suborder BELEMNOPSEINA
Family BELEMNOPSEIDAE

Genus NEOHIBOLITES

Neohibolites ultimus (d'Orbigny)
Plate 45, figures 6–7

Description. Rostrum up to 60 mm in length; symmetrical and club-shaped (lanceolate) in both outline and profile. The position of maximum inflation is in the stem region of the rostrum, and the apex is relatively acute. Transverse sections of the rostrum are weakly dorso-ventrally depressed and elliptical or subcircular in the stem and apical regions, becoming laterally compressed and elliptical in the alveolar region. A short ventral alveolar groove is confined to the anterior third of the rostrum, not extending into the stem region. Two parallel, fine lateral lines (*Doppellinien*) are present for the length of each flank. The alveolus is well calcified.

Occurrence. Lower Cenomanian, *M. mantelli* Zone; widespread but rare.

Family BELEMNITELLIDAE

Genus ACTINOCAMAX

Actinocamax plenus (Blainville)
Plate 44, figures 1–4

Description. Rostrum up to 90 mm in length; symmetrical and lanceolate in outline; weakly lanceolate or even cylindrical with parallel sides in profile. The position of maximum inflation is in the apical third of the rostrum, and the apex is often attenuated. Transverse sections of the rostrum are generally dorso-ventrally depressed and elliptical in the stem and apical regions, becoming pyriform in the alveolar region. The alveolus is lost owing to imperfect calcification, leaving an anteriorly conical fracture sometimes with a small central pit. Consequently, only vestiges of the ventral alveolar fissure may be observed. Lateral lines are present as *Doppellinien*, extending anteriorly as narrow dorso-lateral depressions. The surface of the rostrum is otherwise smooth.

Occurrence. Upper Cenomanian, *M. geslinianum* and *N. juddii* zones; common in the Plenus Marls.

PLATE 44

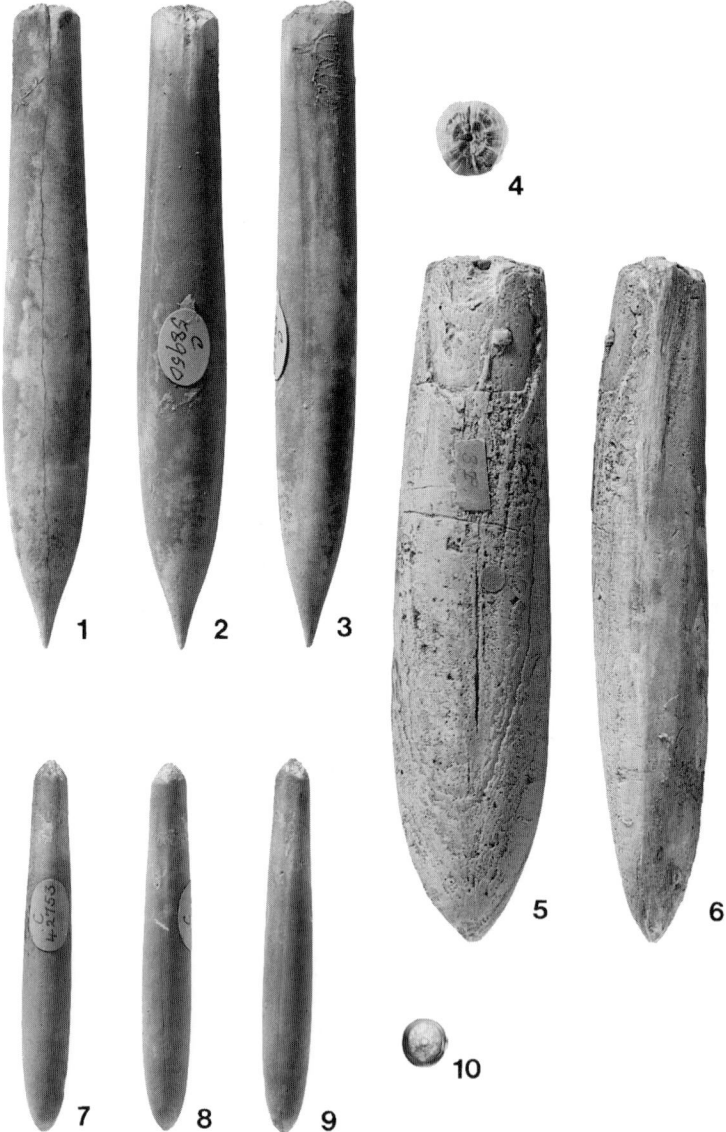

Actinocamax verus Miller
Plate 44, figures 7–10

Description. Rostrum up to 50 mm in length; symmetrical and lanceolate in both outline and profile, the position of maximum inflation being close to the apex. The apex is generally weakly mucronate. Transverse sections of the rostrum are weakly compressed laterally and subcircular in the stem and apical regions, becoming pyriform in the alveolar region. The alveolus and ventral alveolar slit are lost owing to imperfect calcification, leaving an acute, anteriorly conical fracture. Lateral lines are present as *Doppellinien* developing anteriorly into broad dorso-lateral depressions that give the transverse section its pyriform shape. The surface of the rostrum is otherwise smooth.

Occurrence. Coniacian, *M. coranguinum* Zone, to Santonian, *M. testudinarius* Zone; common, especially in the *M. testudinarius* Zone.

Genus GONIOTEUTHIS

Gonioteuthis quadrata (Blainville)
Plate 46, figures 5–8

Description. Rostrum up to 90 mm in length; symmetrical and weakly lanceolate in both outline and profile, in some cases becoming cylindrical. The position of maximum inflation is in the stem region, and the apex of the rostrum is usually mucronate. The transverse sections are slightly compressed laterally and subcircular in the stem and apical regions, becoming pyriform in the alveolar region. A ventral fissure is usually preserved, even though a pseudoalveolus with a quadrate section and a depth of about 15 mm is present in this species. Lateral lines are present as *Doppellinien* in the posterior, developing into relatively narrow, well-defined, dorso-lateral depressions anteriorly. The surface of the rostrum has a granulated appearance.

EXPLANATION OF PLATE 44

Figs 1–4. *Actinocamax plenus* (Blainville). Upper Cenomanian, Eastbourne, Sussex. 1, ventral outline; 2, dorsal outline; 3, left profile; 4, decayed alveolar region; ×1.

Figs 5–6. *Belemnellocamax grossouvrei* (Janet) (plaster cast). Santonian, Fimber, Yorkshire. 5, ventral outline; 6, left profile; ×1.

Figs 7–10. *Actinocamax verus* Miller. Upper Santonian, Gravesend, Kent. 7, ventral outline; 8, left profile; 9, slightly oblique dorsal outline; 10, decayed alveolar region; ×1.

Remarks. Some authors consider this species as the last in a long lineage of Chalk *Gonioteuthis*, characterised by successive deepening of the pseudoalveolus (e.g. Rowe 1904; Jarvis 1980).

Occurrence. Campanian, ?*O. pilula–G. quadrata* zones; widespread but rather rare.

Genus BELEMNITELLA

Belemnitella mucronata (Schlotheim) *sensu lato*
Plate 45, figures 1–5

Description. Rostrum up to 130 mm in length; symmetrical and weakly lanceolate to cylindrical with parallel flanks in outline, symmetrical and cylindrical to weakly cylindri-conical in profile. The position of maximum inflation is in the stem region and the apex of the rostrum is mucronate. Transverse sections are weakly depressed dorso-ventrally and subcircular in the stem and apical regions, becoming slightly compressed laterally and elliptical in the alveolar region. The ventral alveolar fissure is usually preserved, and the Schatsky distance is large (> 4 mm). Lateral lines are found as *Doppellinien* in the posterior of the rostrum, developing into broad dorso-lateral depressions. Vascular impressions are well developed and branch adapically onto the venter from the lateral lines. The alveolus is usually preserved intact, and casts of it possess a dorsal keel (Pl. 45, fig. 5).

Occurrence. Upper Campanian, *B. mucronata* Zone, to ?Early Maastrichtian, *B. lanceolata* Zone; common in the Upper Campanian of Northern Ireland and the southern counties of England; known from drift deposits in north-east England.

Genus BELEMNELLA

Belemnella lanceolata (Schlotheim) *sensu lato*
Plate 46, figures 1–4

Description. Rostrum length up to 130 mm; symmetrical and generally strongly lanceolate in outline, symmetrical and weakly lanceolate to

---

PLATE 45

1

2

3

4

5

6

7

Fossils of the Chalk

cylindrical with parallel sides in profile. The position of maximum inflation is in the apical third of the rostrum, and the apex is generally mucronate. The transverse sections of the rostrum are dorso-ventrally depressed and elliptical in the stem and apical regions, becoming more compressed and pyriform in the alveolar region. The ventral alveolar slit is usually well developed, and the Schatsky distance is small (< 4 mm). The lateral lines are developed as *Doppellinien* in the stem and apex, developing anteriorly into relatively broad dorso-lateral depressions. Vascular impressions are present, although they may be reduced, and branch apically onto the venter from the lateral lines. The alveolus is usually intact and casts of it possess a dorsal keel similar to *Belemnitella*.

Occurrence. Lower Maastrichtian, *B. lanceolata* Zone; common in Norfolk; known from Antrim, Northern Ireland.

Genus BELEMNELLOCAMAX

Belemnellocamax grossouvrei (Janet)
Plate 44, figures 5–6

Description. Rostrum up to 100 mm in length; symmetrical and lanceolate in outline, asymmetrical and arched to symmetrical and cylindrical in profile. The apex is strongly inflated in outline and mucronate, with the position of maximum inflation close to the apex. Transverse sections of the rostrum are strongly depressed dorso-ventrally and lozenge-shaped or elliptical for the length of the rostrum. The ventral fissure is usually lost owing to the imperfect calcification of the alveolar region, and a shallow conical fracture is present in the anterior. Lateral lines are well developed and occur as *Doppellinien* in the posterior, being extended anteriorly as narrow but deep dorso-lateral depressions. The surface of the rostrum is otherwise smooth.

EXPLANATION OF PLATE 46

Figs 1–4. *Belemnella lanceolata* (Schlotheim) *sensu lato*. Lower Maastrichtian, *B. sumensis* Zone within the succession 'White Chalk without *Ostrea lunata* to Grey Beds', near Trimmingham, Norfolk. 1, ventral outline; 2, dorsal outline; 3, right profile; 4, partial longitudinal section, venter to left; ×1.

Figs 5–8. *Gonioteuthis quadrata* (Blainville). Campanian, East Harnham, Salisbury, Wiltshire. 5, ventral outline, showing granulation of surface; 6, pseudoalveolus; 7, dorsal outline; 8, right profile; ×1.

Figs 9–12. *Belemnocamax boweri* Crick. Middle Cenomanian, Louth, Lincolnshire. 9, ventral outline; 10, dorsal outline, showing striations at apex; 11, left profile; 12, pseudoalveolus; ×1.

PLATE 46

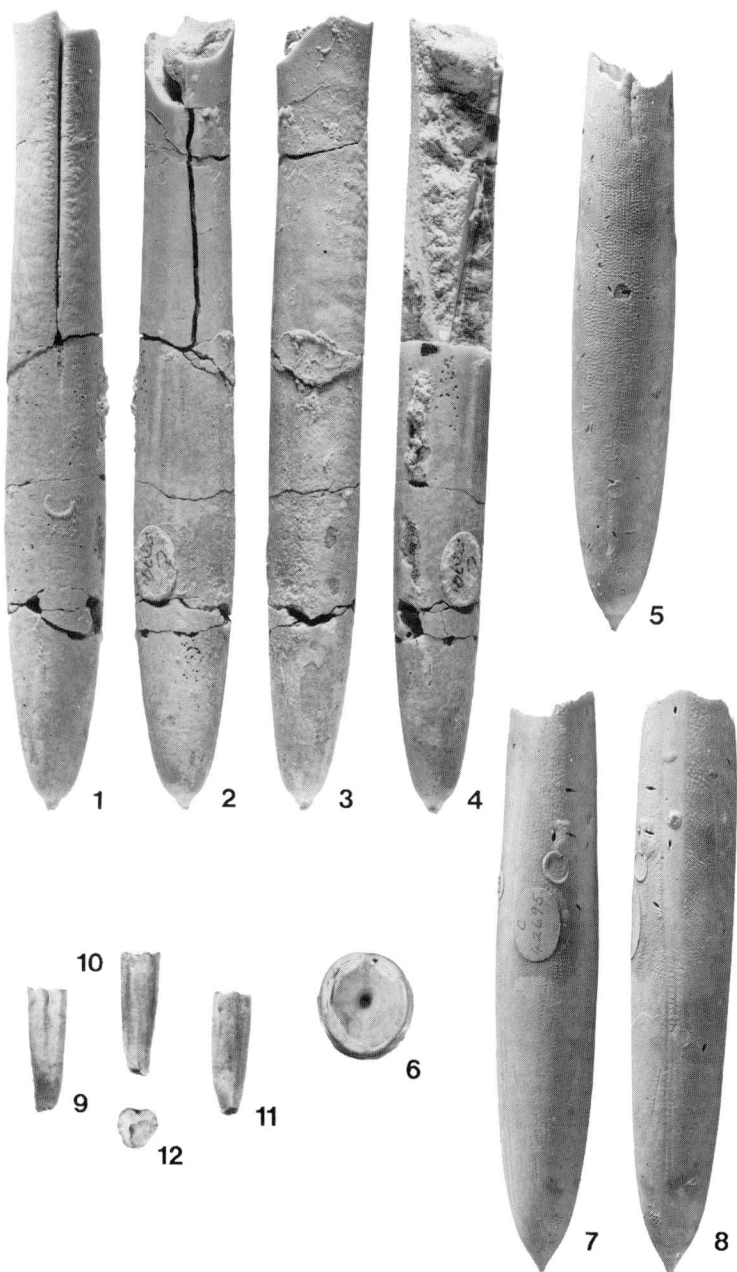

Occurrence. Lower Campanian, *O. pilula* Zone; rare, known only from north-east England.

Genus BELEMNOCAMAX

Belemnocamax boweri Crick
Plate 46, figures 9–12

Description. Rostrum up to 15 mm in length; symmetrical or nearly symmetrical and conical in both outline and profile. The apex is acute. Transverse sections are pyriform or clover-leaf in shape, being more pyriform at the apex. A ventral alveolar fissure is present, extending into the stem region as a shallower groove. Lateral lines are present as incised dorso-lateral depressions, producing the clover-leaf-shaped transverse section, and incised ventro-lateral lines may also be present. Dorsal striations may be developed all along the dorsum, but are commonly best preserved at the apex. The alveolus is lost owing to imperfect calcification, and a shallow pseudoalveolus or weakly conical anterior fracture may be present. The rostrum is otherwise smooth.

Occurrence. Middle Cenomanian, *A. rhotomagense* Zone; rare, known only from Yorkshire, Lincolnshire and Norfolk.

12. ARTHROPODS

by S. F. MORRIS *and* J. S. H. COLLINS

Arthropods that are likely to be encountered in the Chalk belong to two major groups, the Malacostraca and the Cirripedia. The Malacostraca include crabs, lobsters, squat lobsters, callianasids and marine isopods. The British Chalk crabs have been monographed by Wright and Collins (1972), the lobsters by Woods (1925–29), but the other three groups of British Malacostraca have never been properly dealt with in this way. Chalk cirripedes have been monographed by Withers (1935). For terminology, the appropriate sections in the Treatise of Invertebrate Paleontology, Part R, should be consulted (Brooks *et al.* 1969; Newman *et al.* 1969) along with Wright and Collins (1972).

MALACOSTRACA

Three major groups of Malacostraca, 'Macrura', Brachyura, and Isopoda, are found in the Chalk of Britain. A fourth group, Anomura (squat lobsters, mud-shrimps and hermit crabs) have yet to be found in chalk facies, although one species of galatheid squat lobster occurs in the sandy Cenomanian beds of Wilmington and the mud-shrimp *'Callianassa' neocomiensis* is common in the Cenomanian of Colin Glen near Belfast, Northern Ireland.

'Macrura', or 'long-tailed' crustaceans (i.e. lobsters, prawns, shrimps), are largely confined to the Cenomanian and Turonian of Kent and Sussex. Only *Enoploclytia leachii* is found in the Upper Chalk. Nowhere are they abundant.

Brachyura (crabs) are very uncommon in the Chalk, in marked contrast to the arenaceous facies of the Cenomanian of Devon and Wiltshire where 28 species have been recorded, 24 of these from Wilmington, Devon alone. Three species (*Notopocorystes normani*, *Necrocarcinus woodwardii* and *Diaulax oweni*) occur sparsely in the Lower Cenomanian of Kent, Sussex and the Isle of Wight, while from the Upper Chalk only single specimens of four species are known.

An anomalous species of marine isopod is recorded rarely from the Cenomanian chalk of Bedfordshire, Kent and Surrey. It is assigned to the 'lump' genus *Palaega*. Only the posterior half of this animal is ever preserved. Whilst the same body plan is found in several isopod families, it has been suggested that it might be assigned to the Cirolanidae near to the Recent genus *Bathynomus*. Swelling of the branchial chambers of crabs such as *Notopocorystes* has been doubtfully assigned to the parasitic isopod *Bopyrus*.

DESCRIPTIONS

Order DECAPODA
Infraorder BRACHYURA
Family RANINIDAE

Genus NOTOPOCORYSTES

Notopocorystes normani (Bell)
Plate 47, figure 7

Description. Carapace ovoid with gently curved anterior and nearly straight postero-lateral margins; one-fifth longer than wide; very convex transversely. Straight orbito-frontal margin forming 49 per cent of carapace width with narrow, sulcate, bifid rostrum. Outer orbital spine sharp. Two spines on antero-lateral margins of carapace, one at lateral angle where the carapace reaches its widest point, followed by a smaller one behind the branchiocardiac furrow. Four tubercles disposed in a diamond-shaped pattern on each side of the mesogastric lobe, and a median row of tubercles extending to the posterior margin.
Occurrence. Cenomanian; Kent and the Isle of Wight.

Genus HEMIOON

Hemioon circumviator Wright and Collins
Plate 47, figure 1

Description. Carapace longitudinally oval and truncated posteriorly, strongly arched transversely. Orbito-frontal margin forming half of carapace width. Maximum width at anterior third of carapace. Separate regions hardly differentiated over carapace. Fine anteriorly-directed spine on anterolateral margin one-third out from outer orbital angle to the strong epibranchial spine. Rostrum small and bifid.
Occurrence. Upper Turonian, *S. plana* Zone; Surrey.

Family CALAPPIDAE

Genus NECROCARCINUS

Necrocarcinus labeschii (Deslongchamps)
Plate 47, figures 2, 4

Description. Rounded and inflated carapace, slightly wider than long. Well-produced spiny rostrum, triangular and sulcate; may be directed anteriorly or distinctly down-turned. Orbits close together, directed obliquely upwards and with two deep notches on the upper margin. Antero-lateral borders rounded with three or four sharp spines (largest at widest point of carapace). Cervical groove broad and deep, and interrupted at midline. Regions and lobes tumid. Large tubercles on upper surface arranged as one transverse and three longitudinal rows.

Occurrence. Lower Albian, *D. mammillatum* Zone, to Lower Cenomanian, *M. mantelli* Zone; Lower Cenomanian of Wiltshire, Devon, Sussex, Kent, Isle of Wight and Cambridgeshire; also known from the Upper Albian of Wiltshire, Dorset and Cambridge.

Necrocarcinus woodwardii Bell
Plate 47, figure 3

Description. Resembling *N. labeschii* in having a rounded, inflated carapace, but with a greater number of tubercles on the front half of the cephalothorax arranged as two distinct transverse rows, and with a transverse row of three tubercles on the mesogastric lobe. Simple triangular rostrum.

Occurrence. Upper Albian–Lower Cenomanian, *M. mantelli* Zone; Cenomanian of Wiltshire, Devon, Sussex, Isle of Wight and Cambridgeshire; also known from the Upper Albian of Wiltshire, Dorset and Cambridge.

Family DIAULACIDAE

Genus DIAULAX

Diaulax oweni (Bell)
Plate 47, figure 5

Description. Carapace hexagonal in outline; broader than long with greatest breadth occurring in the anterior half. Anterior half of carapace down-turned and posterior half flat. Small, deeply sulcate rostrum. Orbits deep, subcircular and occupying outer quarters of orbito-frontal width. Cervical furrow narrow and shallow, but distinct, ending in a notch on the antero-lateral margin. Ornament of densely crowded flat-topped granules.

Occurrence. Lower Albian, *L. regularis* Subzone, and Middle Cenomanian, *A. rhotomagense* Zone; Cenomanian of Devon, Somerset and Kent.

Family CARPILIIDAE

Genus CALOXANTHUS

Caloxanthus purleyensis (Withers)
Plate 47, figure 6

Description. Carapace transversely oval and slightly wider than long. Strongly arched longitudinally. Frontal margin straight with strong orbital indentations. Cervical groove distinct medially, but obsolete on the hepatic region. Lateral margins entire. Regions weakly differentiated. Surface ornament consists of sparse tubercles with groups of fine, denser tubercles near postero-lateral margins.
Occurrence. Coniacian or basal Santonian; Surrey.

Infraorder ASTACIDEA
Family ERYMIDAE

Genus ENOPLOCLYTIA

Enoploclytia leachii (Mantell)
Plate 48, figure 1

EXPLANATION OF PLATE 47

Fig. 1. *Hemioon circumviator* Wright and Collins. Upper Turonian, Godalming Bypass, Guildford, Surrey; dorsal carapace; ×1.

Figs 2, 4. *Necrocarcinus labeschii* (Deslongchamps). Cenomanian, White Hart Pit, Wilmington, Devon; two carapaces in dorsal view; ×1.

Fig. 3. *Necrocarcinus woodwardii* Bell. Lower Cenomanian, Warminster, Wiltshire; dorsal carapace; ×2.

Fig. 5. *Diaulax oweni* (Bell). Cenomanian, Snowdon Hill, Chard, Somerset; dorsal view of carapace with both chelipeds; ×1.

Fig. 6. *Caloxanthus purleyensis* (Withers). Coniacian or basal Santonian, Purley, Surrey; dorsal view of carapace; ×2.

Fig. 7. *Notopocorystes normani* (Bell). Cenomanian, Dover, Kent; dorsal view of carapace; ×1.

Fig. 8. *Cretiscalpellum glabrum* (Roemer). Campanian, *Gonioteuthis quadrata* Zone, East Harnham, Salisbury, Wiltshire; right lateral view with tergum, carina and upper latus in place; ×3.

Fig. 9. *Arcoscalpellum fossula* (Darwin). Campanian, *Gonioteuthis quadrata* Zone, East Harnham, Salisbury, Wiltshire; left lateral view of capitulum; ×2..

Fig. 10. *Zeugmatolepas mockleri* Withers. Cenomanian, Cambridge; nearly complete capitulum viewed from left side; ×6.

Fig. 11. *Stramentum pulchellum* (Sowerby). ?Cenomanian, Kent; lateral view of complete capitulum with peduncle; ×1·6.

Figs 12–13. *Brachylepas naissanti* (Hébert). Campanian, Thorpe St. Andrew, Norwich, Norfolk; apical and right lateral views of capitulum without opercular valves. 7, ×1·5; 8, ×2.

PLATE 47

Description. Carapace and first pereiopods coarsely ornamented. Rostrum strongly dentate. Gastro-orbital groove wide, deep, short; postcervical groove deep; branchiocardiac groove short. Chelae long and slender; fingers long and slender, twice the length of the palm and without spines.
Occurrence. Turonian–Upper Campanian; Sussex, Kent and Surrey.

Genus PALAEASTACUS

Palaeastacus sussexiensis (Mantell)
Plate 48, figures 2–3

Description. Carapace moderately coarsely ornamented, rostrum dentate. Gastro-orbital groove weak, postcervical groove stronger and separate from branchiocardiac groove. Chelae short, stout and spinose with fingers of equal length to the palm.
Occurrence. Cenomanian–Turonian; Sussex, Kent, Surrey, the Isle of Wight and Cambridgeshire.

Infraorder PALINURA
Family GLYPHEIDAE

Genus GLYPHEA

Glyphea willetti (Woodward)
Plate 48, figures 4–5

Description. Cephalothorax laterally compressed, with a mid-dorsal carina and a row of tubercles on each side of it. Rostrum short. Three strong, tubercle-bearing carinae run from the cervical groove to the anterior margin. Cervical groove very deep. Marginal groove and carina present. Abdominal terga smooth.
Occurrence. Upper Albian–Cenomanian; Sussex, Kent, Hertfordshire and Cambridgeshire.

EXPLANATION OF PLATE 48

Fig. 1. *Enoploclytia leachii* (Mantell). ?Campanian, Arundel, Sussex; dorsal view; ×0·5.

Figs 2–3. *Palaeastacus sussexiensis* (Mantell). 2, ?Turonian, Sussex; left lateral view of carapace; ×1. 3, Cenomanian, Burham, Kent; outer surface of left cheliped; ×1.

Figs 4–5. *Glyphea willetti* (Woodward). 4, Cenomanian, Dover, Kent; right lateral view of carapace; ×1. 5, Cenomanian, near Hitchin, Hertfordshire; lateral view of incomplete specimen; ×0·5.

PLATE 48

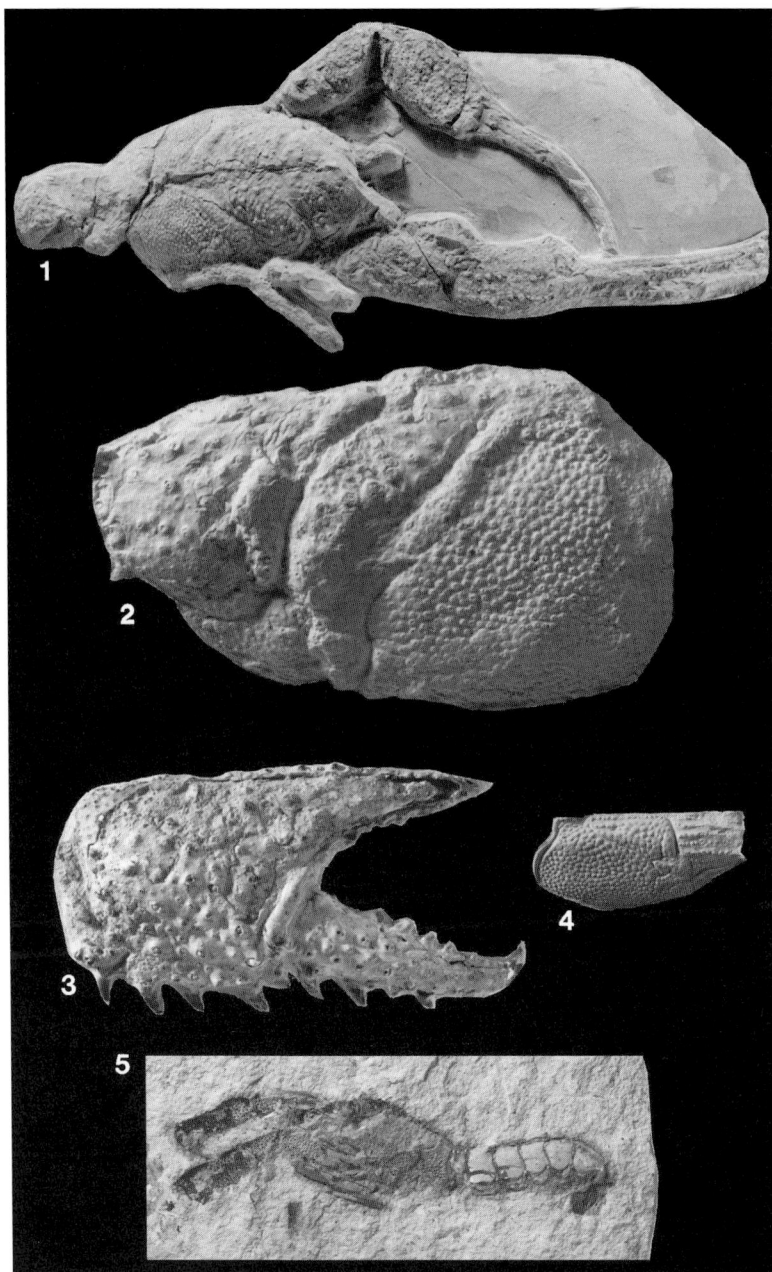

CIRRIPEDES

Fourteen genera of cirripedes are known from the British Chalk. Four range upward from lower deposits and three, *Arcoscalpellum*, *Verruca* and *Calantica s.l.*, persist to the present-day. Of 42 species and subspecies, two, *Cretiscalpellum glabrum* (Roemer) and *Arcoscalpellum angustatum* (Geinitz), range throughout the Chalk, but most are quite rare.

The Stramentidae have strongly armoured peduncles and therefore have less tendency to disintegrate after death. Readily distinguishable scalpellid species include *Zeugmatolepas mockleri* Withers and *Z. cretae* (Steenstrup), whose valves are diminutive, and *Arcoscalpellum maximum* (J. de C. Sowerby) (Upper Santonian–Upper Campanian), whose valves are large and robust. The suborder Verrucomorpha, a group of sessile asymmetrical cirripedes, appears in the British Cenomanian with *Proverruca* (?) *nodosa* Withers, but a more characteristic form is *Verruca prisca* Bosquet, from the Campanian. The suborder Brachylepadomorpha is represented in the Chalk by *Pycnolepas*? *scalaris* Withers (Cenomanian), *Brachylepas naissanti* (Hebert) (Campanian) and *Brachylepas fallax* (Darwin) (Turonian–Campanian). This suborder includes the commonest sessile, symmetrical cirripedes.

The widespread Balanidae of Recent shorelines is not represented in the British Chalk but very rare specimens have been recorded from the Upper Cretaceous of Hungary.

DESCRIPTIONS

Class CIRRIPEDIA
Order THORACICA
Suborder BRACHYLEPADOMORPHA
Family BRACHYLEPADIDAE

Genus BRACHYLEPAS

Brachylepas naissanti (Hébert)
Plate 47, figures 12–13

Description. Shell radially symmetrical with semi-conical carina and rostrum, bearing strong, raised, longitudinal ribs. Tergum with apico-basal ridge that is curved towards scutum. Four rows of imbricating plates, outer whorls of plates having median basal notch. Basis membranous. Valves only moderately thick.
Occurrence. Campanian; Dorset, the Isle of Wight and Norfolk.

Suborder SCALPELLOMORPHA
Family SCALPELLIDAE
Subfamily CALANTICINAE

Genus CRETISCALPELLUM

Cretiscalpellum glabrum (Roemer)
Plate 47, figure 8

Description. Capitulum composed of eighteen valves. Carina comparatively narrow and strongly arched transversely, with strong apico-basal ridge. Tergum broad with sharp carinal angle; scutum with tergo-lateral part moderately broad. These valves and the lower latera have a strong apico-basal ridge.

Occurrence. Upper Albian–Maastrichtian; very widely distributed in Britain throughout the species range, but particularly common in the Turonian of the Alton district, Hampshire.

Subfamily ZEUGMATOLEPADINAE

Genus ZEUGMATOLEPAS

Zeugmatolepas mockleri Withers
Plate 47, figure 10

Description. Capitulum with an erect shape and at least 34 valves including three or more whorls of subtriangular lower latera with V-shaped growth lines. Scutum with umbo situated from just below apex to a little more than one-third the distance from the apex, according to growth stage and stratigraphic position. Carina widening rapidly from apex. Basal margin acutely angular. Tergum subrhomboidal, about half as wide as long, with apex directed towards the carina.

Remarks. *Zeugmatolepas cretae* (Steenstrup) is essentially similar but the umbo of the scutum is more central in position with the apex developed into an acute angle; the carina is relatively wider and the median ridge more obscure; and the tergum is more rounded in appearance. It is first found in the Turonian *T. lata* Zone of Surrey and continues into the Coniacian *M. cortestudinarium* Zone in that county. It also occurs in the Campanian *A. quadratus* Zone of Sussex and Wiltshire and the *B. mucronata* Zone of Norfolk.

Occurrence. Upper Albian–Upper Cenomanian, *N. juddii* Zone; Upper Albian of Kent and near Cambridge; Lower Cenomanian of Sussex, Kent and Hampshire; Upper Cenomanian (*N. juddii* Zone) of Surrey and Berkshire; Cenomanian (undifferentiated) of Kent and Wiltshire.

Subfamily ACROSCALPELLINAE

Genus ARCOSCALPELLUM

Arcoscalpellum fossula (Darwin)
Plate 47, figure 9

Description. Capitulum with 15 closely fitting valves including four pairs of latera. Peduncle furnished with numerous small plates. Carina having the tectum almost flat and bordered on each side by a wide, protuberant, flat-topped ridge; basal margin obtusely angular but often rounded.
Occurrence. Santonian–Campanian; Santonian of Kent, Surrey, Hampshire, Suffolk and Norfolk; Campanian of Sussex, Hampshire, Wiltshire, Dorset, Suffolk and Norfolk.

Family STRAMENTIDAE

Genus STRAMENTUM

Stramentum pulchellum (Sowerby)
Plate 47, figure 11

Description. Capitulum composed of a single whorl of nine valves. Broad peduncle, closely armoured with eight vertical rows of intersecting plates; the upper margins of the three median rows of plates either side are markedly convex. Scutum with umbo situated up to one-third of length of valve from the apex and with occludent margin almost straight.
Occurrence. Cenomanian–Turonian; Cenomanian of ?Kent; Turonian of Kent, Surrey and County Antrim, Northern Ireland.

13. ECHINODERMS

by A. B. SMITH *and* C. W. WRIGHT

Echinoderms are a diverse and relatively common element of the Chalk fauna, with a number of species acting as excellent stratigraphical markers. The echinoderm fauna is enormous and only a small proportion of the known species can be illustrated here. Echinoids, because of their large size and fairly robust test of calcareous plates, are the most widely encountered and three genera in particular (*Conulus*, *Echinocorys* and *Micraster*) are amongst the most characteristic fossils of the British Chalk. Most Chalk echinoids lived on the surface of the sea floor, but a few irregular echinoids were shallow infaunal burrowers. Although tests of echinoids are commonly found complete, it is much rarer to find specimens that have their associated spines still attached.

British Cretaceous echinoids were first monographed by T. Wright (1864–82) and are currently being revised and updated by Smith and Wright (1989, 1991, 1994, 1996, 1999, 2000). Only the British fauna of spatangoids and holasteroids now remains to be updated. Although some spatangoid and holasteroid genera have been the subject of detailed study (*Micraster*: Ernst 1970; Stokes 1975; Fourey 1982; David and Fourey 1984; *Echinocorys*: Smiser 1935; *Infulaster* and *Hagenowia*: Gale and Smith 1982) descriptions of the German Chalk fauna (Ernst 1968, 1970, 1971, 1972; Ernst and Schulz 1974; Ernst and Seibertz 1977; Schulz 1984) remain the best source of general information.

Crinoids are most commonly encountered in the Chalk in the form of isolated skeletal elements, though bourgueticrinid cups are not uncommon and can be useful stratigraphically. Two crinoids, *Marsupites* and *Uintacrinus*, are very important stratigraphical markers in the Campanian. Otherwise, isolated crinoid elements are of limited use and cannot often be ascribed to a species. The Chalk crinoids have been thoroughly monographed by Rasmussen (1961) and the Maastrichtian forms decribed in admirable detail by Jagt (1999).

Asteroids and ophiuroids, like crinoids, are commonly encountered only in the form of disarticulated ossicles, though with luck it is still possible to find rare articulated specimens. The British fauna was monographed by Sladen and Spencer (1891–1908). Identification of isolated asteroid marginals is possible. The large marginal ossicles of goniasterids are particularly useful since they can usually be identified to species level on the basis of their characteristic ornamentation. These have been the subject of a number of monographs (Wright and Wright 1940; Rasmussen 1950; Schulz and Weitschat 1975; Gale 1986, 1987; Breton 1992).

DESCRIPTIONS

Phylum ECHINODERMATA
Subphylum PELMATOZOA
Class CRINOIDEA
Subclass ARTICULATA
Order ISOCRINIDA
Suborder ISOCRININA
Family ISOCRINIDAE

Remarks. Isocrinids have a long column, composed of generally pentagonal or pentalobate columnals whose articular faces have five 'petals' bordered by crenellae; periodical columnals bear cirri which attach to oval articulation facets. The theca (cup) is small and the arms long and branching. Normally only individual ossicles or short lengths of column are found in the Chalk.

Genus AUSTINOCRINUS

Remarks. Columnals subpentagonal to circular, with short, narrow petals inside a broad margin with fine, radiating crenellae.
Occurrence. Campanian–Maastrichtian, and possibly in younger deposits.

Austinocrinus bicoronatus (Hagenow)
Plate 49, figures 1–2

Description. Columnals round to weakly subpentagonal; outer surface smooth and non-crenulate between columnals. Articulation facets with narrow sunken areolae bordered by short, stout, broadly V-shaped crenellae. Crenellae becoming larger towards margin, but missing from end of areolae. Several circles of much finer radial crenellae form vermiculate margin to the articulation face.
Occurrence. Upper Campanian, *B. mucronata* Zone, to Lower Maastrichtian, *B. sumensis* Zone; rather rare in quarries around Norwich and on the Norfolk coast at Trimingham.

Genus ISSELICRINUS

Remarks. Articular faces of columnals normally with wide petals inside a ring of short, uniform crenellae around the margin; nodal columnals bear fewer than five cirri.
Occurrence. Maastrichtian–Miocene.

Isselicrinus buchii (Roemer)
Plate 49, figures 7–8

Description. Columnals subpentagonal to subcircular with smooth outer face and non-crenulate sutures. Articular face with 13 or 14 crenellae forming the outer margin of the petals and usually extending to the margin of the articulation face. Areolae broad and smooth. Crenellae along the ridges between petals are generally absent.
Occurrence. Lower Maastrichtian; Trimingham, Norfolk.

Genus NIELSENICRINUS

Remarks. Cretaceous isocrinids with arms having a synarthrial articulation at II Br1–2.
Occurrence. Valanginian–Maastrichtian.

Nielsenicrinus cretaceus (Leymerie)
Pl. 49, figure 13

Description. Columnals strongly pentalobate, alternately thick and thin; mostly smooth, though with weak median ridge on most proximal columnals.
Occurrence. Aptian–Cenomanian; only common in the Middle–Upper Cenomanian of south-east England.

Genus ISOCRINUS?

Remarks. Articular faces of columnals have elliptical petals surrounded by crenellae; nodal columnals have five cirrus sockets at mid height on the outer face. The columnal morphology of *Isocrinus* is highly conservative and similar to that seen in several other genera, and the genus is defined by characters of the theca and arms. Consequently Chalk species based on columnal morphology alone are only tentatively assigned to this genus and may prove to belong to other genera of Isocrinidae.
Occurrence. Triassic–Recent.

Isocrinus? *granosus* Valette
Plate 49, figures 3–4

Description. Columnals subpentagonal with a slightly swollen outer face and a low rounded median ridge of rather coarse granules. Sutures distinctly crenulate.
Occurrence. Middle Turonian, *C. woolgari* Zone, to Upper Turonian, *S. neptuni* Zone; isolated columnals are widespread and abundant in southern England.

PLATE 49

Isocrinus? *legeri* (Repelin)
Plate 49, figures 5–6

Description. Columnals pentagonal in outline with flat outer face except for a ridge across each interradial angle. Outer surface with fine pustular granulation.
Occurrence. Lower Albian–Lower Cenomanian, *M. mantelli* Zone; Kent and the sandy Cenomanian of Devon, probably elsewhere.

Isocrinus? *minutus* (Valette)
Plate 49, figures 9–10

Description. Slender subpentagonal columnals; outer face with a median row of fine, close spines and a narrow ridge or row of granules above and below.
Occurrence. Coniacian, *R. petrocoriense* Zone, to Upper Santonian, *P. polyopsis* Zone; rather rare in Kent and Surrey.

Suborder BOURGUETICRINIDA
Family BOURGUETICRINIDAE

Genus BOURGUETICRINUS

Remarks. Small crinoids with a root of branched cirri and a rather short column of normally smooth and relatively tall ossicles; columnals

---

EXPLANATION OF PLATE 49

Figs 1–2. *Austinocrinus bicoronatus* (Hagenow). Upper Campanian, *B. mucronata* Zone, Trowse, Norfolk. 1, articular face of columnal; 2, side view of part of column; ×3.

Figs 3–4. *Isocrinus*? *granosus* Valette. Upper Turonian, *S. plana* Zone, Dover, Kent. 3, articular face of columnal; 4, side view of part of column; ×4·5.

Figs 5–6. *Isocrinus*? *legeri* (Repelin). Cenomanian, Folkestone, Kent. 5, articular face of columnal; 6, side view of part of column; ×5.

Figs 7–8. *Isselicrinus buchii* (Roemer). Lower Maastrichtian, *B. lanceolata* Zone, Trimingham, Norfolk. 7, articular face of columnal; 8, side view of part of column; ×5.

Figs 9–10. *Isocrinus*? *minutus* (Valette). Santonian, *M. coranguinum* Zone, Thanet coast, Kent. 9, articular face of columnal; 10, side view of part of column; x10.

Figs 11–12. *Roveacrinus alatus* Douglas. Coniacian, *M. cortestudinarium* Zone, Seaford Read, Sussex. 11, side, and 12, top (ventral) views of theca; ×10.

Fig. 13. *Nielsenicrinus cretaceus* (Leymerie). Middle or Upper Cenomanian, Folkestone, Kent; stem with cirri, cup and basal part of arms; ×2.

Fig. 14. *Applinocrinus cretaceus* (Bather). Lower Campanian, *G. quadrata* Zone, Durrington, Sussex; side view of theca; ×20.

normally with elliptical articular faces with the long axis of the ellipse at different angles on the two ends of the ossicle. The theca is generally fused, composed of an undivided proximale, a ring of basals (except in *B. elegans*) and an alternating ring of radials. Columnals and thecae of *Bourgueticrinus* are among the commonest fossils in the Upper Chalk of southern England and many species are useful stratigraphic indicators.

Bourgueticrinus brydonei Rasmussen
Plate 50, figures 11–12

Description. Theca smooth and conical, widest at or near the base of the proximale, then tapering to a truncated top.
Occurrence. Upper Campanian, *B. mucronata* Zone, and/or Lower Maastrichtian, *B. lanceolata* Zone; Trimingham, Norfolk.

Bourgueticrinus cylindricus M'Coy
Plate 50, figures 2–3

Description. Theca high, subcylindrical to pear-shaped; basals large, and radials very small and largely hidden in side view, giving a low, rounded, upper end; proximale generally with elliptical lower articular face.
Occurrence. Upper Turonian, *S. plana* Zone, to Upper Campanian, *B. mucronata* Zone; moderately common in southern England throughout its range.

Bourgueticrinus elegans Griffith and Brydone
Plate 50, figures 4–5

Description. Theca with club- or pear-shaped proximale, topped by a slightly narrower low collar of radials, without trace of basals. Articular face of proximale is much smaller than its lower end. Columnals are long, slender and subcylindrical, with small elliptical articular faces.
Occurrence. Lower Santonian, *M. coranguinum* Zone, to Lower Campanian, *G. quadrata* Zone; widespread in southern England and highly characteristic of *O. pilula* and *G. quadrata* zones, rarer below.

Bourgueticrinus ellipticus Miller
Plate 50, figure 6

Description. Theca large, pear- or club-shaped with wide, flat top; sutures inconspicuous.
Occurrence. Upper Turonian, *S. plana* Zone, to Lower Campanian, *G. quadrata* Zone; widespread in southern England in the *M. coranguinum* Zone but scarcer below and above.

Bourgueticrinus fritillus Griffith and Brydone
Plate 50, figures 8–9

Description. Theca with proximale widest at base and top, narrower in between.
Occurrence. Lower Campanian, *O. pilula–G. quadrata* zones; widespread in southern England.

Bourgueticrinus hureae (Valette)
Plate 50, figure 7

Description. Theca more or less cylindrical; basals and radials of similar size and strongly tumid, separated by depressed sutures.
Occurrence. Lower Santonian, *M. coranguinum* Zone, to Lower Campanian, *G. quadrata* Zone; rather sparse at all horizons but widespread in southern England.

Bourgueticrinus papilliformis Griffith and Brydone
Plate 50, figure 10

Description. Theca slender, spindle- or pear-shaped with high basals and the radials forming a very narrow collar on top.
Occurrence. Lower Santonian, *M. coranguinum* Zone, to Lower Campanian, *G. quadrata* Zone; widespread and common in southern England, particularly characteristic of the *U. socialis* and *M. testudinarius* zones.

Order COMATULIDA

Remarks. Adult comatulids ('feather stars') are small, free-swimming crinoids with feathery arms springing from a flat-topped discoid, hemispherical or conical centrodorsal plate on the underside of which are few to many cirri with which the crinoid may cling to the bottom. Normally in the Chalk only centrodorsals are found, occasionally with basals and radials attached. They occur sporadically in most zones and regions.

Genus GLENOTREMITES

Remarks. Centrodorsal with prominent stellate scar on aboral surface and five deep, radial pits surrounding the central cavity on the ventral surface. Large cirrus sockets with weakly crenelate margins developed around the margin, tending to be in vertical rows, between ten and 20 in total.
Occurrence. Albian–Santonian.

Glenotremites paradoxus Goldfuss
Plate 51, figures 5–6

PLATE 50

Description. Centrodorsal curved, with broad base and tending to be hemispherical, about half as high as wide.
Occurrence. Lower Cenomanian, *M. mantelli* Zone, to Coniacian, *M. cortestudinarium* Zone; Kent, Sussex, Isle of Wight, Dorset, Devon and Yorkshire, least rare in Kent.

Glenotremites rotundus (Carpenter)
Plate 51, figures 3–4

Description. Centrodorsal tends to be rounded subconical in profile. Cirrus sockets smaller, more numerous and regular, and more densely packed than in *G. paradoxus*.
Occurrence. Upper Albian, *M. inflatum* Zone, Lower Cenomanian, *M. mantelli* Zone; Isle of Wight, Wiltshire, Dorset and Devon.

Genus PLACOMETRA

Remarks. Small, without radial pits on oral surface of centrodorsal; underneath flat or arched, without cirri; the flat sides have very few, large cirrus sockets.

Placometra laticirra Carpenter
Plate 51, figures 1–2

Description. Underneath of centrodorsal almost smooth. Cirrus sockets tall and with radiating crenellae round their edges.

EXPLANATION OF PLATE 50

Fig. 1. *Uintacrinus socialis* Grinnell. Upper Santonian, *U. socialis* Zone, Bridlington, Yorkshire; theca obliquely viewed from below; ×1·5.
Figs 2–3. *Bourgueticrinus cylindricus* M'Coy. Lower Santonian, *M. coranguinum* Zone, Salisbury, Wiltshire. 2, oral, and 3, side views of theca; ×4.
Figs 4–5. *Bourgueticrinus elegans* Griffith and Brydone. Lower Campanian, *G. quadrata* Zone, Seaford Head, Sussex. 4, oral, and 5, lateral views of theca; ×3.
Fig. 6. *Bourgueticrinus ellipticus* Miller. Lower Santonian, *M. coranguinum* Zone, Salisbury, Wiltshire; side view of theca; ×5.
Fig. 7. *Bourgueticrinus hureae* (Valette). Santonian, Gravesend, Kent; side view of theca with associated characteristic bourgueticrinid columnal; ×5.
Figs 8–9. *Bourgueticrinus fritillus* Griffith and Brydone. Lower Campanian, *G. quadrata* Zone, Salisbury, Wiltshire. 8, oral, and 9, side views of theca; ×4.
Fig. 10. *Bourgueticrinus papilliformis* Griffith and Brydone. Santonian, *U. socialis* Zone, Devizes, Wiltshire; side view of theca; ×2.
Figs 11–12. *Bourgueticrinus brydonei* Rasmussen. Lower Maastrichtian, Trimingham, Norfolk. 11, oral, and 12, side views of theca; ×4.

Occurrence. Upper Turonian, *S. plana* Zone, to Lower Santonian, *M. coranguinum* Zone; rare in Surrey, Hampshire, Wiltshire and Dorset.

Order ROVEACRINIDA

Family ROVEACRINIDAE

Remarks. Minute, free-swimming crinoids with conical or spherical theca containing a small double cavity and normally with spines, ridges or flanges. Roveacrinids are occasionally common in the Chalk and may be washed out with foraminifera or picked off air-weathered surfaces.

Genus ROVEACRINUS

Remarks. Theca commonly high and conical; the radials bear a vertical ridge or wing.
Occurrence. Albian–Santonian.

Roveacrinus alatus Douglas
Plate 49, figures 11–12

Description. Radials with large vertical wing-like flanges, commonly extending well below the body of the theca. Theca strongly stellate in outline.
Occurrence. Cenomanian, zone unknown, to Coniacian, *M. cortestudinarium* Zone; rare in Sussex, Devon and Cambridgeshire.

Family SACCOCOMIDAE

Genus APPLINOCRINUS

Remarks. Minute; thecal plates thin, surrounding a large single cavity; arms very slender.
Occurrence. Kimmeridgian–Campanian.

Applinocrinus cretaceus (Bather)
Plate 49, figure 14

EXPLANATION OF PLATE 51

Figs 1–2. *Placometra laticirra* Carpenter. Upper Chalk, Wylye, Wiltshire. 1, side, and 2, underneath views of centrodorsal; ×10.

Figs 3–4. *Glenotremites rotundus* (Carpenter). Haldon, Devon, ?Cenomanian. 3, side, and 4, underneath views of centrodorsal; ×9.

Figs 5–6. *Glenotremites paradoxus* Goldfuss. Upper Turonian, *S. plana* Zone, Culver Cliff, Isle of Wight. 5, side, and 6, underneath views of centrodorsal plate with radials; ×9.

Fig. 7. *Marsupites testudinarius* (Schlotheim). Upper Santonian, *M. testudinarius* Zone, Kent; lateral view of cup; ×1·5.

PLATE 51

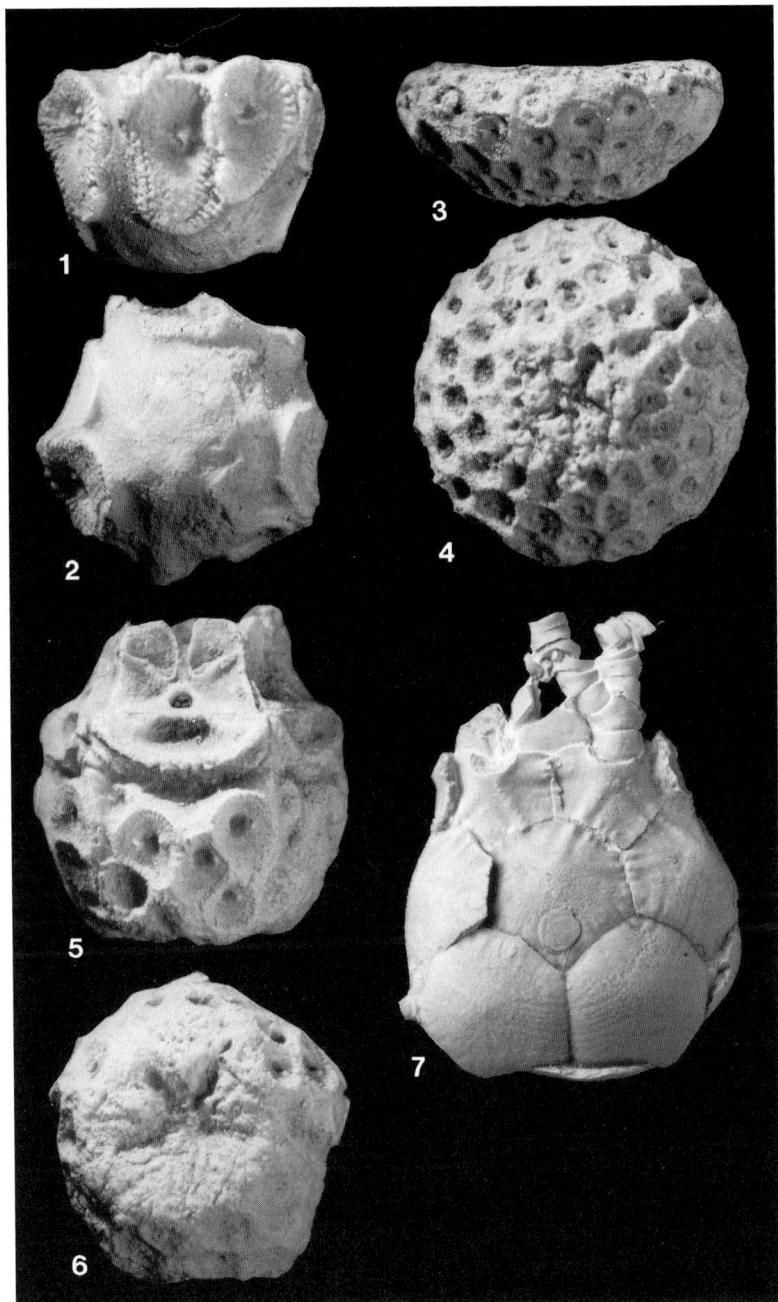

Description. Theca subconical with inflated sides and more or less pointed distally; upper surface has five protruding radial double ridges.
Occurrence. Lower Campanian, *G. quadrata* Zone, in a narrow band near the base of the zone; Sussex and Hampshire; now also known from the Upper Campanian–uppermost Maastrichtian of the Netherlands.

Order UINTACRINIDA

Remarks. Free-swimming crinoids consisting of a large sac-like theca and long arms.

Family UINTACRINIDAE

Genus UINTACRINUS

Remarks. Theca composed of many rather small plates, thick in the middle, thinner at the edges; the lower plates of the arms form part of the thecal sac and there is a gradual transition from theca to free arms.
Occurrence. *Uintacrinus* appears suddenly in vast numbers in the Upper Santonian of Europe, Australia and North America and disappears by the end of the stage. Its ancestry remains cryptic.

Uintacrinus socialis Grinnell
Plate 50, figure 1

Description. Ossicles of the theca with regular edges and almost smooth. The only other known species, *U. anglicus* Rasmussen from the top of the *M. testudinarius* Zone, has the theca ossicles with irregular edges and folds on the surface.
Occurrence. Upper Santonian, *U. socialis* Zone; abundant wherever the zone occurs, from Kent to Dorset and north to Yorkshire.

Family MARSUPITIDAE

Genus MARSUPITES

Remarks. Theca is a sac composed of 16 large, thin, almost equal plates; the arms are sharply distinguished from the theca and articulated to its upper edge.
Occurrence. *Marsupites* appears suddenly in the Upper Santonian, slightly later than *Uintacrinus*, in Europe, North Africa, Madagascar, India, Australia and North America.

Marsupites testudinarius (Schlotheim)
Plate 51, figure 7

Description. The large pentagonal thecal plates vary from almost smooth to strongly ornamented with granulated radiating ridges.

Occurrence. Upper Santonian, *M. testudinarius* Zone; common wherever its zone occurs, from Kent to Dorset and Yorkshire.

Subphylum ELEUTHEROZOA
Class ASTEROIDEA
Order VALVATIDA
Suborder TUMULOSINA
Family STAURANDERASTERIDAE

Remarks. Marginals and ossicles of upper surface are notched at the corners so that they are shield- or cross-shaped. The upper surface has a swollen central ossicle and a ring of five large radial ossicles.

Genus ASPIDASTER

Remarks. Arms swollen, separated from disc by a narrow neck.

Occurrence. Bathonian–Danian.

Aspidaster bulbiferus (Forbes)
Plate 52, figure 11

Description. Major ossicles have an even spread of fine pits, except for a smooth band around the edge.

Occurrence. Lower Turonian, *I. labiatus* Zone, to Upper Campanian, *B. mucronata* Zone; found in all areas in most zones, particularly in the *M. coranguinum* Zone in the south-east.

Suborder GRANULOSINA
Family GONIASTERIDAE
Subfamily GONIASTERINAE

Genus METOPASTER

Remarks. Flat, pentagonal form with arms not or only slightly produced; upper marginals differentiated into rectangular intermediates, 2–6 a side, and triangular ultimates. Surface of marginals with distinct central area, with or without well-separated pits, and narrow, smooth or finely pitted edge.

Occurrence. Cenomanian–Eocene, ?Recent (the living genus *Tosia* may be synonymous).

Metopaster hunteri (Forbes)
Plate 52, figure 3

Description. Pentagonal, with three upper marginals in each half radius, low in profile but rather swollen. Ultimates scarcely larger than other marginals. Upper marginals swollen aborally and with characteristic ornament of rugosities and pits.

Occurrence. Coniacian–Lower Campanian, but only common in the Lower Santonian, *M. coranguinum* Zone; Kent, Surrey, Hampshire and Sussex.

Metopaster parkinsoni (Forbes)
Plate 52, figures 1–2

Description. Intermediate upper marginals normally four in number, low in profile but rather swollen; ultimates long and triangular with evenly rounded surface. Raised central area on marginals with pits but no rugosities.

Occurrence. Rare in Cenomanian but common from the Middle Turonian, *T. lata* Zone, to Lower Campanian, *G. quadrata* Zone; fairly common at most horizons in most areas.

Metopaster thoracifer (Geinitz)
Plate 52, figure 12

Description. Two intermediate upper marginals only; upper marginals low in profile, ultimate supermarginals triangular, low, longer than broad with one or more prominent horn-like projections; all marginals with uniform pits over well-defined central area.

Occurrence. Cenomanian–Middle Turonian, *T. lata* Zone; in England, known only from cliff-top exposures in calcarenites of the *I. labiatus* and *T. lata* zones west of Branscombe, Devon, where it is common, and as rare strays to the immediate east.

Metopaster tumidus Spencer
Plate 52, figure 4

Description. Four supermarginals in each half radius, these being high with the upper surface raised in a broad swelling to which pits tend to be confined; ultimate marginals short and blocky, not produced in a point.

Occurrence. Upper Campanian, *B. mucronata* Zone, to Lower Maastrichtian; rather common in the Isle of Wight, Dorset and Norfolk.

Metopaster uncatus (Forbes)
Plate 52, figure 6

Description. Ultimate upper marginals long and pointed; intermediates two or three in each half radius; no pits, but upper marginals may have a low ridge and swelling near the inner edge.

Occurrence. Middle Turonian, *T. lata* Zone, to Upper Campanian, *B. mucronata* Zone; rare at base and top of range, otherwise common and widespread.

Genus PARAMETOPASTER

Remarks. Like *Metopaster* but marginals without differentiated central and marginal zones.

Occurrence. Cenomanian–Maastrichtian.

Parametopaster undulatus (Spencer)
Plate 52, figure 13

Description. Small, with three supermarginals in each half radius. Median marginals low and broad, often with slightly irregular surface towards margins of outer face. Surface covered in close-packed pits. Ultimate marginals short.

Occurrence. Lower Campanian, *G. quadrata* Zone, to Lower Maastrichtian, *B. lanceolata* Zone; rather rare in Hampshire, Dorset and Norfolk.

Genus RECURVASTER

Remarks. Flat, pentagonal form with arms distinctly produced and upturned; upper marginals numerous, those towards the distal tip meeting along the centre of the arm. Surface of marginals as in *Metopaster*.

Occurrence. Campanian–Maastrichtian.

Recurvaster cf. *radiatus* (Spencer)
Plate 52, figure 7

Description. Supermarginals smooth, inferomarginals with reduced central area.

Occurrence. Upper Campanian, *B. mucronata* Zone; common in the Weybourne Chalk of Norfolk.

PLATE 52

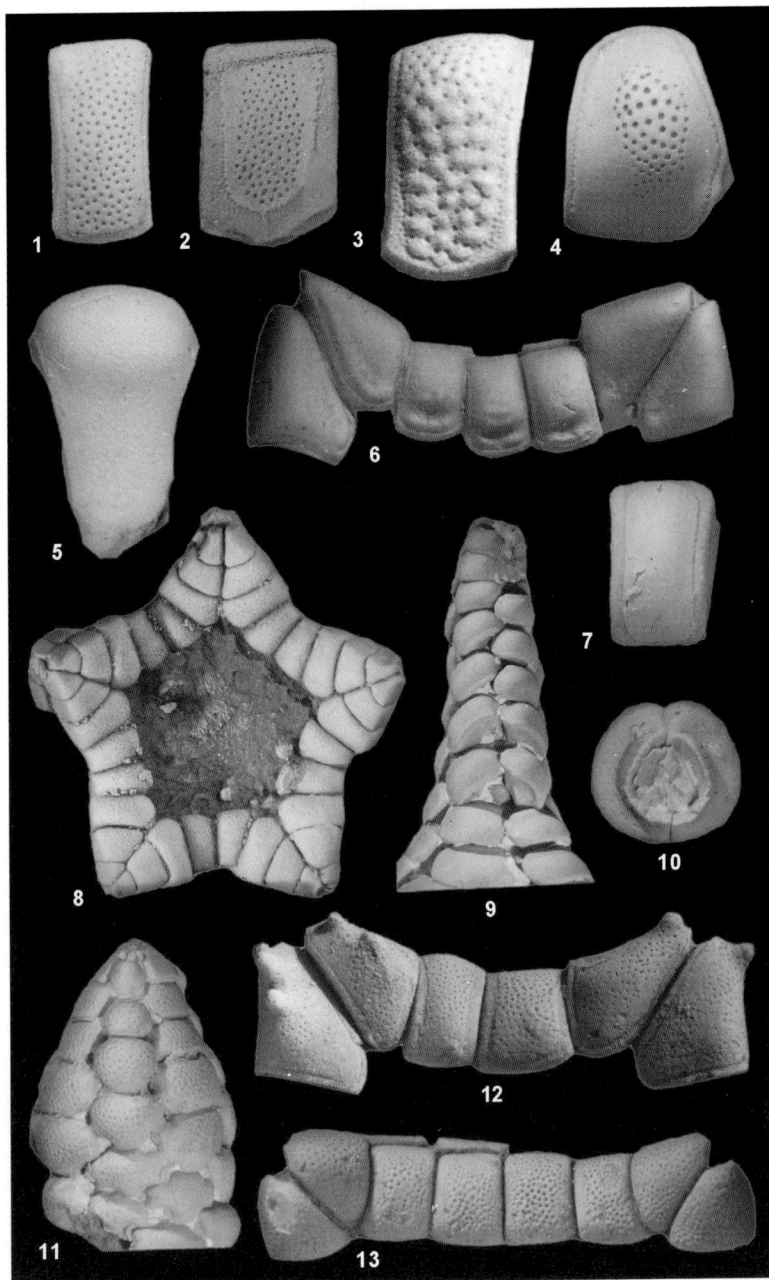

Subfamily uncertain

Genus CRATERASTER

Crateraster quinqueloba (Goldfuss)
Plate 52, figure 8

Description. Arms slightly to moderately produced, with upper marginals in contact on the midline; upper marginals with well-demarcated and flat upper and outer faces set at right angles, with lipped crater-like pits and rugosities on all but the earliest forms.

Occurrence. Upper Albian, ?*S. dispar* Zone, to Upper Campanian, *B. mucronata* Zone; the commonest Chalk starfish, in all areas and in most zones.

Family PYCINASTERIDAE

Genus PYCINASTER

Remarks. Rather large with long, stout arms; marginals high, alternating.
Occurrence. Lower Jurassic–Miocene.

EXPLANATION OF PLATE 52

Figs 1–2. *Metopaster parkinsoni* (Forbes); outer face of median superomarginal ossicles. 1, Santonian, *M. coranguinum* Zone; Isle of Wight; ×3. 2, Upper Campanian, *B. mucronata* Zone, Isle of Wight; ×3.

Fig. 3. *Metopaster hunteri* (Forbes). Lower Santonian, *M. coranguinum* Zone, Bromley, Kent; outer face of superomarginal ossicle; ×4.

Fig 4. *Metopaster tumidus* Spencer. Upper Campanian, *B. mucronata* Zone, locality unknown; outer face of superomarginal ossicle; ×1·5.

Fig. 5. *Pycinaster magnificus* Spencer. Lower Campanian, *G. quadrata* Zone, East Harnham, Salisbury, Wiltshire; outer face of superomarginal ossicle; ×2.

Fig. 6. *Metopaster uncatus* (Forbes). Lower Santonian, *M. coranguinum* Zone, Bromley, Kent; upper side, segment of marginal frame; ×2.

Fig. 7. *Recurvaster* cf. *radiatus* (Spencer). Upper Campanian, *B. mucronata* Zone, Weybourne, Norfolk; outer face of superomarginal ossicle; ×2·5.

Fig. 8. *Crateraster quinqueloba* (Goldfuss). Lower Santonian, *M. coranguinum* Zone, Micheldever, Hampshire; upper side; ×1·5.

Fig. 9. *Pycinaster angustatus* (Forbes). Santonian, *M. coranguinum* Zone, locality unknown, Kent; upper surface of one arm; ×1.

Fig. 10. *Ophiomusium subcylindricum* (Hagenow). Upper Campanian, *B. mucronata* Zone, Sheringham, Norfolk; end view of fragment of arm showing two lateral arm plates surrounding central vertebra; ×3.

Fig. 11. *Aspidaster bulbiferus* (Forbes). Lower Santonian, *M. coranguinum* Zone, Newlands Corner, Surrey; upper side of an arm; ×2.

Fig. 12. *Metopaster thoracifer* (Geinitz). Lower Turonian, *I. labiatus* Zone, west of Branscombe, Devon; side of a specimen reconstructed from dissociated marginals of different individuals; ×2.

Fig. 13. *Parametopaster undulatus* (Spencer). Upper Campanian, *B. mucronata* Zone, Studland, Dorset; upper side, segment of marginal frame; ×1·5.

Pycinaster angustatus (Forbes)
Plate 52, figure 9

Description. Size moderate, upper marginals of interrays not conspicuously swollen, outer surface smooth.
Occurrence. Santonian, *M. coranguinum–M. testudinarius* zones; moderately common in all areas.

Pycinaster magnificus Spencer
Plate 52, figure 5

Description. Marginals very large, those of interrays conspicuously swollen; outer surface with very shallow pits.
Occurrence. Campanian, *O. pilula–B. mucronata* zones; rather common in all areas where these zones are exposed.

Class OPHIUROIDEA
Order OPHIURIDA
Suborder PHRYNOPHIURINA
Family OPHIURIDAE

Remarks. Individual arm plates and less commonly fragments of arms or complete specimens of brittle-stars belonging to several genera of this family occur sporadically throughout the Chalk but are commonest in parts of the Cenomanian and in the Campanian.

Subfamily OPHIOLEPIDINAE

Genus OPHIOMUSIUM

Remarks. Upper and lower arm plates vestigial so that, at least in the outer part of the arms, the side plates meet above and below.
Occurrence. Jurassic–Recent.

Ophiomusium subcylindricum (Hagenow)
Plate 52, figure 10

Description. Readily recognized by the side plates of the arm being semicircular in end view, with the outer part well separated from neighbours; plates smooth.
Occurrence. Upper Campanian, *B. mucronata* Zone; rather rare in Norfolk.

Class ECHINOIDEA
Order CIDAROIDA
Family CIDARIDAE

Remarks. Regular, more or less globular echinoids with narrow ambulacra and wide interambulacra composed of few, large plates each bearing a single large perforate tubercle. Ambulacra made up of simple plates throughout. Perignathic girdle composed of apophyses only. Complete tests are rare in the Chalk but dissociated plates and the stout short or long spines are common.
Occurrence. Upper Jurassic–Recent.

Subfamily STEREOCIDARINAE

Remarks. Uppermost interambulacral plates with no or only rudimentary tubercle; horizontal sutures between plates commonly grooved.
Occurrence. Aptian–Recent.

Genus TEMNOCIDARIS
Subgenus STEREOCIDARIS

Remarks. Upper plates distinctly taller than wide; never more than a single plate in each interambulacral column with a rudimentary tubercle.
Occurrence. Aptian–Recent.

Temnocidaris (*Stereocidaris*) *carteri* (Forbes)
Plate 53, figure 11

Description. Test very small and rather high and inflated. Interambulacra with just four plates to a column, the uppermost in each column with rudimentary tubercle. Spines unknown.
Occurrence. Middle or Upper Cenomanian; rare, Cambridgeshire.

Temnocidaris (*Stereocidaris*) *intermedia* (Wiltshire, *in* Wright)
Plate 53, figure 4

Description. Test uniformly rounded. Interambulacra with five or six plates to a column, with well-developed tubercles on all but one adapical plate in each zone. Interambulacral plates not increasing in height adapically and with fine, dense granulation. Spines long and cylindrical with fine thorned ribs.
Occurrence. Upper Turonian, *S. plana* Zone, to Lower Santonian, *M. coranguinum* Zone; fairly common in southern England.

Temnocidaris (*Stereocidaris*) *sceptrifera* (Mantell)
Plates 53, figures 6–7

Description. Test equally flattened above and below or slightly subconical above. Interambulacra with five or six plates to a column, the uppermost in each very high with rudimentary tubercle and surrounding bald area, the latter commonly extended upwards. Spines stout, moderately long, inflated and widest at the inner third, then narrowing gradually, with longitudinal rows of fine thorns.

Occurrence. Upper Turonian, *S. plana* Zone, to Upper Campanian, *G. quadrata* Zone; widespread, common from *M. cortestudinarium–M. testudinarius* zones.

Genus PRIONOCIDARIS

Remarks. Interambulacral plates wide with primary tubercle occupying almost entire height of plate; more than six plates in a column.
Occurrence. Albian–Recent.

Prionocidaris vendocinensis (Agassiz and Desor)
Plate 53, figure 5

Description. Test very high, evenly rounded in profile, with up to nine plates in each interambulacral column; uppermost plates with well-developed tubercles. Primary tubercles on interambulacral plates not separated by miliary granules. Spines very long, tapering extremely gradually, with rather few, weak ridges bearing distant prominent thorns.

Occurrence. Upper Turonian, *S. plana* Zone, to Lower Santonian, *M. coranguinum* Zone; widespread and common in the *M. coranguinum* Zone in southern England, rather rare below and elsewhere.

Genus PHALACROCIDARIS

Remarks. Aboral interambulacral plates lacking a primary tubercle and distinctly wider than tall.
Occurrence. Cenomanian–Recent.

Phalacrocidaris merceyi (Cotteau)
Plate 53, figure 10

Description. Test subconical, with uppermost 2–5 plates lacking properly developed tubercles. Ambital interambulacral plates as wide as tall and primary tubercles well separated from one another. Spines very long and slender, with rather few, weak, serrated ridges.

Occurrence. Upper Turonian, *S. plana* Zone, to Lower Santonian, *M. coranguinum* Zone; widespread and common in the *M. coranguinum* Zone in southern England, rather rare below and elsewhere.

Genus HIRUDOCIDARIS

Remarks. Like *Temnocidaris* but with fusiform spines that are ribbed.
Occurrence. Ceonmanian–Campanian.

Hirudocidaris hirudo (Sorignet)
Plate 53, figure 9

Description. Test rather high and inflated. Interambulacra with five or six plates to a column, the uppermost in alternate columns with rudimentary tubercle. Spines short and stout, widest at mid-point, not narrowing much to the truncated and crowned end; with dense, fine, more or less definitely granulated ridges.
Occurrence. Middle Cenomanian–Late Campanian, *B. mucronata* Zone; widespread at all horizons but abundant in Lower Turonian of southern England.

Family PSYCHOCIDARIDAE

Remarks. All but adapical primary tubercles imperforate; spines acorn- or club-shaped.
Occurrence. Albian–Recent.

Genus TYLOCIDARIS

Tylocidaris (*Tylocidaris*) *clavigera* (Mantell)
Plate 53, figure 1

Description. Test rather flat above and below; five interambulacral plates in each column. Primary tubercles mostly imperforate, although occasional adapical tubercles may have a small rudimentary perforation. Spines variable, club-shaped, always with more or less rounded top and with distinct ridges with fine thorns.
Occurrence. Upper Turonian, *S. plana* Zone, to Lower Santonian, *M. coranguinum* Zone; widespread in southern England but only common in the Lower Santonian.

Tylocidaris (*Tylocidaris*) *sorigneti* (Desor)
Plate 53, figures 2–3

Description. Test as in *T. clavigera*, but interradial granular zones narrower. Spines short and clavate with coarser serrated ribs and greatly reduced handle; ending in a conical tip.

PLATE 53

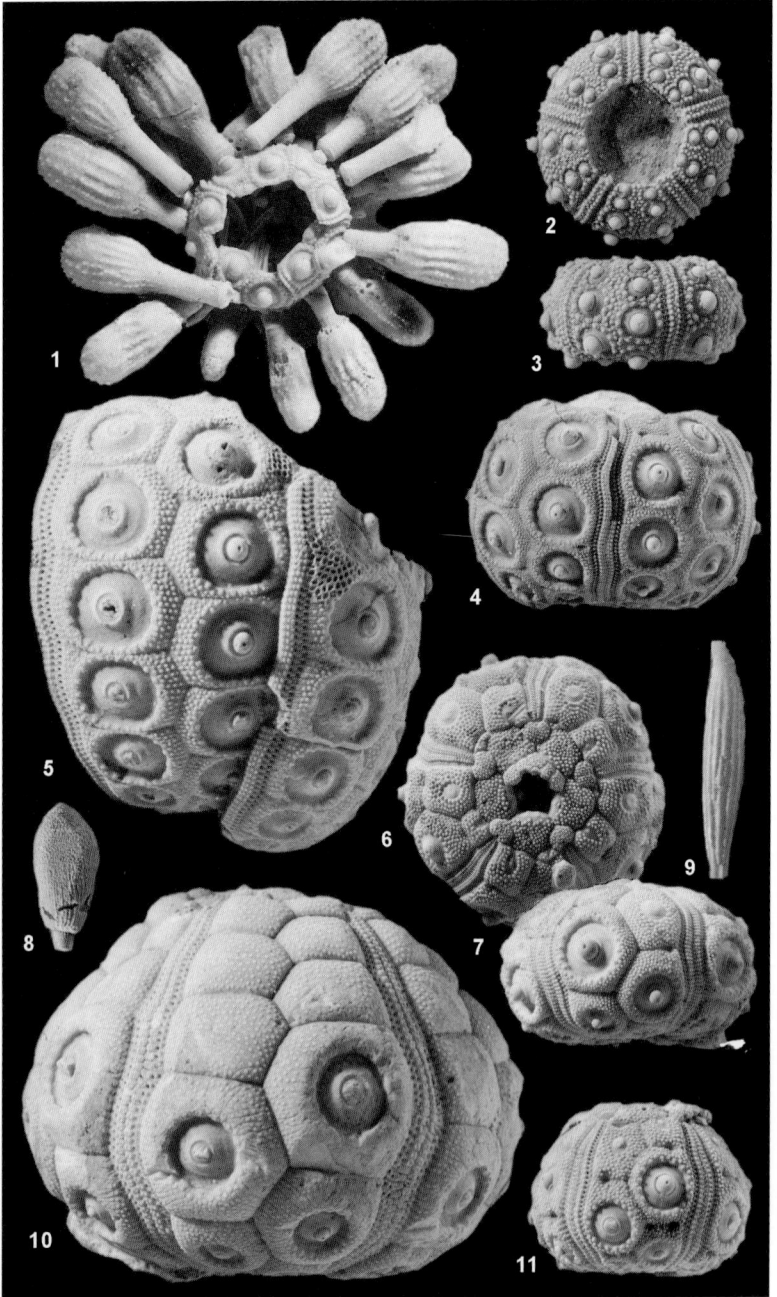

Occurrence. Lower Turonian, *I. labiatus* Zone; Devon and Dorsct.

Tylocidaris (*Oedematocidaris*) *asperula* (Roemer)
Plate 53, figure 8

Description. Test as in *T. sorigneti*. Spines glandiform and covered in dense, fine, granular ornament.
Occurrence. Middle–Upper Cenomanian; locally common from south-east England to Yorkshire.

Cohort ECHINACEA
Order CALYCINA
Family SALENIIDAE

Remarks. Small, regular echinoids with few interambulacral plates, bearing large imperforate tubercles; the apical disc is large and cap-like, with the periproct offset from centre, and with an extra (suranal) plate inside the ring of ocular and genital plates.

Genus HYPOSALENIA

Remarks. Periproct offset directly backwards to face between two ambulacra.
Occurrence. Jurassic–Palaeocene.

EXPLANATION OF PLATE 53

Fig. 1. *Tylocidaris* (*Tylocidaris*) *clavigera* (Mantell). Upper Chalk, no locality data; test with spines; ×1.

Figs 2–3. *Tylocidaris* (*Tylocidaris*) *sorigneti* (Desor). Lower Turonian, Branscombe, Devon. 2, oral, and 3, lateral views of test; ×2.

Fig. 4. *Temnocidaris* (*Stereocidaris*) *intermedia* (Wiltshire). Upper Chalk, no locality details; lateral view of test; ×1.

Fig. 5. *Prionocidaris vendocinensis* (Agassiz and Desor). Santonian, Gravesend, Kent; lateral view of partially dissociated test; ×1.

Figs 6–7. *Temnocidaris* (*Stereocidaris*) *sceptrifera* (Mantell). Lower Santonian, *M. coranguinum* Zone, Beddingham, Sussex. 6, apical, and 7, lateral views of test; ×1.

Fig. 8. *Tylocidaris* (*Oedematocidaris*) *asperula* (Roemer). Middle Cenomanian, Southerham, Sussex; spine; ×1.

Fig. 9. *Hirudocidaris hirudo* (Sorignet). Lower Santonian, *M. coranguinum* Zone, Bromley, Kent; spine; ×1.

Fig. 10. *Phalacrocidaris merceyi* (Cotteau). Lower Santonian, *M. coranguinum* Zone, Bromley, Kent; lateral view of test; ×1.

Fig. 11. *Temnocidaris* (*Stereocidaris*) *carteri* (Forbes). Cenomanian, Cambridge; lateral view of test; ×4.

Hyposalenia clathrata (Woodward)
Plate 54, figures 7–8

Description. Normally small but specimens up to 26 mm diameter are known from Kent. Apical disc about four-fifths of total diameter, subconical with lobed outline, deeply but variably indented by oval pits or grooves radiating from the genital pores.

Occurrence. Upper Albian, *S. dispar* Zone, to Lower Cenomanian, *M. dixoni* Zone; sparse but widespread in southern England and East Anglia.

Genus SALENIA

Remarks. Periproct offset towards the latero-posterior, facing towards an ambulacrum. Apical disc thickened and elevated above remainder of test; generally smooth.

Occurrence. Hauterivian–Recent.

Salenia petalifera (Defrance, *in* Desmarest)
Plate 54, figures 1–2

Description. Rather large, evenly rounded in profile. Apical disc half to two-thirds of total diameter, slightly convex to low subconical, with the edges of the periproct raised; elongated pits cross the sutures and the plates vary from nearly smooth to being covered with fine, irregular ridges. Ambulacra strictly bigeminate throughout (i.e. two pairs of ambulacral pore-pairs to each primary tubercle).

Occurrence. Cenomanian, *M. mantelli–C. guerangeri* zones; sparse, throughout all exposures.

Salenia magnifica Wright
Plate 54, figures 3–4

Description. Large and rather tall test; raised apical disc with only relatively small sutural pitting developed. Ambulacra composed of alternately simple and bigeminate plates.

Occurrence. Upper Santonian, *M. testudinarius* Zone, to Lower Maastrichtian; fairly common throughout southern England in the Upper Santonian and Campanian.

Genus SALENOCIDARIS

Remarks. Apical disc hardly raised above remainder of test and covered in dense pustular ornament. Ambulacral plating biserial adorally or throughout.

Occurrence. Albian–Recent.

Salenocidaris granulosa (Woodward)
Plate 54, figures 5–6

Description. Small, rather tall to depressed. Apical disc very wide, without distinct edge, covered with fine to rather coarse rugosities.
Occurrence. Lower Turonian, *I. labiatus* Zone, to Upper Campanian, *B. mucronata* Zone; widespread at most horizons, commonest in the Turonian and Lower Santonian.

Order PHYMOSOMATOIDA
Family DIPLOPODIIDAE

Remarks. Circular, normally flat test with large, pentagonal apical disc opening; primary tubercles perforate and crenulate, smaller and more numerous than in cidarids or saleniids. Ambulacral plating trigeminate adorally; pores arranged biserially on the upper surface in adults.
Occurrence. Upper Jurassic (Oxfordian) – Cenomanian.

Genus TETRAGRAMMA

Remarks. Two or more primary tubercles on each interambulacral plate. Rows of pore pairs doubled in upper part of ambulacra.

Tetragramma variolare (Brongniart)
Plate 55, figures 1–2

Description. Height and inflation variable, though generally rather depressed; 2–4 primary tubercles on each interambulacral plate.
Occurrence. Upper Albian, *M. inflatum* Zone, to Upper Cenomanian, *C. guerangeri* Zone; common in the sandy Cenomanian of Devon, rather rare in the Lower Chalk of Kent and in the Chalk Basement Beds of Dorset, Wiltshire and Somerset.

Family PHYMOSOMATIDAE

Remarks. Primary tubercules imperforate and crenulate. Ambulacral plating polygeminate, with characteristic plate compounding style.
Occurrence. Middle Jurassic (Callovian) – Eocene.

Genus PHYMOSOMA

Remarks. Pore pairs in doubled rows on upper side and near the mouth; peristome hardly invaginated; tuberculation relatively coarse.
Occurrence. Callovian–Palaeocene.

PLATE 54

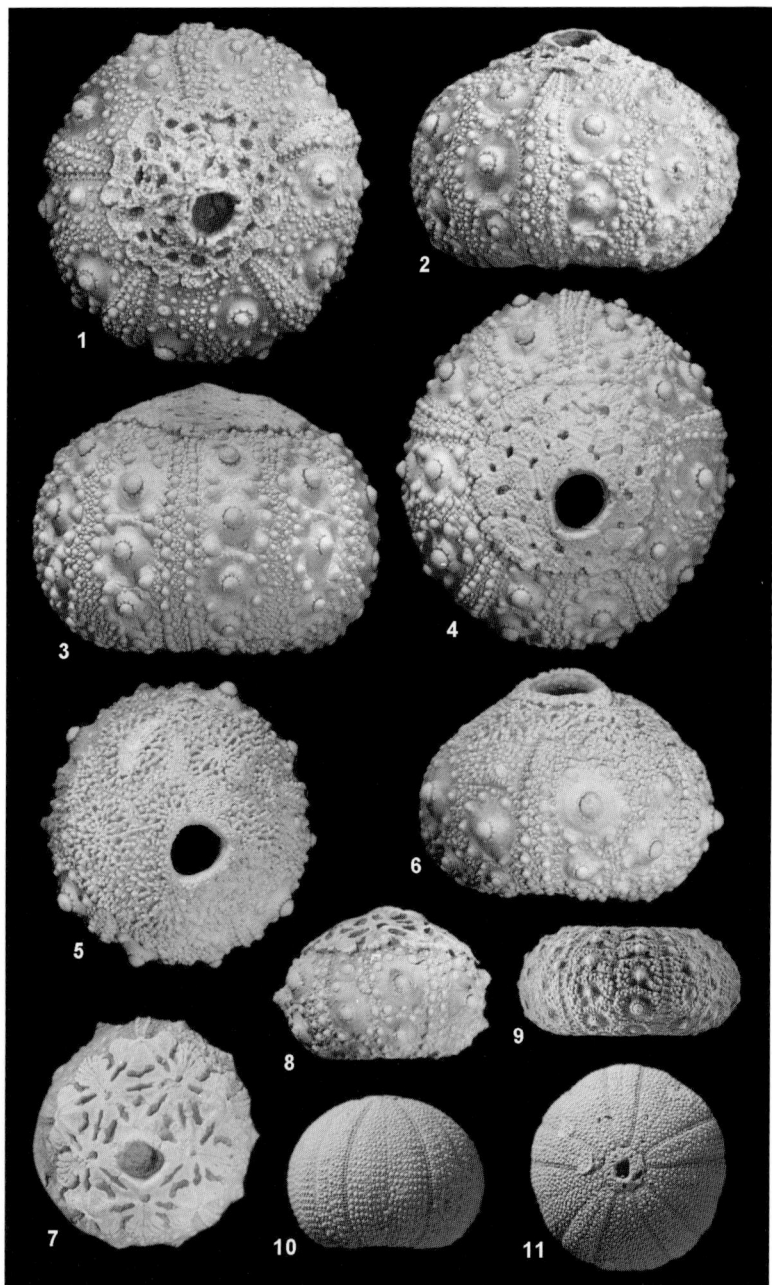

Phymosoma koenigi (Mantell)
Plate 55, figures 5–6

Description. Rather large and flattened in profile. Differs from the contemporary *P. corollare* (Leske) mainly in having secondary tubercles on interambulacral plates above the ambitus that are almost as big as the primaries. Contemporary *Gauthieria* species differ in having pore pairs in single rows throughout and deeply sunken peristomes.

Occurrence. Coniacian, *M. coranguinum* Zone, to Lower Campanian, *O. pilula* Zone; fairly common in the *M. coranguinum* Zone, less so above and below; widespread but rare in the north.

Superorder CAMARODONTA
Order TEMNOPLEUROIDA

Remarks. Small regular echinoids with strongly ridged or grooved test.

Family ZEUGLOPLEURIDAE

Genus ECHINOCYPHUS

Remarks. Apical disc rather large and pentagonal with plates only loosely bound to test and thus almost always lost. Adapical horizontal sutures with paired V-shaped sutures. Interambulacral tubercles imperforate and crenulate.

Occurrence. Cenomanian–Turonian.

Echinocyphus rotatus Cotteau
Plate 54, figure 9

EXPLANATION OF PLATE 54

Figs 1–2. *Salenia petalifera* (Defrance, *in* Desmarest). Lower Cenomanian, *M. mantelli* Zone, Burham, Kent. 1, apical, and 2, lateral views of test; ×1·3.

Figs 3–4. *Salenia magnifica* Wright. Upper Santonian, *M. testudinarius* Zone, Thanet coast, Kent. 3, lateral, and 4, apical views of test; ×1·3.

Figs 5–6. *Salenocidaris granulosa* (Woodward). Lower Santonian, *M. coranguinum* Zone, Kent. 5, apical, and 6, lateral views of test; ×2.

Figs 7–8. *Hyposalenia clathrata* (Woodward) forma *umbrella*. Middle Cenomanian, *A. rhotomagense* Zone, Dover, Kent. 7, apical, and 8, lateral views of test; ×2.

Fig. 9. *Echinocyphus rotatus* Cotteau. Middle Cenomanian, *A. rhotomagense* Zone, Dover, Kent; lateral view of test; ×2.

Figs 10–11. *Cottaldia benettiae* (König). Lower Cenomanian, Warminster, Wiltshire. 10, lateral, and 11, apical views of test; ×2.

PLATE 55

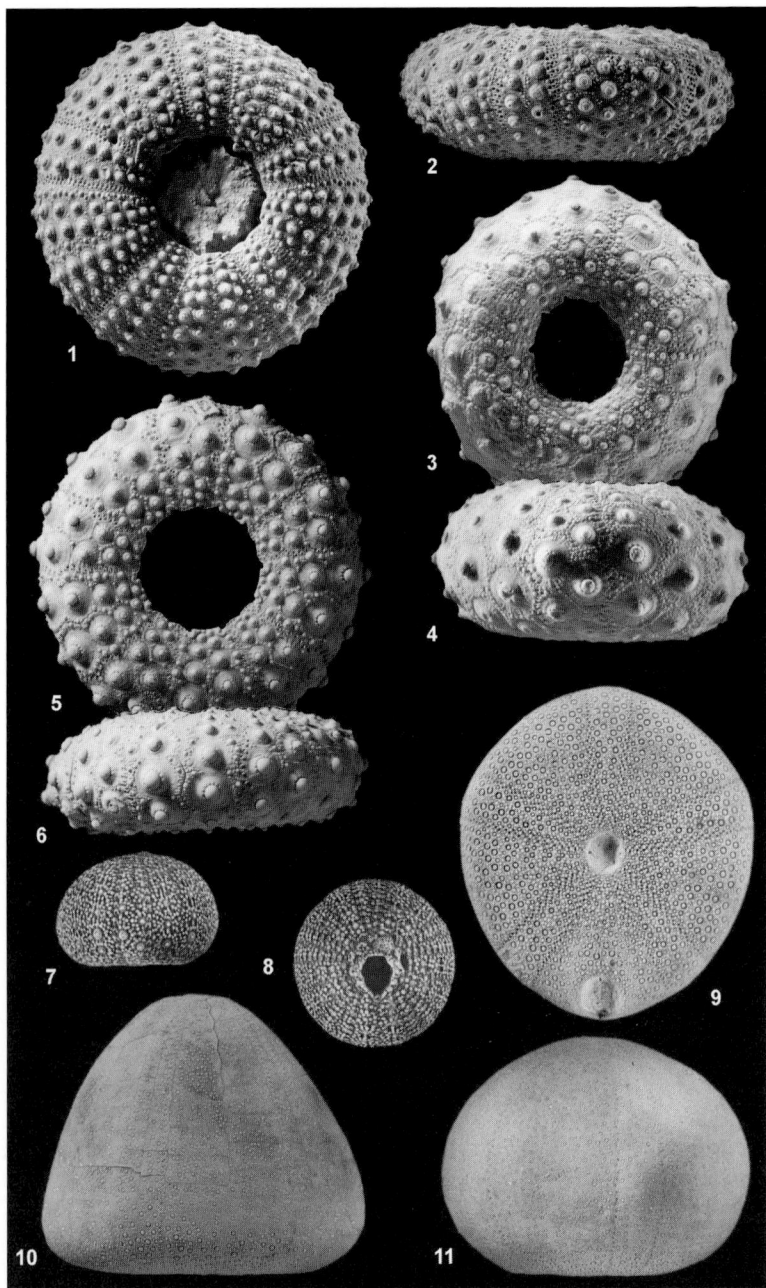

Description. Small, flat-based and rather depressed test. Interambulacral plates with single primary tubercle surrounded by dense granulation. Ambulacral tuberculation irregularly developed; typically with just a single column of tubercles in each zone.
Occurrence. Upper Albian, *S. dispar* Zone, to Middle Cenomanian, *A. rhotomagense* Zone; common in sandy Cenomanian in Devon but rare in Chalk facies, occurring sparsely in southern England and East Anglia.

Genus ZEUGLOPLEURUS

Remarks. Small, depressed or globular test. Interambulacral plates with ridges or rows of granules radiating from the primary tubercles. Apical disc plates firmly bound to the test and with notched periproctal margin.
Occurrence. Cenomanian–Maastrichtian.

Zeuglopleurus rowei Gregory, *in* Rowe
Plate 55, figures 7–8

Description. Very small, inflated, and with very large periproct; tubercles small but those at margin and below larger than those above.
Occurrence. Santonian, *M. coranguinum–M. testudinarius* zones; rare but widespread, known in Kent, Surrey, Sussex, Isle of Wight and Yorkshire.

Order uncertain

Genus COTTALDIA

Remarks. Globular test with many small, uniform tubercles covering ambulacral and interambulacral plates. Apical disc small and usually preserved in life position. Pore-pairs uniserial throughout.
Occurrence. Albian–Cenomanian.

EXPLANATION OF PLATE 55

Figs 1–2. *Tetragramma variolare* (Brongniart). Cenomanian, ?*M. dixoni* Zone, Folkestone, Kent. 1, oral, and 2, lateral views of test; ×1·3.

Figs 3–4. *Gauthieria radiata* (Sorignet). Lower Turonian, *I. labiatus* Zone, Dover, Kent. 3, oral, and 4, lateral views of test; ×1·3.

Figs 5–6. *Phymosoma koenigi* (Mantell). ?Santonian, Gravesend, Kent. 5, oral, and 6, lateral views of test; ×1·3.

Figs 7–8. *Zeuglopleurus rowei* Gregory. Upper Santonian, *U. socialis* Zone, Westgate, Kent. 7, lateral, and 8, apical views of test; ×3.

Figs 9–10. *Conulus albogalerus* Leske. Lower Santonian, *M. coranguinum* Zone, Northfleet, Kent. 9, oral, and 10, lateral views of test; ×1·3.

Fig. 11. *Conulus subrotundus* Mantell. Lower Turonian, *I. labiatus* Zone, Burham, Kent; lateral view of test; ×1·3.

Cottaldia benettiae (König)
Plate 54, figures 10–11

Description. Test up to 15 mm in diameter. Globular in profile and with single row of primary tubercles on each interambulacral plate.
Occurrence. Cenomanian, *M. mantelli–C. guerangeri* zones; uncommon in the Cenomanian of south-east England, more common in the sandy Lower Cenomanian facies of Devon and Wiltshire.

Cohort IRREGULARIA
Order HOLECTYPOIDA

Remarks. Irregular echinoids with anus outside the apical system; outline pentamerally or bilaterally symmetrical. Peristome large and central, with buccal notches.
Occurrence. Lower Jurassic (Toarcian) – Upper Cretaceous (Maastrichtian).

Family DISCOIDIDAE

Genus DISCOIDES

Remarks. Small, bun-shaped to subconical test, circular to subpentagonal in outline and covered with dense, minute tubercles. Only one of the genital plates in the apical system is perforated by madrepores. Periproct on the flat base. Thick ridges developed on interior of oral surface.
Occurrence. Albian–Coniacian.

Discoides subuculus (Leske)
Plate 56, figures 6–8

Description. Depressed with convex sides, to almost conical. Peristome large; periproct occupies more than half the space between peristome and margin. Granulation on plates dense and irregular.
Occurrence. Upper Albian, *S. dispar* Zone, to Upper Cenomanian, *C. guerangeri* Zone; common in Chalk Basement Beds and lower part of the Cenomanian in the south, absent in the north.

Genus CAMEROGALERUS

Remarks. Bun-shaped to subconical test, circular to subpentagonal in outline and covered in dense, minute tubercles. All five of the genital plates in the apical system are perforated by madrepores. Thin, blade-like ridges developed on interior of oral surface.
Occurrence. Albian–Maastrichtian.

Camerogalerus minimus (Agassiz)
Plate 56, figures 4–5

Description. Small rounded to subconical test. All five genital plates of the apical system are perforated by gonopores. Sides and margin are more evenly rounded and the peristome and periproct much smaller than in *Discoides subuculus*. Granulation on plates obviously aligned.

Occurrence. Lower Turonian, *I. labiatus* Zone, to Coniacian, *M. cortestudinarium* Zone; widespread and common in the south in the lower part of the Turonian, rarer in the north; very rare in the *S. plana* Zone from Kent to Devon and in the *M. cortestudinarium* Zone of Devon.

Camerogalerus cylindricus (Lamarck)
Plate 56, figs 1–3

Description. Rather large; circular or subpentagonal in outline; young tests are bun-shaped with flat base, older ones with increasingly vertical sides and low, rounded top. Often appears to be smooth but well-preserved tests are covered in minute tubercles.

Occurrence. Cenomanian, *M. mantelli–M. geslinianum* zones; generally distributed, sometimes common.

Order ECHINONEOIDA
Family CONULIDAE

Remarks. Outline almost but not quite circular or subpentagonal to oval. Peristome small, central and lacking buccal notches. Periproct subambital, on or near the margin. Tubercles minute; ambulacra contain some plates of reduced size.

Occurrence. Valanginian–Maastrichtian.

Genus CONULUS

Conulus albogalerus Leske
Plate 55, figures 9–10

Description. Flat-based, almost circular or subpentagonal in outline with ambitus almost at the base, higher than wide, rounded subconical to tall and nearly conical in profile.

Occurrence. Upper Turonian, *S. plana* Zone, to Upper Campanian, *B. mucronata* Zone; rare in *S. plana* and *M. cortestudinarium* zones, common in Santonian and Lower Campanian in the south-east and East Anglia, less common in Dorset, absent in Yorkshire.

PLATE 56

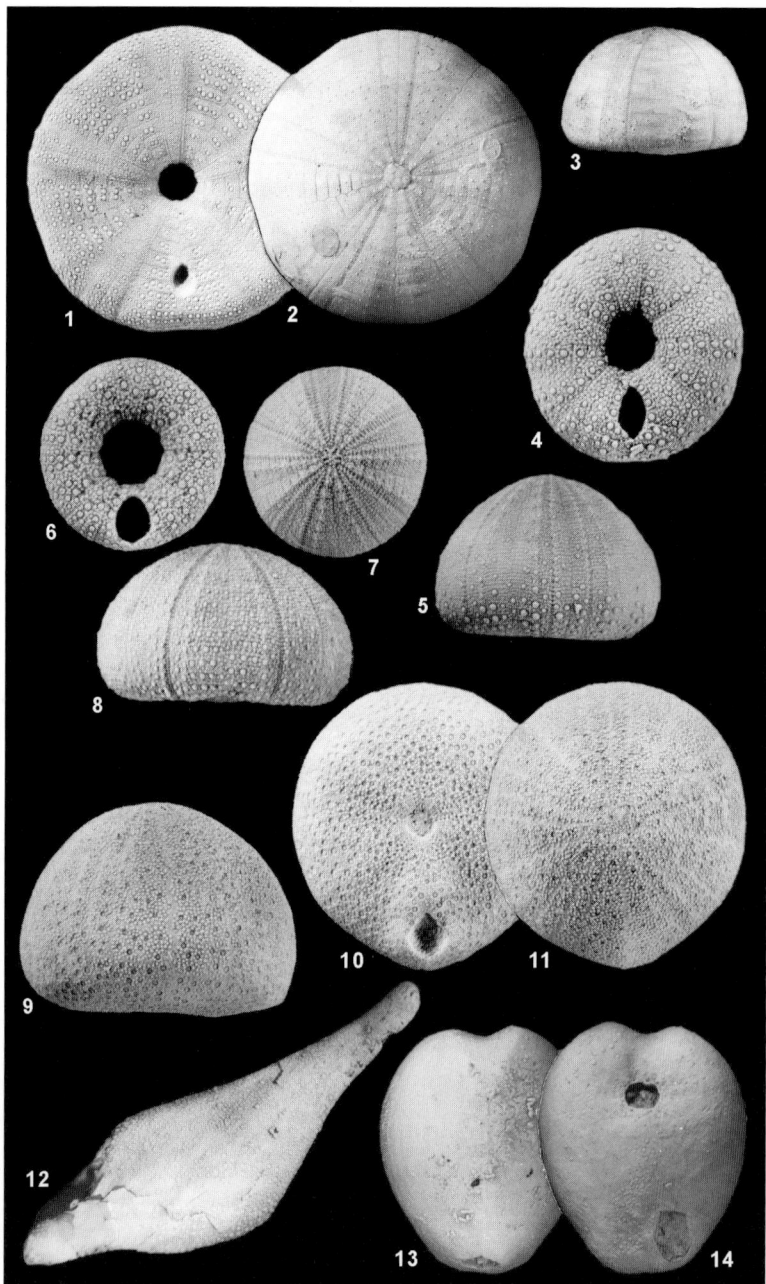

Conulus subrotundus Mantell
Plate 55, figure 11

Description. Like *C. albogalerus* but much more inflated in profile, with ambitus positioned well above the base.
Occurrence. Lower Turonian, *I. labiatus* Zone, to Lower Campanian, *U. socialis* Zone; common in Lower–Middle Turonian of southern England, becoming rarer in the *S. plana* and *M. cortestudinarium* zones and common once more in the *U. socialis* Zone of Kent.

Family GALERITIDAE

Remarks. Like Conulidae but ambulacra consist solely of simple uniform plates.
Occurrence. Middle Jurassic–Lower Eocene.

Genus GALERITES

Galerites hemisphaericus (Desor)
Plate 56, figures 9–11

Description. More or less hemispherical with flat base. Peristome very small with slight inturned rim. Posterior interambulacrum slightly raised on oral surface. Ambulacral pores uniserial, becoming weakly offset into triads close to peristome.
Occurrence. Upper Campanian, *B. mucronata* Zone; rather rare in Norfolk.

EXPLANATION OF PLATE 56

Figs 1–3. *Camerogalerus cylindricus* (Lamarck). 1–2, Middle Cenomanian, *H. subglobosus* Zone, Burham, Kent. 1, oral, and 2, apical views; ×1. 3, Middle Cenomanian, *H. subglobosus* Zone, Chardstock, Somerset; lateral view of another specimen; ×1.
Figs 4–5. *Camerogalerus minimus* (Agassiz). Lower Turonian, *I. labiatus* Zone, Dover, Kent. 4, oral, and 5, lateral views of test; ×3.
Figs 6–8. *Discoides subuculus* (Leske). Lower Cenomanian, *N. carcitanense* Zone, Warminster, Wiltshire. 6, lower surface; ×2. 7, upper surface; ×2. 8, lateral view; ×3.
Figs 9–11. *Galerites hemisphaericus* (Desor). Upper Campanian, *B. mucronata* Zone, Thorpe St Andrew, Norfolk. 9, lateral, 10, oral, and 11, aboral views; ×2.
Fig. 12. *Hagenowia rostrata* (Forbes). Lower Santonian, *M. coranguinum* Zone, Broadstairs, Kent; lateral view of test; ×2.
Figs 13–14. *Cardiaster truncatus* (Goldfuss). Lower Turonian, *I. labiatus* Zone, Dover, Kent. 13, apical, and 14, oral views of test; ×2.

Superorder ATELOSTOMATA
Order HOLASTEROIDA
Family HOLASTERIDAE

Remarks. Bilaterally symmetrical, shape variable, generally globular, helmet- or heart-shaped. The paired ambulacra are flush with the test, the unpaired one commonly in a groove. Test smooth with sparse, minute tubercles. Apical system elongated. Peristome near the anterior end of the underside; periproct marginal.

Genus HOLASTER

Holaster nodulosus (Goldfuss)
Plate 57, figures 1–4

Description. Longer than wide, slightly heart-shaped with shallow anterior groove, flat-based, with semicircular transverse cross-section; test thin.
Occurrence. Cenomanian, *M. mantelli–C. guerangeri* zones; common in the sandy Cenomanian of Devon, less so in the Chalk facies but generally distributed in southern England.

Holaster subglobosus (Leske)
Plate 59, figures 3–4

Description. Thick-shelled with convex base and broadly rounded margin; tubercles very small and feeble. Occurs in five distinct shape-forms, nearly spheroid, heart-shaped, broad and depressed, elongate with subconical apex, and short and high with steep anterior end.
Occurrence. Middle Cenomanian, *A. rhotomagense* Zone, to Upper Cenomanian, *C. guerangeri* Zone; common and widespread in all areas.

Genus STERNOTAXIS

Remarks. As *Holaster* but interambulacral plates behind the peristome in a single row, not alternating.
Occurrence. Cenomanian–Campanian.

Sternotaxis gregoryi (Lambert)
Plate 57, figures 10–11

Description. Rather large, oval, flat-based but with well-rounded margin, with test evenly arched in all directions. Periproct positioned very low on posterior face. Aboral surface densely covered in large tubercles.
Occurrence. Lower Cenomanian, *M. dixoni* Zone; locally common in Kent, sparse elsewhere.

Sternotaxis trecensis (Leymerie)
Plate 57, figures 8–9

Description. Like *Holaster nodulosus* but with much higher test, with the lower part of the sides vertical above a sharp margin. Plastron with three plates uniserially arranged behind the peristome.

Occurrence. Middle–Upper Cenomanian; one of the commonest fossils in the Cenomanian chalks of southern England.

Sternotaxis plana (Mantell)
Plate 58, figures 5–7

Description. Normally longer than wide, high, with slight anterior groove, convex base and well-rounded margin. No large tubercles developed aborally. Periproct relatively high on posterior face.

Occurrence. Lower Turonian, *I. labiatus* Zone, to Coniacian, *M. cortestudinarium* Zone; common in *S. plana* Zone and sometimes in *T. lata* Zone, otherwise rare; widespread throughout Chalk areas.

Sternotaxis placenta (Agassiz)
Plate 58, figures 8–9

Description. May be very large; flat-based with sharply curved margin, lower part of sides vertical; hardly any anterior groove; test very thin. Distinctly broader than *S. gregoryi* and *S. trecensis*. Periproct on inward-sloping posterior face.

Occurrence. Lower Turonian, *I. labiatus* Zone, to Lower Campanian, *O. pilula* Zone; rather common in *S. plana* and *M. cortestudinarium* zones, rare above and below; widespread in all areas.

Genus CARDIASTER

Remarks. Heart-shaped with deep, sharp-edged anterior groove; apical disc subcentral; plates of the plastron alternate behind the peristome; there is a distinct fasciole (narrow band of very dense minute granules) around the margin. Anterior sulcus extending to the peristome.

Occurrence. Cenomanian–Maastrichtian.

Cardiaster truncatus (Goldfuss)
Plate 56, figures 13–14

Description. Small (generally less than 2 cm long), narrow, with rear end oblique, the base projecting.

Occurrence. Upper Cenomanian, *N. juddii* Zone, to Middle Turonian, *T.*

PLATE 57

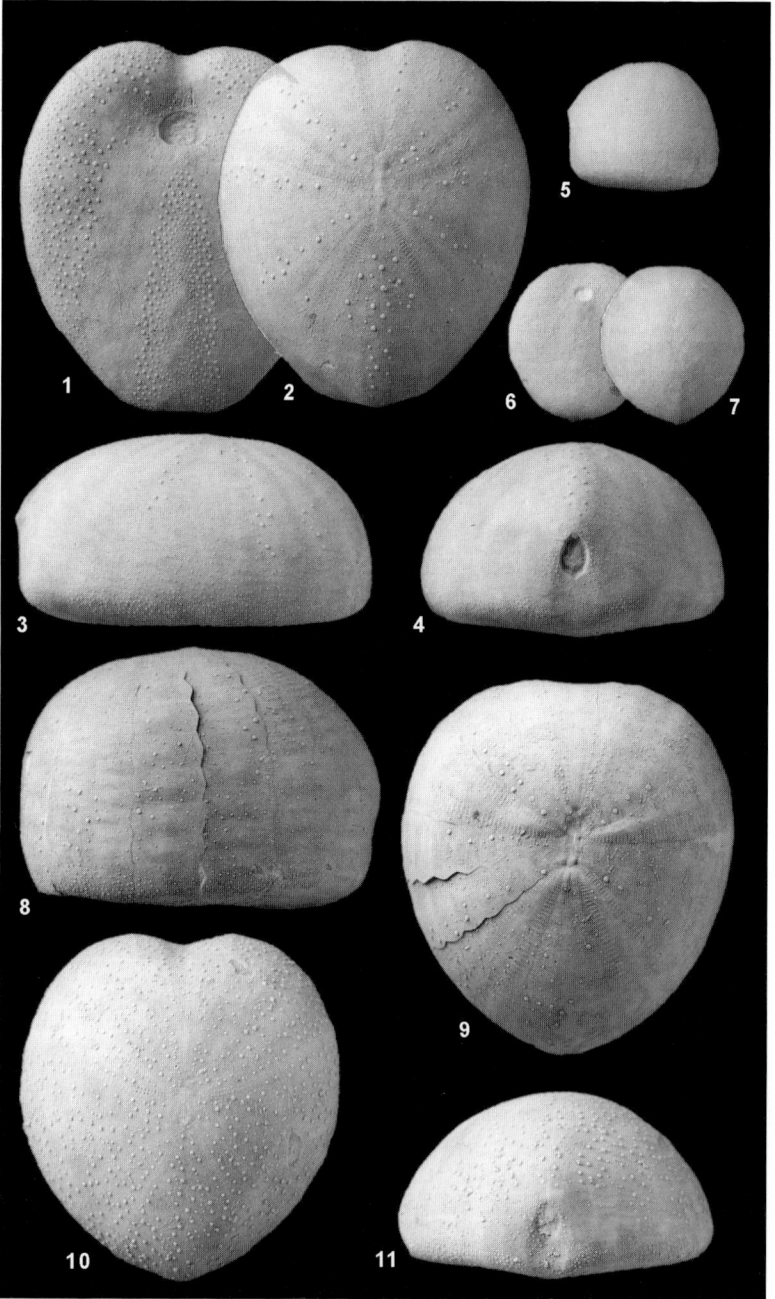

lata Zone; common in the Upper Cenomanian of Devon and in the *I. labiatus* Zone of southern England, otherwise rare; absent in Yorkshire.

Cardiaster sp.
Plate 58, figures 1–2

Description. Rather flat-topped in profile, but usually with apex slightly pinched and margins to frontal groove raised. Anterior sulcus shorter than typical for *C. cretaceus* and apical disc displaced far towards the anterior. No enlarged aboral tubercles.

Remarks. Intermediate between *Cardiaster* and *Infulaster*, having the biserial plastron plating of a *Cardiaster*, but with the apical disc positioned close to the anterior.

Occurrence. Middle Turonian, *T. lata* Zone, to Upper Turonian, *S. plana* Zone; Norfolk and Lincolnshire, absent from the south of England.

Cardiaster granulosus (Goldfuss)
Plate 58, figures 3–4

Description. Rather large, normally a little longer than wide; tubercles relatively large, particularly along the posterior keel and the sharp edges of the anterior groove.

Occurrence. Upper Campanian, *B. mucronata* Zone, to Lower Maastrichtian, *B. lanceolata* Zone; widespread.

Genus ECHINOCORYS

Remarks. Ovate in outline with no frontal groove. Periproct just below margin; no fascioles.

Occurrence. Cenomanian–Palaeocene.

Echinocorys scutata Leske
Plate 59, figures 1–2; Text-figure 13.1

Description. Size and shape very variable. Here treated as a single variable species but many authors apply numerous specific or subspecific names.

EXPLANATION OF PLATE 57

Figs 1–4. *Holaster nodulosus* (Goldfuss). Cenomanian, Wendover, Buckinghamshire. 1, oral, 2, aboral, 3, lateral, and 4, posterior views; ×1.

Figs 5–7. *Offaster pilula* (Lamarck). Lower Campanian, *G. quadrata* Zone, Winchester, Wiltshire. 5, lateral, 6, oral, and 7, aboral views; ×1.

Figs 8–9. *Sternotaxis trecensis* (Leymerie). Middle or Upper Cenomanian, Dover, Kent; lateral and apical views; ×1.

Figs 10–11. *Sternotaxis gregoryi* (Lambert). Lower Cenomanian, *M. mantelli* Zone, Dover, Kent; apical and posterior views; ×1.

PLATE 58

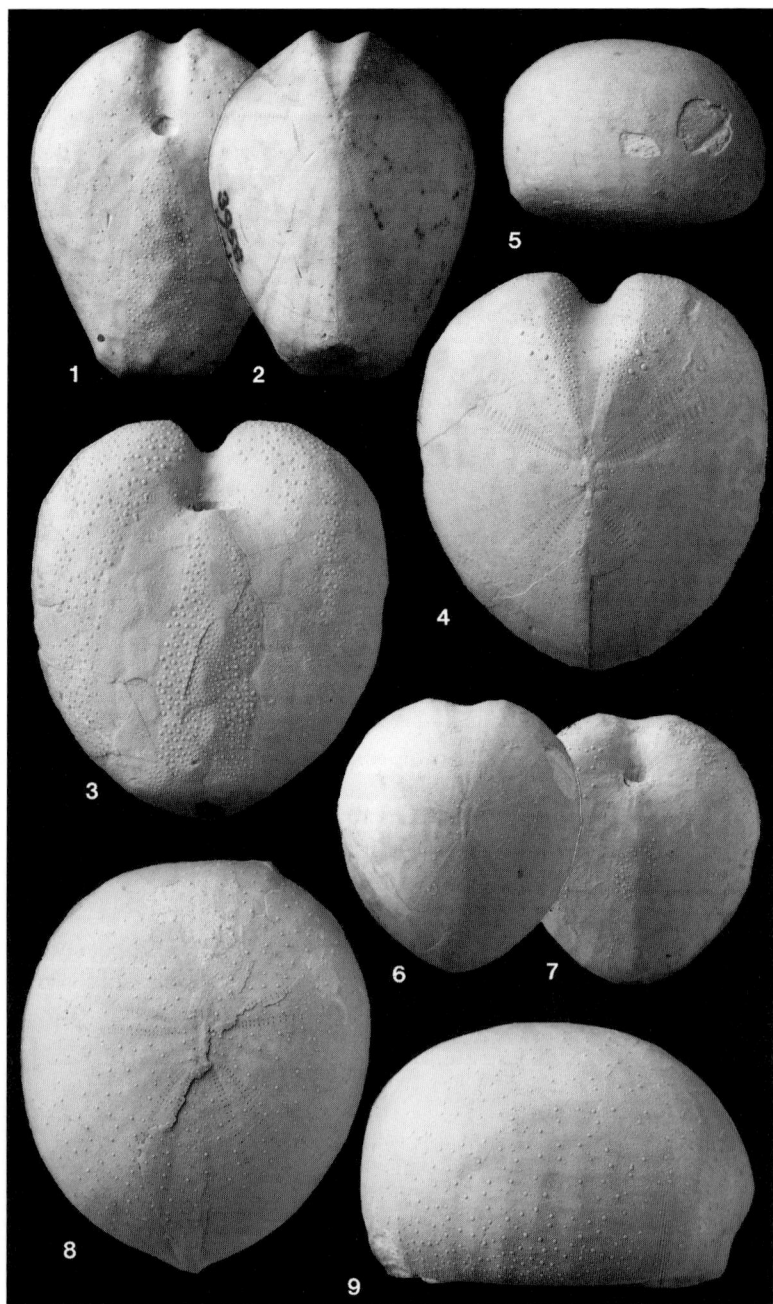

Text-figure 13.1 shows a selection of forms, some of which are of great stratigraphical value, being confined to narrow horizons, others less so, with long ranges.

Occurrence. Middle Turonian, *T. lata* Zone, to Maastrichtian, *B. lanceolata* Zone (to Palaeocene elsewhere); rare in the *T. lata* Zone of south-east England and East Anglia, otherwise common and generally distributed.

Genus OFFASTER

Remarks. Small and subglobular with no frontal groove. Interambulacral plates high; periproct above the margin. A marginal fasciole is present. Plastron plates biserial.

Occurrence. Santonian–Danian.

Offaster pilula (Lamarck)
Plate 57, figures 5–7

Description. Small and globular to more or less flat-based, shape variable, but always tall.

Occurrence. Upper Santonian, *M. testudinarius* Zone, to Lower Campanian, *G. quadrata* Zone; common in the lower beds of the Campanian, rare above and below; widespread in Norfolk and southern England.

Genus HAGENOWIA

Remarks. Very small, thin-shelled, with the test drawn forwards and upwards into a rostrum, at the top of which is the anterior part of the apical system; the rear part is near the base of the rostrum.

Occurrence. Upper Coniacian–Upper Campanian.

Hagenowia rostrata (Forbes)
Plate 56, figure 12

Description. Rostrum relatively short and stout, pointing obliquely forwards. More extreme forms with more slender rostra, previously recorded as *H. rostrata*, have now been distinguished as separate species.

EXPLANATION OF PLATE 58

Figs 1–2. *Cardiaster* sp. nov. Upper Turonian, *S. plana* Zone, Kiplingcoates, Yorkshire; oral and aboral views; ×1.

Figs 3–4. *Cardiaster granulosus* (Goldfuss). Upper Campanian, *B. mucronata* Zone, Mousehold, Norfolk. 3, oral, and 4, aboral views; ×1.

Figs 5–7. *Sternotaxis plana* (Mantell). Upper Turonian, *S. plana* Zone, Chatham, Kent. 5, lateral, 6, apical, and 7, oral views; ×1.

Figs 8–9. *Sternotaxis placenta* (Agassiz). Upper Turonian, *S. plana* Zone, Dover, Kent. 8, apical, and 9, lateral views; ×1.

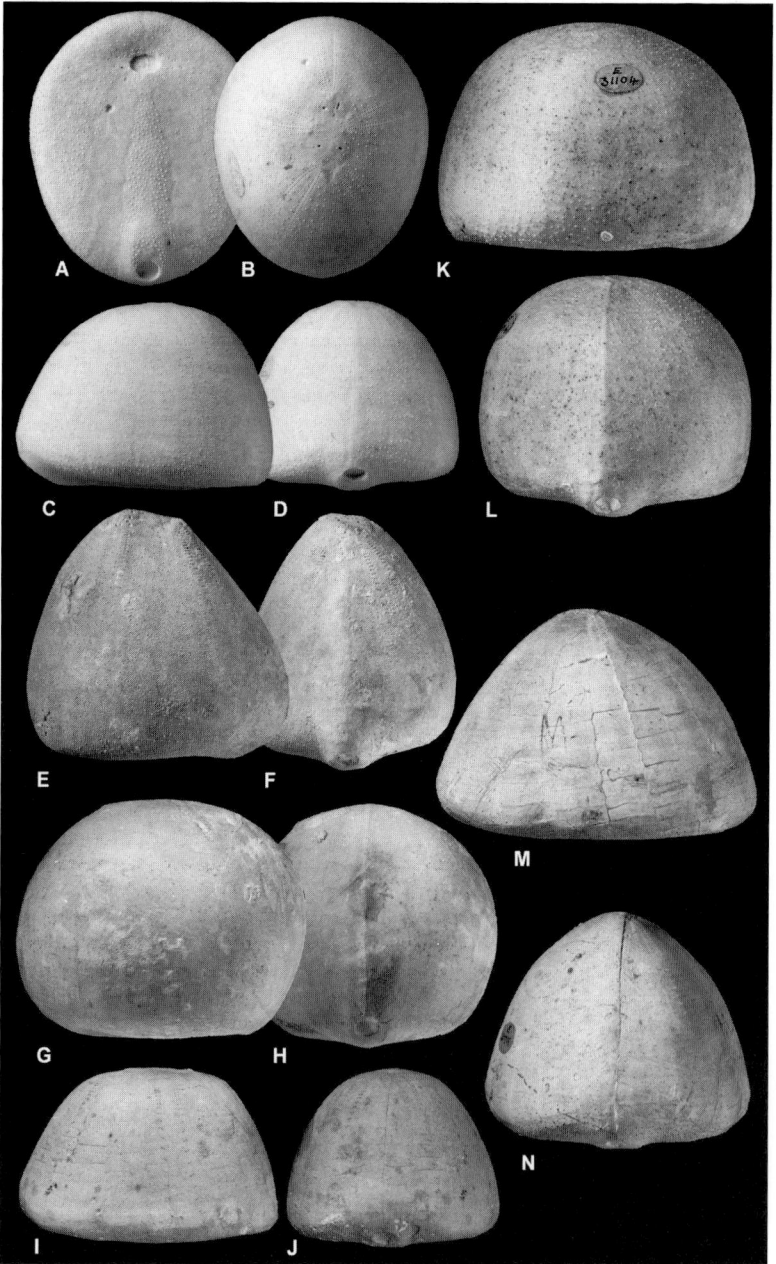

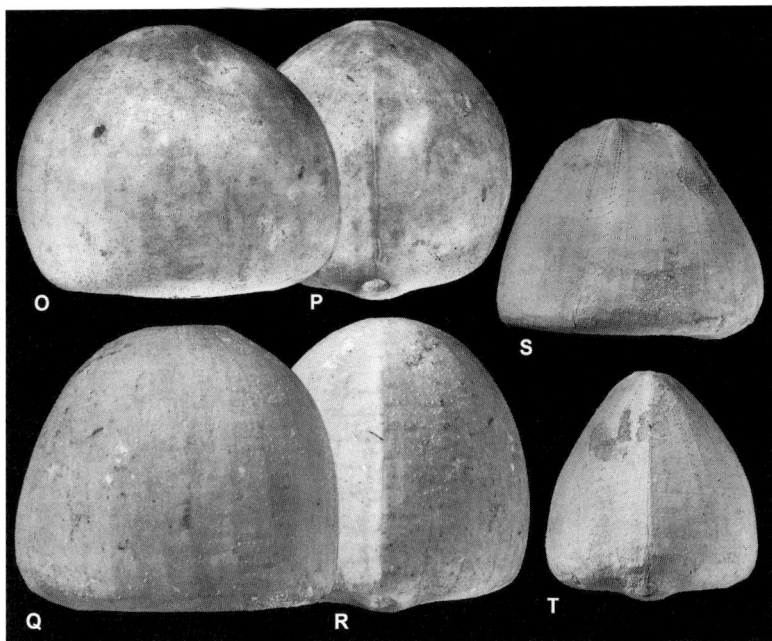

TEXT-FIG. 13.1. Shape varieties of *Echinocorys scutata* (Leske) from the Upper Chalk. A–D, Santonian, *G. quadrata* Zone, Paulsgrove, Hampshire. E–F, variety '*subconicula*' Brydone, Campanian, *B. mucronata* Zone, Norwich, Norfolk. G–H, variety '*gravesi*', Coniacian, *M. cortestudinarium* Zone, Chatham, Kent. I–J, variety '*depressula*' Brydone, Santonian, *M. testudinarius* Zone, Margate, Kent. K–L, Santonian, *M. coranguinum* Zone, Northfleet, Kent. M–N, variety '*pyramidalis*' Portlock, Campanian, *B. mucronata* Zone, Norwich, Norfolk. O–P, variety '*subglobosus*' Goldfuss, Santonian, *M. coranguinum* Zone, Micheldever, Wiltshire. Q–R, variety '*marginatus*' Goldfuss, Santonian, *M. coranguinum* Zone, Northfleet, Kent. S–T, variety '*elevatus*' Griffith and Brydone, no locality details. E–T show each specimen in lateral and posterior profile; all ×0·67.

Occurrence. Upper Coniacian and Lower Santonian, *M. coranguinum* Zone; common in narrow bands in Kent, rare elsewhere in southern England and very rare in Yorkshire (the abundant *Hagenowia* with slender rostrum of the Yorkshire coast is *H. anterior* Ernst and Schultz).

Order SPATANGOIDA

Remarks. Commonly heart-shaped with the paired ambulacra in grooves on the upper surface. Apical disc compact, never elongate.

PLATE 60

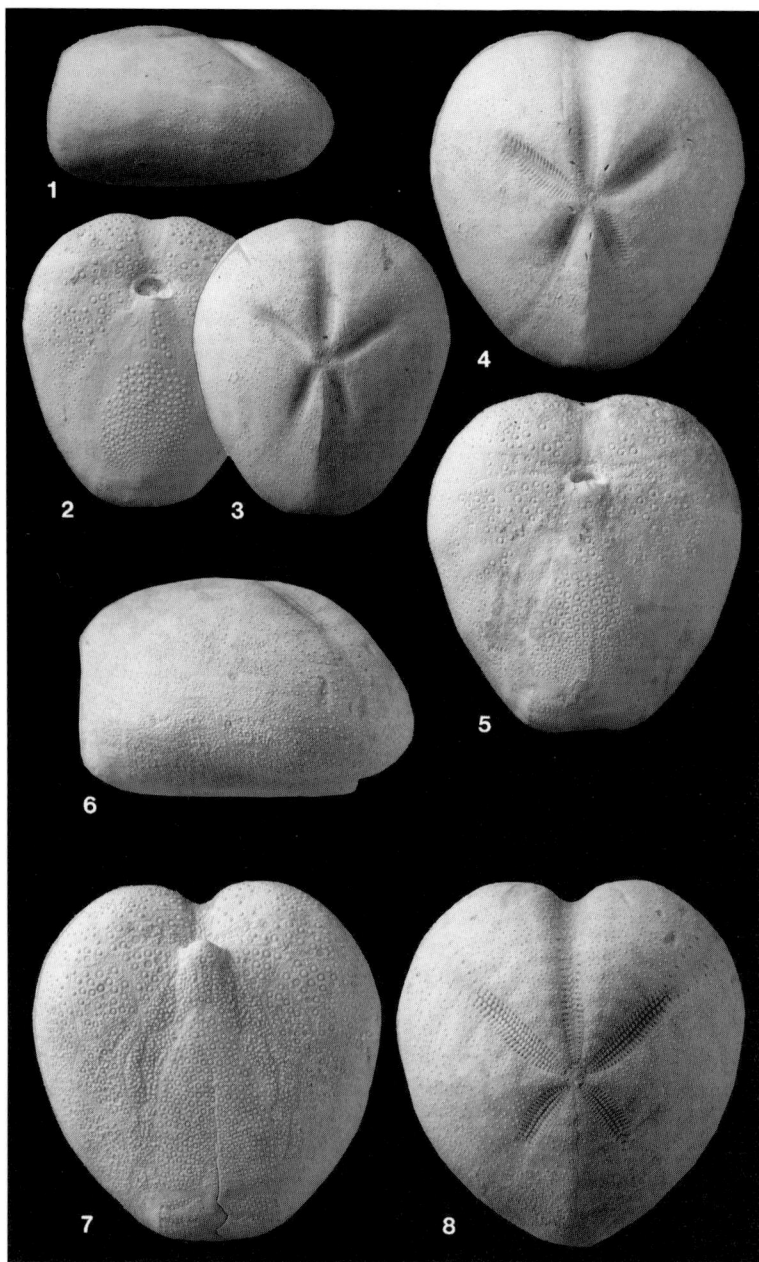

posterior part raised. Ambulacral petals rather shallow but with deep groove between the rows of pore pairs. The two ambulacral areas on either side of the plastron behind the mouth are very heavily granulated; the mouth is near the margin, with a prominent lip.

Occurrence. Santonian, *M. coranguinum–M. testudinarius* zones; generally distributed, common in the south, rarer in the north.

Micraster (*Gibbaster*) *gibbus* (Lamarck)
Plate 59, figures 5–7

Description. Like *M. coranguinum* in petal, peristome and plastron appearance, but easily distinguished by its strongly conical profile and low periproct.

Occurrence. Santonian, *M. coranguinum–M. testudinarius* zones; generally distributed, common in the south, rarer in the north.

EXPLANATION OF PLATE 60

Figs 1–3. *Micraster leskei* Desmoulins. Turonian, *T. gracilis* Zone, no locality details; lateral, oral and apical views; ×1.

Figs 4–5. *Micraster cortestudinarium* (Goldfuss). Coniacian, Arundel, Sussex; apical and oral views; ×1.

Figs 6–8. *Micraster coranguinum* (Leske). Lower Santonian, *M. coranguinum* Zone, no locality data; oral and apical views; ×1.

14. FISHES

by A. E. LONGBOTTOM and C. PATTERSON

Remains of fishes are widely distributed in the Chalk but are nowhere very common (except perhaps minute sharks' teeth and scales, recovered by acid digestion of chalk). Nowadays only fragmentary remains are usually found, such as teeth, fragments of jaws or other skull bones, vertebrae and scales. The magnificent complete fishes seen in some museums are from the cabinets of the 'gentlemen collectors' of the nineteenth century, when innumerable Chalk pits were handworked by quarrymen who knew that good fishes could be sold to collectors. Smith Woodward's monograph (1902–12) illustrates many of those fossils, and is still the most complete account of Chalk fishes. Today, with fewer Chalk pits and mechanical working, the flow of Chalk fishes has virtually stopped, apart from occasional lucky finds by casual collectors.

In this chapter fishes are divided into Chondrichthyes (sharks, skates, rays, chimaeras, with cartilage skeleton) and Osteichthyes (bony fishes). Because of their relatively fragile skeletons, fossil sharks and their relatives usually occur as isolated teeth and so the diagnostic features tend to be confined to dental characters. Many Chalk bony fishes are identifiable only when more or less intact, and the guide given here concentrates on identifiable fragments.

Because of the rarity of fishes, and because most of the old collections were made without proper details of horizon, the stratigraphic range of many Chalk fishes is known only in a very sketchy way.

DESCRIPTIONS

Class CHONDRICHTHYES

Remarks. Sharks' teeth are the most commonly found fish remains in the Chalk. This is not surprising since one individual may produce many thousands of teeth during its lifetime. In a shark's jaw there are many rows of teeth and during life the teeth within any one row are replaced, with new teeth erupting on the inside of the jaw and old teeth being shed on the outside. In this chapter only the more commonly found sharks' teeth are described and illustrated. It should be noted that very small teeth (less than 3 mm wide) belonging to dogfish and catsharks (such as *Squalus*, *Triakis* and *Scyliorhinus*) may be very common but are rarely seen except as a result of bulk sampling. Most of the terms used in the descriptions are explained in Text-figure 14.1. The terms lingual (towards the inside: tongue) and labial (towards the outside: lips) are used instead of lateral and medial because the orientation of a tooth surface relative to the axis of the fish will change according to the position of the tooth upon the jaw. Anterior refers to a position towards the symphysis (front) of the jaw and posterior refers to a position towards the angle of the jaw. The terms height and width refer to complete teeth (crown plus root) unless otherwise stated and are used as shown in Text-figure 14.1.

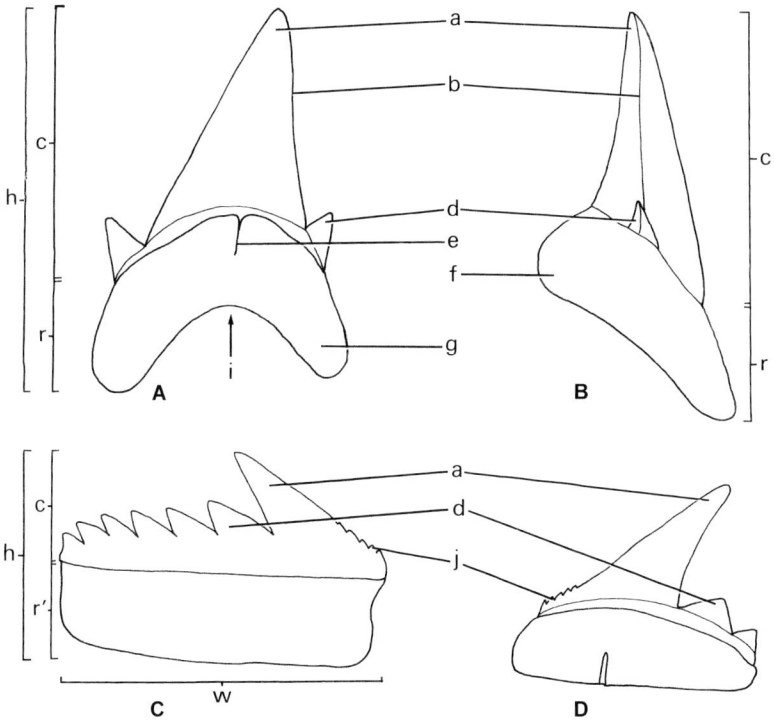

TEXT-FIG. 14.1. Diagrams of sharks' teeth to illustrate the descriptive terms used and to show tooth variation. A, lamnid shark tooth, lingual face. B, lamnid shark tooth, side view. C, hexanchid shark lower tooth, labial view. D, hexanchid shark upper tooth, lingual view. a, main cusp; b, cutting edge; c, crown; d, accessory cusps; e, nutritive cleft; f, lingual protrusion; g, root-half (long, rounded); h, tooth height; i, angle between root-halves (rounded); j, serrations on anterior cutting edge; r, root (deep, forked); r', root (deep, unforked); w, tooth width.

Individual sharks' teeth are often difficult to identify to species or even genus for the various reasons given below. Most modern identifications, especially in the lamnoids, are based on complete associated dentitions or large samples from known horizons and localities. Sharks' teeth are fairly rare in the English Chalk and most of the revised identifications used here are based on finds from other parts of the world.

In sharks there is often a difference in the shapes of anterior and posterior teeth. This change in shape is usually gradual along the jaw but in some genera the change is abrupt, a phenomenon known as heterodonty (e.g. in *Heterodontus*). Sometimes there is a marked difference between teeth of the upper and the lower jaws; for instance, in *Hexanchus* the crowns of the lower teeth are saw-like with many cusps whereas those of the upper teeth have only one or two cusps (Text-fig. 14.1). There is also sometimes a difference in the teeth in males and females (see descriptions of *Hexanchus microdon*). Obviously these differences in tooth shape do provide problems of identification of individual fossil teeth.

Order SYNECHODONTIFORMES
Family PALAEOSPINACIDAE

Genus SYNECHODUS

Remarks. An extinct genus. The anterior and posterior teeth differ considerably, although the change of shape is gradual along the jaw. The anterior teeth are taller with well-defined cusps and can usually be identified to species. The more posterior teeth are low-crowned, wide, with less distinguishable cusps, and are very difficult to separate into species.

Synechodus dubrisiensis (Mackie)
Plate 61, figure 9; Text-figure 14.2F

Description. A species with fairly small teeth (anterior teeth up to 5 mm wide, posterior teeth 10 mm wide). The anterior teeth have a fairly prominent main cusp with two or three pairs of accessory cusps on either side. The antero-posterior teeth have a tall main cusp and four or five pairs of accessory cusps. The most posterior teeth have a lower main cusp and the accessory cusps become less well defined. The base of the crown is continuous and marked by fine reticulating wrinkles, and the cusps are strongly striated vertically in both anterior and posterior teeth. The anterior teeth are about 6 mm high; most posterior teeth about 3 mm high. The root is shallow and marked by several nutritive furrows on the labial margin.
Occurrence. Cenomanian; Kent, Sussex, Oxfordshire, Cambridgeshire.

Synechodus nitidus Woodward
Text-figure 14.2D–E

Description. The anterior teeth differ from those of *S. dubrisiensis* in that they are larger (up to 11 mm wide), with a tall, slender, unstriated main cusp, usually slightly arched lingually, and 2–4 accessory cusps on each side. Posterior teeth have a lower crown, which is feebly marked with fine vertical wrinkles at the base. Anterior teeth are 7–8 mm high, and posterior teeth about 3 mm high. The root is shallow, marked by several nutritive furrows on the labial margin and with a slightly prominent lingual protrusion.
Occurrence. Cenomanian; Kent, Sussex.

Genus PARAORTHACODUS

Paraorthacodus recurvus (Trautschold)
Text-figure 14.2G–H

Description. The teeth are up to 10 mm high and 12 mm wide. The large, slender main cusp has sharp cutting edges, a flattened labial face, which is smooth or slightly wrinkled at the base, and a convex lingual face with only a few fine vertical wrinkles at the base. The accessory cusps are separately inserted on the root, and are long, slender and pointed, with fine vertical wrinkles at the base. The root/crown junction is straight or only slightly arched. The lingual protrusion on the root is very prominent. The basal edge of the labial face of the root is marked with several nutritive furrows. The root is pierced by several foramina of which the central one is most obvious and forms a cleft on the lingual face of the root. The crown curves sinusoidally when viewed from the side.
Occurrence. Santonian-Campanian, Norfolk.

Order HEXANCHIFORMES
Family HEXANCHIDAE

Genus HEXANCHUS

Remarks. The six-gilled sharks, or cow sharks, show a notable difference in shape between the teeth of the upper and lower jaws (Text-fig. 14.1). The lower teeth are comb-like with many cusps and can be distinguished in different species. The upper teeth have one large main cusp and one, two or no accessory cusps on one side and are not easily identifiable to species.

Hexanchus microdon (Agassiz)
Plate 61, figure 2; Text-figure 14.2M–N

Description. The lower teeth are flattened labio-lingually and consist of a series of pointed cusps fixed on a deep, unforked root. All the cusps are inclined to one side. The main cusp is appreciably larger than the accessory cusps, which diminish regularly in size posteriorly. The main cusp has fine but prominent serrations on its anterior edge extending from the base to about midway up the cusp. There are usually six or seven unserrated accessory cusps. The root is deep (equal to or slightly more than the height of the main cusp). This is a small species of *Hexanchus* with the teeth only about 12 mm wide.

Remarks. Most British Chalk hexanchid teeth have in the past been identified as *H. microdon*. More recently many of these were described as belonging to a different species, *H. gracilis* (Davis). It is now thought that the form *H. mirodon* represents teeth from males and *H. gracilis* those of females. The lower teeth of '*H. gracilis*' (Text-fig. 14.2N) differ from those of *H. microdon* (Text-fig. 2M) in that the main cusp is barely longer than the succeeding one and all the cusps diminish regularly in size posteriorly. There are usually four or five unserrated accessory cusps. The

---

PLATE 61

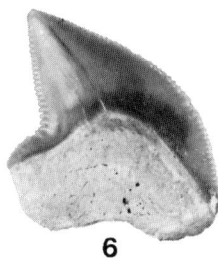

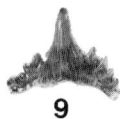

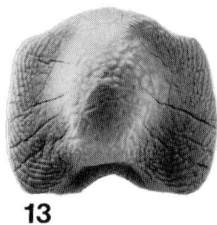

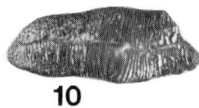

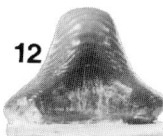

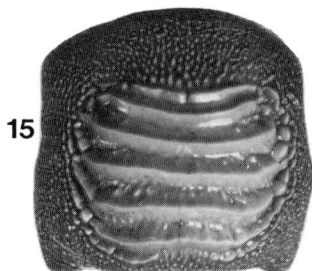

main cusp is serrated on the anterior edge but these serrations are often less prominent than in *H. microdon*. The root tends to be less deep in the females and the teeth are about 11 mm wide. *H. microdon* is a fairly rare Chalk fossil.

Occurrence. Cenomanian–Maastrichtian; Kent, Surrey, Sussex, Hampshire, Wiltshire, Bedfordshire, Norfolk.

Order SQUATINIFORMES
Family SQUATINIDAE

Genus SQUATINA

Remarks. This includes the living angel sharks; skate-like sharks with wing-shaped pectoral fins.

Squatina cranei Woodward
Text-figure 14.2K–L

Description. A species with teeth up to 6 mm high and 8 mm wide. The posterior teeth have a single, slender, tall main cusp on a broadly expanded base, without accessory cusps. The root is shallow, flattened basally and subtriangular in shape when viewed from below. One prominent central nutritive foramen and often others, less prominent, are present on the basal surface of the root. The lingual protrusion on the root is very pronounced. A small boss of enameloid hangs over the root on the labial face. Anterior teeth are similar but not as wide.

Occurrence. Cenomanian–Campanian; Kent, Sussex, Wiltshire, Norfolk.

Order HETERODONTIFORMES
Family HETERODONTIDAE

Genus HETERODONTUS

Remarks. Living examples include the Port Jackson or Bullhead sharks. They are noted for the great difference in shape between the anterior and posterior teeth. They are among the few living sharks with fin spines. The spines do not have barbs on the posterior face.

Heterodontus canaliculatus (Egerton)
Plate 61, figure 10; Plate 63, figure 5; Text-fig. 14.2O–P

Description. The anterior teeth (Text-fig. 14.2O–P) have a low triangular main cusp and one or sometimes two pairs of accessory cusps. The crown is smooth and overhangs the root on all faces except lingually where the root has a prominent lingual protrusion. The root is forked with short root halves. The lingual prominence has a large central nutritive cleft.

The posterior teeth (Pl. 61, fig. 10) are low crowned and 6–8 mm wide. When viewed from above the crown is elongated with rounded ends and with a keel, slightly sigmoidal in shape, which runs widthways along the centre of the crown. The occlusal surface is ornamented with a conspicuous reticulation of ridges that diverge from the keel. The ornamentation is often reticular on the labial side of the crown but formed of parallel ridges at right angles to the keel on the lingual side. The crown overhangs the root all around but more so on the labial side. The root is shallow and flat underneath and has a conspicuous enclosed central canal that crosses the root slightly obliquely from the lingual to labial side. On the labial side the canal opening lies in a depression and on the lingual side the root is protuberant on either side of the opening. These canals line up when the teeth lie naturally on the jaw within each tooth family. The posterior teeth lie in close contact on the jaw and form a continuous pavement of crushing teeth.

Remarks. Dorsal fin spines have been described under the name *Spinax major* (Agassiz). Later discoveries of associated jaws and spines showed that some of these belonged to *H. canaliculatus*. The spines (Pl. 63, fig. 5) are curved and ornamented with fine striations, which run the length of the spine. The posterior face is flat or channelled and, unlike those of hybodonts, does not possess barbs.

Occurrence. Cenomanian–Campanian; Kent, Surrey, Sussex, Wiltshire.

Order LAMNIFORMES

Remarks. The next four families are the lamnoid sharks (including the living porbeagle and mackerel sharks). The species and genera are difficult to distinguish from isolated teeth and the species boundaries are rather arbitrary in some cases. The teeth show a lot of variation between anterior and posterior positions on the jaw and in some species between upper and lower teeth. Isolated or broken teeth are therefore very difficult to identify.

Family ANACORACIDAE

Genus SQUALICORAX

Remarks. An extinct genus, limited to the Cretaceous. These sharks are noted for the possession of teeth with distinctly serrated cutting edges.

Squalicorax falcatus (Agassiz)
Text-figure 14.2CC

Description. A small- to medium-sized *Squalicorax* with teeth up to 12 mm high and 15 mm wide. The crown is subtriangular with the anterior edge straight or only slightly convex. The posterior edge is usually vertical and then notched, almost forming an accessory cusp. The cutting edges are distinctly serrated from base to tip. The root is large and slightly forked with a rounded, obtuse angle between the halves. There is no nutritive cleft.

Occurrence. Cenomanian–Campanian; Kent, Surrey, Sussex, Hampshire, Dorset, Wiltshire, Essex, Norfolk; a fairly common fossil.

Squalicorax kaupi (Agassiz)
Plate 61, figure 6; Text-figure 14.2DD

Description. A common Chalk fossil like *S. falcatus* but with larger and relatively taller teeth (up to 21 mm high and 16 mm wide). The anterior cutting edge is usually very convex. The posterior edge is vertical and slightly convex with the notch fairly indistinct in most teeth and only rarely forming an accessory cusp. The cutting edges are distinctly serrated. The root is large, deep and slightly forked with an obtuse angle between the halves. There is no nutritive cleft.

Occurrence. Coniacian–Campanian; Kent, Surrey, Sussex, Hampshire, Norfolk.

Genus PSEUDOCORAX

Pseudocorax laevis Leriche
Text-figure 14.2BB

Description. The teeth are fairly small (10–11 mm high, 9 mm wide) with more slender crowns than in *Squalicorax* and with smooth cutting edges. The anterior edge of the crown is straight and slightly notched at the base. The posterior edge is vertical and prominently notched at the base, thus forming a fairly distinctive, broad accessory cusp. The root is slightly forked with an acute angle between the two halves and a slight central nutritive cleft on the lingual face.

Remarks. Most of the teeth from the British Chalk previously named *P. affinis* (Agassiz) are here considered to be *P. laevis* because of their lack of serrations.

Occurrence. Santonian–Campanian; Sussex, Norfolk.

Family MITSUKURINIDAE

Remarks. This includes the living goblin or elfin shark, which has an elongate, pointed snout above the jaws.

Genus SCAPANORHYNCHUS

Scapanorhynchus rhaphiodon (Agassiz)
Plate 61, figure 3; Text-figure 14.2w–x

Description. The teeth have a long, slender main cusp with sharp cutting edges. The posterior teeth have a single pair of accessory cusps; the anterior teeth have none. The teeth are up to 30 mm high. The lingual face of the main cusp has very prominent striations, which are closely arranged at the base and diverge towards the tip. The labial face of the crown is smooth in all teeth. The root is forked with both halves elongated. The root halves are separated by an acute angle and each half is long and pointed. The root has a prominent lingual protrusion and this is deeply cut by a nutritive cleft. The most anterior teeth lack accessory cusps and are sigmoidally bent when viewed from the side. The more posterior teeth have slightly shorter crowns, a single pair of slender pointed accessory cusps, and shorter, more rounded root-halves.

Occurrence. Cenomanian–Campanian; Kent, Surrey, Sussex, Hampshire, Wiltshire, Devon, Norfolk; a common fossil in the Chalk.

Scapanorhynchus subulatus (Agassiz)
Text-figure 14.2y

Description. The teeth are smaller (about 14 mm high) than in *S. rhaphiodon*, with a slightly more triangular crown, especially in posterior teeth. The crown is smooth on both faces. One pair of accessory cusps is usually present. These are long and slender in anterior teeth but broader and more triangular in posterior teeth. The root is forked, with both halves shorter and more rounded than in *S. rhaphiodon*, and divided by a rounded angle. The root has a fairly prominent lingual protrusion cut by a central nutritive cleft.

Occurrence. Cenomanian–Campanian; Kent, Sussex, Wiltshire, Essex, Norfolk; fairly common.

Genus PARANOMOTODON

Paranomotodon angustidens (Reuss)
Plate 61, figure 8; Text-figure 14.2q

Description. The teeth are up to 14 mm high. The crown has a tall, slender, main cusp expanded at the base but without accessory cusps. The crown is smooth. The root is forked with the halves fairly long in anterior teeth but less so in posterior teeth. The tooth is compressed labio-lingually and the

Fishes 307

crown is slightly sigmoidal when viewed from the side. The nutritive cleft is distinguishable in anterior teeth only.

Occurrence. Cenomanian–Turonian; Kent, Surrey, Sussex, Cambridgeshire.

Family CRETOXYRHINIDAE

Genus CRETOXYRHINA

Cretoxyrhina mantelli (Agassiz)
Plate 61, figure 5; Text-figure 14.2R

Description. The teeth are large, about 33 mm high and 28 mm wide. Anterior and posterior teeth have a triangular main cusp, the posterior more so than the anterior. The anterior teeth are fairly robust with a convex lingual face on the main cusp; the posterior teeth are more compressed labio-lingually with a less convex lingual face. Both faces are

---

TEXT-FIG. 14.2. A, *Scyliorhinus antiquus* (Agassiz), Cenomanian, Burham, Kent; lingual view; ×4. B–C, *Scyliorhinus dubius* (Woodward), Cenomanian, Dover, Kent. B, lingual view; C, side view; ×6. D, *Synechodus nitidus* Woodward, Middle Cenomanian, *H. subglobosus* Zone, Wouldham, Kent; anterior tooth, labial view; ×2. E, *Synechodus nitidus* Woodward, Middle Cenomanian, *H. subglobosus* Zone, Snodland, Kent; posterior tooth, labial view; ×2. F, *Synechodus dubrisiensis* (Mackie), Cenomanian, Lewes, Sussex; worn anterior tooth, labial view; ×4. G–H, *Paraorthacodus recurvus* (Trautschold), Campanian, Norwich, Norfolk. G, labial view; H, side view; ×2. I–J, *Polyacrodus illingworthi* (Dixon), Cenomanian, Southeram, Sussex; posterior tooth. I, lingual view; J, oral view; ×2. K–L, *Squatina cranei* Woodward, Santonian or Campanian, Lewes, Sussex. K, labial view; L, basal view; ×3. M–N, *Hexanchus microdon* (Agassiz). M, Campanian, Norwich, Norfolk; lower tooth, labial view; ×2. N, '*Hexanchus gracilis*' (Davis), Chalk; England; lower tooth, labial view; ×2. O–P, *Heterodontus canaliculatus* (Egerton), Upper Turonian–Campanian, Guildford, Surrey; anterior tooth. O, lingual view; P, side view; ×3·5. Q, *Paranomotodon angustidens* (Reuss), Turonian, *T. lata* Zone, Whyteleafe, Surrey; labial view; ×2. R, *Cretoxyrhina mantelli* (Agassiz), Chalk, Kent; lingual view; ×1·5. S, *Archaeolamna kopingensis* (Davis), Campanian, Norwich, Norfolk; lingual view; ×1. T, *Cretodus crassidens* (Dixon), Chalk, Houghton, Sussex; labial view, no root; ×1. U–V, *Cretodus semiplicata* (Agassiz), Chalk, Charing, Kent. U, side view; V, labial view; ×1. W–X, *Scapanorhynchus rhaphiodon* (Agassiz), Chalk, Kent. W, lingual view; X, side view; ×1·5. Y, *Scapanorhynchus subulatus* (Agassiz), Cenomanian, Dover, Kent; lingual view; ×2. Z, *Cretalamna appendiculata* (Agassiz), Cenomanian or Turonian, Burham, Kent; lingual view; ×1·5. AA, *Dwardius woodwardi* (Herman), Chalk, Maidstone, Kent; lingual view; ×1·5. BB, *Pseudocorax laevis* Leriche, Campanian, Norwich, Norfolk; lingual view; ×2. CC, *Squalicorax falcatus* (Agassiz), Chalk, Sussex; labial view; ×2. DD, *Squalicorax kaupi* (Agassiz), Campanian, *G. quadrata* zone, Downend Quarry, Fareham, Hampshire; lingual view; ×2.

smooth and the crowns do not have serrated cutting edges. At the base the crown extends laterally over a wide root but rarely forms accessory cusps. The root is wide in posterior teeth, and forked with short, fairly divergent root-halves. The lingual protrusion is slight and the lingual face of the root is often flat, especially in posterior teeth. A small nutritive cleft is present on the lingual face in some teeth.

Occurrence. Turonian–Campanian; Kent, Surrey, Sussex, Hampshire, Wiltshire, Essex, Norfolk; common.

Genus CRETODUS

Cretodus semiplicata (Agassiz)
Text-fig. 14.2U–V

Description. The teeth are very large and robust with a large, sub-triangular main cusp and a single pair of broad, triangular, accessory cusps, sometimes incompletely separated from the main cusp. The root is robust and has a prominent lingual protrusion. Two types, *C. semiplicata* and *C. sulcata* (Geinitz), have been described as different species and there are differences between them. *C. semiplicata* teeth (Text-fig. 14.2U–V) are smaller (about 30 mm high), with fine, regular wrinkles at the base of the crown on the labial and lingual faces. Specimens of *C. sulcata* (Pl. 61, fig. 1) are larger (up to 50 mm high), and have a more slender main cusp with a more convex lingual face and larger accessory cusps. The crown has fine, regular wrinkles at the base on both faces. Herman (1975) believed that these differences are ontogenetic.

Occurrence. Cenomanian–Turonian; Kent, Surrey, Sussex.

Cretodus crassidens (Dixon)
Text-figure 14.2T

Description. Teeth of *C. crassidens* are very similar to those of *C. semiplicata* but differ mainly in having ill-defined or no accessory cusps. The teeth are large (up to 50 mm high) and robust. The main cusp is narrower than in *C. semiplicata* but widens at the base. The accessory cusps are not easily distinguishable from the main cusp but the crown has marked plications on the cutting edge at the base, suggestive of accessory cusps. The labial and lingual faces have a few coarse wrinkles centrally and fine, short, regular wrinkles all across the base. The root has a prominent lingual protrusion.

Remarks. *C. crassidens* is rarer than *C. semiplicata* and considered to be merely an extreme form of it by some.

Occurrence. Turonian; Kent, Sussex.

Genus ARCHAEOLAMNA

Archaeolamna kopingensis (Davis)
Text-figure 14.2s

Description. The teeth are medium-sized to large (up to 30 mm high). The main cusp is taller and more slender than in *Cretodus* species, inclined to one side, and with one pair of large, triangular accessory cusps. The crown is smooth on both faces and flattened labio-lingually. The root is wide and forked with short root halves and a slight lingual protrusion without a nutritive cleft.

Remarks. Teeth previously described as *Plicatolamna arcuata* are now synonymised with this species.

Occurrence. Coniacian–Campanian; Kent, Norfolk.

Genus CRETALAMNA

Cretalamna appendiculata (Agassiz)
Plate 61, figure 4; Text-figure 14.2z

Description. Teeth range from 15 to 25 mm high. The anterior teeth have a slender, upright and subtriangular main cusp, the posterior a triangular main cusp, slightly inclined posteriorly. All teeth usually have a single pair of large, triangular accessory cusps. The main cusp is flattened labio-lingually, with a flat labial face and a slightly convex lingual face without any striations. The lingual protrusion is prominent across the width of the root just below the crown-root junction, then the base of the root is flat. The root is forked but the halves are very short, square, and separated by a rounded angle. A slight nutritive cleft is visible in some teeth.

Remarks. The original spelling of the generic name takes priority and is used here. In the past many lamnid teeth from the Chalk were identified as *C. appendiculata*. However, Herman (1975) recognised another fairly common species among these teeth, which he named *Cretolamna woodwardi*. This has now been reclassified by Siverson (1999) as a new genus, *Dwardius* (described below).

Occurrence. Cenomanian–Lower Maastrichtian; Kent, Surrey, Sussex, Hampshire, Isle of Wight, Devon, Wiltshire, Cambridgeshire, Norfolk; fairly common.

Family *incertae sedis*

Genus DWARDIUS

Dwardius woodwardi (Herman)
Text-figure 14.2AA

Description. The teeth are generally larger (20–30 mm high) than in *C. appendiculata*. The anterior teeth have a tall, slender and upright main cusp; posterior teeth a triangular main cusp, slightly inclined posteriorly. A single pair of large, triangular accessory cusps is usually present. The main cusp has a more convex lingual face than in *C. appendiculata*. The root has a prominent lingual protrusion in the middle of the lingual face; it is forked with long, pointed or rounded halves separated by a rounded acute or obtuse angle. A slight nutritive cleft is visible in some teeth.

Occurrence. Cenomanian–Turonian; Kent, Surrey, Sussex.

Genus LEPTOSTYRAX

Leptostyrax macrorhiza (Cope)
Text-figure 14.3

Description. Anterior teeth are narrow and slender, but widening at the base, with a slightly sigmoid profile. The single pair of accessory cusps are long and slender and also have a sigmoid profile, and in this species are alongside the main cusp in side view or set slightly behind it. The labial face is flat but often with a deep indentation at the base of the crown. The lingual face is convex, usually smooth or with faint striations at the base of the crown, whereas the labial face has strong vertical folds in the enameloid at the base reaching up to one-third of the height of the crown. Folds or striations are also present on the lower half of all faces on the accessory cusps. There is a very pronounced lingual protrusion, and the root-halves are long and separated by an acute angle.

The lateral teeth are relatively shorter and wider. They often develop a second pair of accessory cusps. The labial face has strong folds at the base

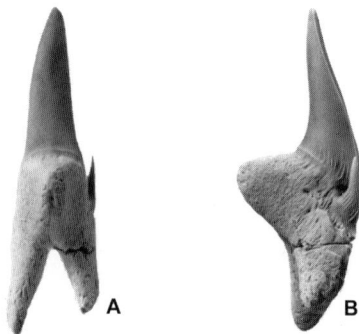

TEXT-FIG. 14.3. *Leptostyrax macrorhiza* (Cope), Chalk, England. A, lingual view; B, side view; ×2.

of the crown but the lingual face is smooth. The lingual protrusion is less pronounced in the lateral teeth and the root-halves are more rounded but still divided by an acute angle.

Occurrence. Unknown, rare.

Order CARCHARHINIFORMES
Family SCYLIORHINIDAE

Genus SCYLIORHINUS

Remarks. This includes the living dogfish. The fossil teeth are very small and are usually only found by bulk processing and sieving.

Scyliorhinus antiquus (Agassiz)
Text-figure 14.2A

Description. The teeth are up to 2·5 mm wide and 3 mm high. The main cusp is long and slender, and one pair of short, stout, accessory cusps is present. The lingual face in anterior teeth is finely striated; both faces are striated in more posterior teeth. The root is short and forked, with rounded ends to the root-halves. The root has a very pronounced lingual protrusion, but is flattened basally. A distinct central nutritive cleft is present. The crown covers most of the root on the labial face.

Occurrence. Cenomanian; Kent.

Scyliorhinus dubius (Woodward)
Text-figure 14.2B–C

Description. The teeth are only up to 1·5 mm high. The main cusp is long and slender, and one pair of long, slender, accessory cusps (up to one-half the height of the main cusp) is present. The faces of the crown are mostly smooth but some teeth have feeble striations at the base of the crown on the labial face. The main cusp and the accessory cusps curve slightly lingually when viewed from the side. The root is short and forked, with each half expanded at the ends. There is a pronounced lingual protrusion with a large nutritive cleft. The crown covers most of the root on the labial face.

Occurrence. Cenomanian; Kent.

Order HYBODONTIDEA
Family POLYACRODONTIDAE

Genus POLYACRODUS

Polyacrodus illingworthi (Dixon)
Text-figure 14.2I–J

Description. The teeth are large, low-crowned, and up to 30 mm wide. The main cusp and accessory cusps are not clearly defined, and are short, stout, and obtusely pointed. The surface of the tooth is marked by fine ridges radiating from the apex of the cusps or from an acute keel that runs across the width of the crown. The root is large, deep, not crimped or forked, and lacks a nutritive cleft.

Occurrence. Cenomanian; Kent, Surrey, Sussex.

Family PTYCHODONTIDAE

Genus PTYCHODUS

Remarks. An extinct genus of shark possibly related to the hybodontids on the basis of internal root morphology of the teeth. They are known only by teeth, jaws and vertebrae. They have a crushing pavement dentition with the teeth aligned closely in rows running labio-lingually over the jaws (Text-fig. 14.4). There are paired lateral rows on each side of a median row. In the lower jaw the median row consists of very small teeth, and in the upper jaw the largest teeth form the median row. The teeth diminish in size in the lateral rows from the row next to the median row to the most lateral row. There are different numbers of rows in different species. Where more complete dentitions are known, estimates are 15 rows in *P. decurrens* Agassiz and 17 rows in *P. mortoni* Mantell.

The crown surface is ornamented with a series of subparallel ridges running across the central area surrounded by a marginal area of granular ornamentation. The number, size, and the relative area of the crown surface covered by these ridges and the marginal area are characters used to differentiate the species. The other main characters used are the degree of elevation of the central coronal area, and the profile of the crown when viewed from the side. The roots are all shallow and simple, and the crown overhangs the root slightly on all sides. The crown has a slight depression on the lingual side and usually slopes regularly from the apex to the labial margin. The labial margin of each tooth fits into the lingual depression of the next tooth in each row. The teeth in the pair of rows next to the median row tend to be square or rectangular and are usually the most characteristic of the species. The more lateral rows have smaller, more asymmetrical teeth, often becoming lozenge-shaped or oval, and are very difficult to identify because they look the same in different species. The following descriptions are mostly based on teeth from the large median row or the first lateral rows. In the figures the labial margin is at the top and the side views are all taken looking at the lingual face.

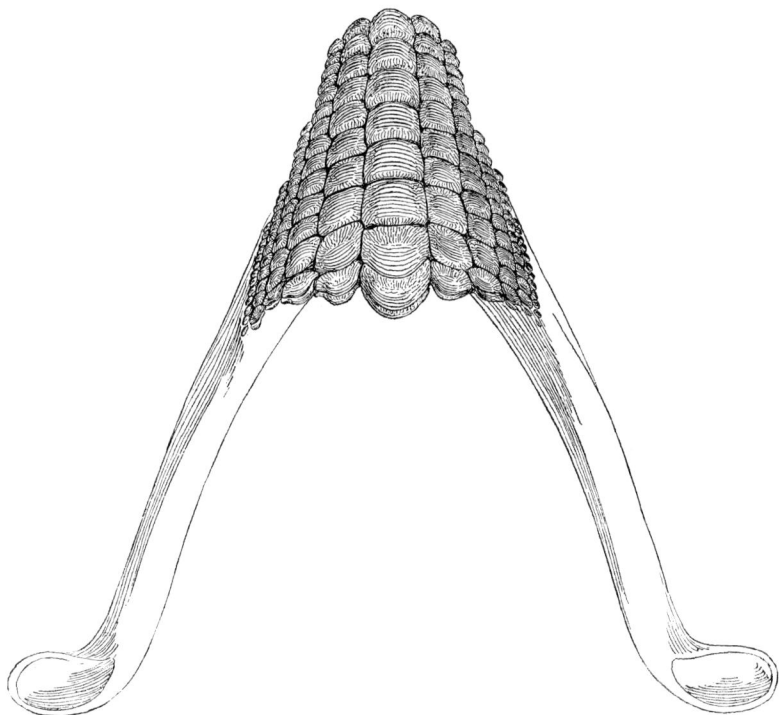

TEXT-FIG. 14.4. Reconstruction of a jaw of *Ptychodus decurrens* Agassiz, natural
size (from Woodward 1904, p. 134).

Ptychodus ranges world-wide but is restricted to the Upper Cretaceous.
The teeth are probably the most distinctive of the Chalk fish fossils, and
many species have been described (see Agassiz 1833–45; Dixon 1850;
Dibley 1911).

Ptychodus mammillaris Agassiz
Plate 61, figures 11–12

Description. The largest teeth are up to 45 mm wide. The central coronal
area is much elevated, especially in smaller teeth, and more or less sharply
defined from the marginal area. The tip of this central hump is slightly
flattened. The raised area is crossed by 8–11 regular, prominent, parallel
ridges, which continue down the sides to the marginal area. These ridges
are rarely curved at the ends but often end with coarse tubercles, which

become finer and merge with the marginal ornamentation. The flat marginal area is relatively wide and ornamented with fine granules concentric to the margins of the tooth. The marginal area is often crossed by radiating shallow puckers.

Occurrence. Turonian–Coniacian; Kent, Surrey, Sussex, Isle of Wight, Dorset, Devon, Wiltshire, Hertfordshire, Bedfordshire, Lincolnshire, Yorkshire.

Ptychodus rugosus Dixon
Plate 61, figures 13–14

Description. The teeth are very similar to those of *P. mammillaris* with a much elevated central area and a well-defined marginal area. This central hump is crossed by 3–5 ridges, which pass into a rugose ornamentation on the labial part of the tooth. The marginal area has concentric rugose ornamentation. Unlike *P. mammillaris* the ornamentation tends to be fainter and less well defined, with the transverse ridges being incomplete or interrupted. The central raised hump also has a more rounded tip.

Occurrence. Coniacian–Santonian; Kent, Surrey, Sussex, Hampshire, Essex.

Ptychodus latissimus Agassiz
Plate 61, figure 15

Description. The teeth are up to 60 mm wide. The central area is gently raised or flattened and crossed by 6–8 coarse, sharp, transverse ridges. The ridges are straight or slightly curved at the ends but never form complete whorls. The marginal area is narrow or wide with coarse granulations, which sometimes continue from the ends of the ridges.

Remarks. A variety, *P. dixoni*, has slightly finer ridges than the normal form and a narrow, coarsely granulated margin. Another variety, *P. paucisulcatus*, has only four or five very prominent transverse ridges with rare or slight curving of the ends, often with large granules between them. The marginal area is wide and coarsely granulated.

Occurrence. Upper Cenomanian–Lower Santonian; the variety *P. paucisulcatus* has only been found in the *M. cortestudinarium* and *M. coranguinum* zones; Kent, Surrey, Sussex, Hampshire, Wiltshire, Buckinghamshire, Hertfordshire, Suffolk, Norfolk.

Ptychodus decurrens Agassiz
Plate 62, figure 1

Description. The teeth are up to 60 mm wide. The central area is raised and gently arched in profile or flattened (var. *depressus*). The raised area is crossed by numerous (9–16) fine ridges, these being parallel and straight or very slightly curved at the ends. In some specimens they are recurved into whorls (var. *multistriatus*). The marginal area is usually narrow with lines of fine granulations more or less at right angles to the margins. The transverse ridges pass gradually into the marginal ornamentation and sometimes branch at the ends.

Remarks. A very varied species.

Occurrence. Cenomanian–Turonian; Kent, Surrey, Hertfordshire, Cambridgeshire, Norfolk.

Ptychodus oweni Dixon
Plate 62, figure 2

Description. The teeth are about 40 mm wide with a moderately elevated crown, which is regularly arched in profile. The central area is not clearly distinguishable from the marginal area. The surface is ornamented with fine but well-defined ridges, mostly incomplete or branched and tending to radiate from the centre rather than be transverse. The marginal area is restricted and ornamented with coarse, elongate granulations.

Remarks. There are specimens intermediate between this and *P. decurrens* so the separation of this into a distinct species is questionable.

Occurrence. Middle Cenomanian; Kent, Sussex.

Ptychodus polygyrus Agassiz
Plate 62, figure 3

Description. The teeth are up to 70 mm wide. They tend to be slightly wider than long and have 7–11 distinct transverse ridges crossing the crown. The ridges often curve at the ends, one or two sometimes forming complete whorls. The marginal area is narrow and has coarse granular ornamentation often continuing concentrically around the crown from the ends of the ridges. The crown is not much elevated and is flattened on top.

Occurrence. Coniacian–Campanian; Kent, Surrey, Hampshire, Wiltshire, Middlesex, Essex.

Ptychodus concentricus Agassiz
Plate 62, figure 4

Description. The teeth are almost triangular in profile with a markedly elevated central area. The central area is well defined and crossed by many fine but prominent transverse ridges, which curve to form complete

PLATE 62

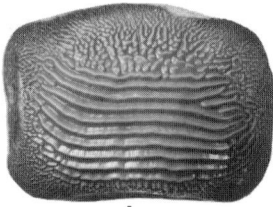

1

A · B

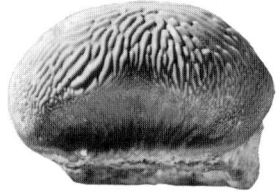

2

A · B

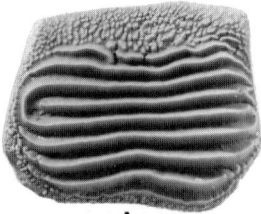
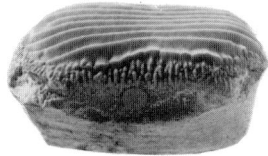

3

A · B

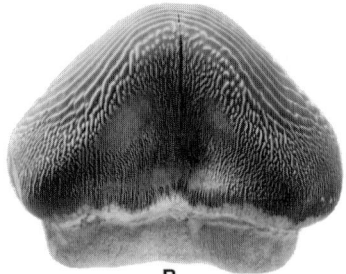

4

A · B

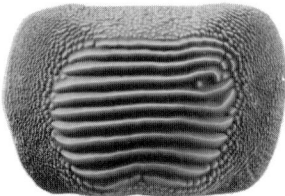

5

A · B

concentric whorls. The marginal area is moderately wide in some teeth and ornamented with fine granulations, which tend to be concentric near the transverse lines, becoming radiate towards the crown margin.

Occurrence. Middle Cenomanian; Kent; this species is not very common in the British Chalk but is common on the continent.

Ptychodus marginalis Agassiz
Plate 62, figure 5

Description. Differs from *P. polygyrus* in having slightly finer and more numerous transverse ridges, up to 14 in number, many of which curve round at the ends and continue concentrically on the marginal area as coarse to fine granulations. The marginal area is extensive and is distinct from the central hump. The teeth are smaller than in *P. polygyrus* and are about 30–53 mm wide.

Occurrence. Turonian–Coniacian; Kent, Sussex, Berkshire.

Class OSTEICHTHYES

Remarks. Bony fish remains found in the Chalk today are commonly disarticulated parts such as scales, teeth, fin-rays, vertebrae, jaws and other skull bones. Otoliths or 'ear-stones', which are common in many late Mesozoic and Cenozoic clays, are composed of aragonite and are virtually unknown in the Chalk. The fragments of bony fishes found in the Chalk are often not diagnostic, since only more or less complete fishes can be identified with any assurance. These are so rare and the fauna is so diverse that no real guidance to their identification can be given here. Smith Woodward's (1902–12) monograph is still the best guide, and if that is unavailable locally those fortunate enough to find a complete fish are advised to take it to a major museum. A fairly common Chalk fossil easily mistaken for a complete fish is the worm tube named *Terebella lewesiensis*. This worm built its tube from fish bones, and the elongate tube, up to 60 cm long, looks vaguely like the skeleton of an eel (and was described

---

EXPLANATION OF PLATE 62

All figures: A, oral view; B, side view.
Fig. 1. *Ptychodus decurrens* Agassiz. Chalk, Maidstone, Kent; ×1.
Fig. 2. *Ptychodus oweni* Dixon. Chalk, Lewes, Sussex; ×1.
Fig. 3. *Ptychodus polygyrus* Agassiz. Santonian, *M. coranguinum* Zone, Banstead, Surrey; ×1.
Fig. 4. *Ptychodus concentricus* Agassiz. Cenomanian, *H. subglobosus* Zone, Lee's Pit, Holborough, Kent; ×1.
Fig. 5. *Ptychodus marginalis* Agassiz. Chalk, Halling, Maidstone, Kent; ×1.

as one by Gideon Mantell in 1822). Yet on detailed examination, the 'skeleton' turns out to have no head or tail, no vertebral column or fins, but to be a jumble of bones and scales, often comprising fragments of different genera. Guides to the identification of bony fish fragments are given below under the headings Scales, Teeth, and Other Remains.

SCALES

Four main types of scale may be encountered: cycloid, ctenoid, coelacanth, and ganoid (listed in sequence from the commonest to the rarest).

Cycloid and ctenoid scales are characteristic of teleosts, the dominant group of bony fishes today (there are over 20,000 living species). These scales are thin, almost translucent plates of bone whose most obvious feature is concentric ridges (circuli) marking the growth of the scale from a central focus. In many teleosts the spacing of the circuli varies with the season in which they were laid down, so that annual growth rings can be recognized, as in a tree, but in Chalk fishes the circuli are generally evenly spaced. The anterior or embedded part of the scale, which is covered in life by the scale in front, usually has a number of grooves (Text-fig. 14.5B) or uncalcified strips (Text-fig. 14.5A), the radii, converging on the focus of the scale.

Cycloid and ctenoid scales differ in the form of the posterior, exposed part of the scale. In cycloid scales (Text-fig. 14.5A) this is smooth, but in ctenoid scales (Text-fig. 14.5B) the exposed circuli develop minute comb-like teeth, the ctenii. Ctenoid scales are characteristic of higher teleosts, especially of those with spines in their fins, the Acanthopterygii (e.g. perch, mackerel). In the Chalk seas the only acanthopterygians belong to the most primitive living group, the Beryciformes. Living beryciforms are marine fishes of two principal types, the holocentroids or soldierfishes, abundant in tropical reefs, and the trachichthyoids, deep-water, oceanic fishes. Both groups existed in the Chalk sea. Cycloid scales characterise the more primitive teleosts (e.g. herring, pike) and so are typical of most Chalk teleosts.

Some Chalk teleosts have scales that are specialised as scutes. One of the V-shaped scutes of *Dercetis* is illustrated (Text-fig. 14.5C). Dercetids are a Late Cretaceous group of eel-shaped fishes with two or more rows of these scutes along the flank.

Coelacanths are represented in the Chalk by two species of *Macropoma*, a Late Cretaceous genus which was thought to be the last of the coelacanths until the living *Latimeria* was discovered in 1939. *Macropoma* scales (Text-fig. 14.5D) are very thin, ovoid plates of fibrous bone, easily recognised by the enamelled tubercles or denticles covering the portion that was exposed in life.

Ganoid scales (Text-fig. 14.5E) are thick, rhomboidal plates of bone with an external layer of smooth, shiny ganoin. Typically each scale has a peg on the upper edge by means of which it articulates with its neighbour in the scale row. Ganoid scales characterise many Palaeozoic fishes and a few surviving genera like the North American gars (*Lepisosteus*) and the African bichirs (*Polypterus*). Four or five genera of fishes with ganoid scales are recorded in the Chalk, but all are extremely rare.

TEETH

Over 40 genera of bony fishes are recorded in the Chalk; about a dozen of these have a distinctive dentition, and the rest have small or nondescript teeth.

Among Chalk teleost teeth, the most likely identifiable finds are teeth of 'enchodonts', an informal collective name for about five genera of predatory fishes, probably related to living deep-sea predators like *Alepocephalus*, the lancetfish. Three genera are illustrated: *Enchodus*, *Apateodus*, and *Cimolichthys*. The characteristic teeth in each of these are the large fangs carried on the palate and the tip of the lower jaw (dentary). In *Apateodus* (Pl. 63, fig. 3) there are two enlarged fangs on the palatine and two on the dentary, and the teeth are compressed, with two sharp edges, and are smooth apart from fine striations near the base. In *Cimolichthys* (Text-fig. 14.5G) there are several fangs on the palatine and on the dentary, and the palatal teeth have a barb at the tip. In *Enchodus* (Text-fig. 14.5F) there is only one fang on the palatine and one on the dentary; the palatine fang is compressed, with two sharp edges, whereas the dentary fang has one sharp edge and one rounded. The bases of *Enchodus* fangs may be striated.

Pachyrhizodus (Text-fig. 14.5H) is a member of an extinct, Cretaceous group of teleosts, the pachyrhizodontoids. In *Pachyrhizodus* the teeth are slightly curved, round in section, and have a characteristic swollen base where they are sheathed in bone attaching them to the jaw. *Tomognathus* (Text-fig. 14.5I) is a primitive bony fish (not a teleost), known only from the Chalk, which has teeth rather like those of *Pachyrhizodus* but more slender, more strongly curved, and with a translucent cap and a few striations near the cap.

Protosphyraena is a large fish with a long rostrum like a swordfish. It was the last survivor and only Late Cretaceous member of the pachy-cormids, a Mesozoic group of primitive bony fishes. In *Protosphyraena*

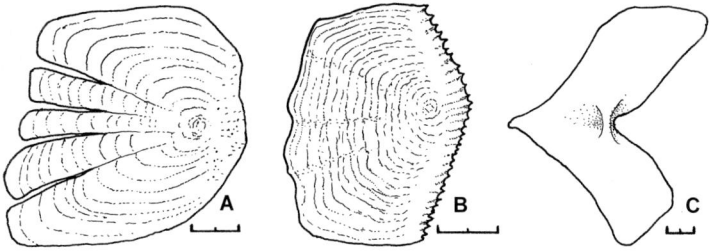

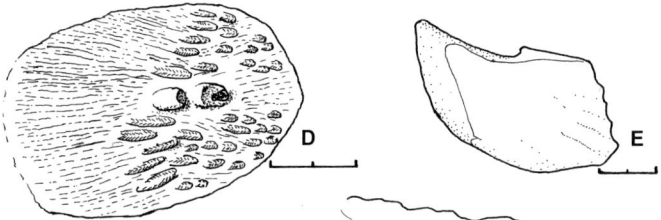

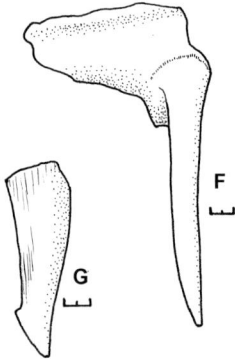

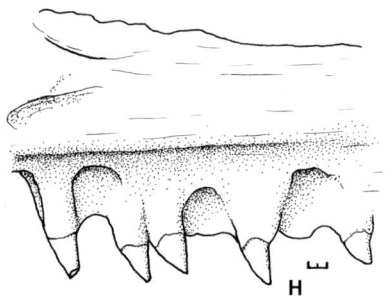

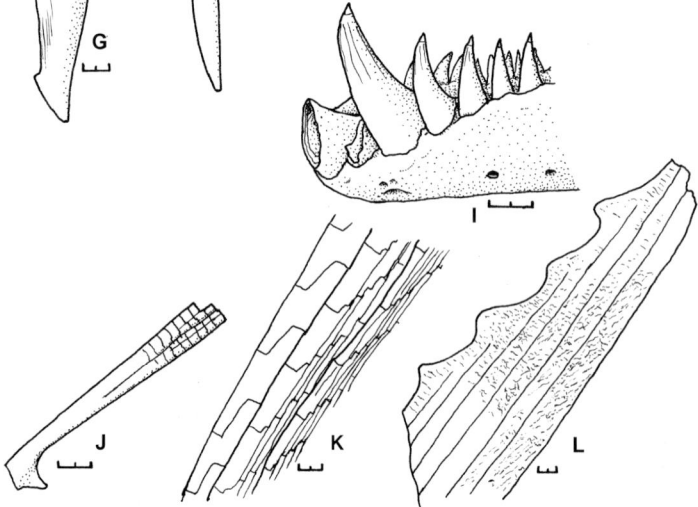

the teeth (Pl. 63, fig. 6) are very like those of the enchodont *Apateodus*, but they are set in deep sockets in the jaw, not fused to the bone as they are in *Apateodus*. This means that *Protosphyraena* teeth have undifferentiated bases, whereas in *Apateodus* the base is slightly swollen and striated. In *Protosphyraena* teeth the surface enameloid is very thin and usually shows characteristic longitudinal post-mortem cracks.

All the teeth mentioned so far are fang-like and obviously belong to predators. There are two groups of bony fishes with crushing dentitions, presumed to be bottom-feeders, whose teeth may be found in the Chalk: pycnodonts and plethodonts. Pycnodonts are found in rocks ranging in age from Late Triassic to Eocene. There are over 15 genera and about five of these are recorded in the Chalk. Pycnodont crushing dentitions consist of a median, symmetrical tooth-plate on the vomer in the upper jaw, and paired tooth-plates on the splenials in the lower jaw. Pycnodont genera are distinguished, in part, by the arrangement of the teeth on these bones, and isolated teeth are rarely identifiable. The pycnodont vomer illustrated (Pl. 63, fig. 1) is assigned to *Phacodus*, a nominal genus whose only species is from the Chalk; the fossil shows the usual form of pycnodont vomers, with a median row of broad, bean-like teeth, and ovoid or rounded teeth in

---

TEXT-FIG. 14.5. A, *Osmeroides lewesiensis* (Mantell) (Osmeroididae), Chalk, Lewes, Sussex; scale, external view. B, *Hoplopteryx lewesiensis* (Mantell) (Trachichthyidae), Chalk, Cuxton, Kent; scale, external view. C, *Dercetis maximus* Smith Woodward (Dercetidae), Upper Chalk, *M. coranguinum* Zone, Grays, Essex; scute, internal view. D, *Macropoma mantelli* Agassiz (Coelacanthidae), Upper Chalk (Upper Turonian–Campanian), Guildford, Surrey; scale, external view; the two denticles in the middle of the scale are broken away and represented only by their bases. E, *Lophiostomus dixoni* Egerton (?Caturidae), Chalk, Lewes, Sussex; scale, external view. F, *Enchodus lewesiensis* (Mantell) ('Enchodontoidei'), Chalk, Sussex; right palatine, external view. G, *Cimolichthys levesiensis* Leidy ('Enchodontoidei'), Chalk, Sussex; tooth from the palatine or ectopterygoid. H, *Pachyrhizodus basalis* Dixon (Pachyrhizodontidae), Cenomanian, Bluebell Hill, Burham, Kent; part of right maxilla, internal view. I, *Tomognathus mordax* Dixon (Tomognathidae), Cenomanian, *H. subglobosus* Zone, Amberley, near Arundel, Sussex; anterior end of left lower jaw, external view; the foremost tooth is broken, and the second is undergoing replacement. J, *Hoplopteryx lewesiensis* (Mantell) (Trachichthyidae), Upper Chalk, near Rochester, Kent; base of a soft ray from dorsal fin. K, *Pachyrhizodus* sp. (Pachyrhizodontidae), Chalk, Kent; fragment of leading edge of tail fin showing zigzag jointing strengthening the outermost fin-rays, and repeated bifurcation in its neighbours. L, *Protosphyraena* sp. (Pachycormidae), Cenomanian or Turonian, Burham, Kent; fragment of leading edge of pectoral fin; the wrinkled ornament is broken and worn away over part of the surface. Scale bars represent 2 mm; head of fish is to the left in all except F and G.

PLATE 63

1

2

3

4

5

6

rows on either side. The splenials illustrated (Pl. 63, fig. 2) represent another genus, *Polygyrodus*, whose only species is from the Chalk. Here there are no bean-like teeth (they occur in most pycnodont splenials) and the teeth are irregularly arranged and roughly circular, with a central cone, a flat rim around the base of the cone, and rough, ornamented surfaces.

The other type of Chalk crushing dentition is named *Plethodus*. The plethodonts are Cretaceous teleosts, ranging in Britain from the Albian Gault to the Upper Chalk. There are several nominal genera in the family, but *Plethodus* is the only one recorded from the Chalk. The dentition consists of two plates with smooth grinding surfaces formed of fused small teeth, which are sometimes still recognisable around the margin of the plate. The tooth-plates are circular, pentagonal or rectangular in form; the upper plate is slightly concave and was carried by the parasphenoid, on the floor of the braincase, and the lower plate is convex and was carried by the basihyal, in the 'tongue'.

OTHER REMAINS

Collected under this heading are comments on miscellaneous parts of bony fishes which may be found in the Chalk: vertebrae, fin-rays and 'coprolites'.

Isolated bony fish vertebrae are not, in general, identifiable to genus or even family. Indeed, it is often difficult to decide whether a vertebra is from a shark or a teleost, since in primitive teleosts the bony neural and haemal arches and spines are not fused to the centrum, so that they fall away after death, leaving isolated centra which can look very like shark centra (Pl. 61, fig. 7).

Fin-rays in bony fishes are of two principal types, 'soft rays' and spines. Fin spines are found in the front portion of the dorsal, anal and pelvic fin

---

EXPLANATION OF PLATE 63

Fig. 1. *Phacodus punctatus* Dixon. Campanian, *G. quadrata* Zone, Whaddon, near Salisbury, Wiltshire; incomplete dentition of vomer, oral view; ×1.

Fig. 2. *Polygyrodus cretaceus* (Agassiz). Turonian, *T. lata* Zone, Cuxton, Kent; lower jaws showing splenial dentition, oral view; ×0·7.

Fig. 3. *Apateodus striatus* Smith Woodward. Cenomanian, near Maidstone, Kent; right ectopterygoid in lateral view; ×1.

Fig. 4. Heteropolar coprolite or enterospira (fossil intestine), probably from a shark; Middle Cenomanian, *H. subglobosus* Zone, Dover, Kent; ×1·5.

Fig. 5. *Heterodontus canaliculatus* (Egerton). Chalk, Sussex; dorsal fin-spine; ×1.

Fig. 6. *Protosphyraena* sp. Lower Cenomanian, *S. varians* Zone, Hauxton, near Cambridge; tooth, probably from the tip of the left lower jaw, in lateral view; ×1.

of acanthopterygians, the higher teleosts represented in the Chalk by beryciforms. They are awl-like bony rods, with a complex articular surface at the base. 'Soft rays' are so called because of their flexibility, a consequence of repeated jointing or segmentation, breaking each bony ray into a series of short cylinders. Soft rays usually bifurcate several times, so that towards the tip they are subdivided into a number of slender, jointed filaments. Soft rays (Text-fig. 14.5J) compose all the fins of primitive bony fishes, and the tail, pectoral and posterior parts of the dorsal, anal and pelvic fins in acanthopterygians. In the leading edges of the fins, especially the tail, the transverse joints in the soft rays are often staggered, producing a serrated effect (Text-fig. 14.5K). This strengthening of the leading edge of the fin reaches a peak of specialisation in the pectoral fins of the pachycormid *Protosphyraena* (Text-fig. 14.5L). These are long, narrow and scythe-shaped, with the fin-rays unjointed, unbifurcated, and closely pressed together. The rays end successively on the leading edge of the fin where they are covered by an enamel-like substance, projecting in a series of saw teeth.

Objects from the Chalk like that shown in Plate 63, figure 4 were at first interpreted as fossil fir-cones. Gideon Mantell proposed that they had an animal origin, and Dean Buckland showed that similar fossils from the Lias of Lyme Regis, Dorset, matched cement casts of the large intestine of living sharks, where the spiral valve imparts a characteristic structure of concentric cones. Buckland christened these fossils coprolites, or fossilised faeces. The Chalk coprolites are conventionally assigned to the coelacanth *Macropoma*, though Smith Woodward remarked that they could as well come from sharks or chimaeroids. More recently there has been argument about whether coprolites are actually fossil faeces (expelled from the fish during life) or fossilised intestinal contents, like the cement casts that Buckland made from sharks; such fossils are now called enterospirae. The source of Chalk coprolites or enterospirae is not known, but the spiral valve is a feature of primitive fishes-sharks, chimaeroids, and nonteleostean bony fishes. Enterospirae have not been found *in situ* in Chalk coelacanths, and sharks are the most likely source of these Chalk fossils.

15. REPTILES

by ANGELA C. MILNER

Reptile remains from the Chalk are few and fragmentary but they include representatives of all the marine groups that occur in the Cretaceous. Terrestrial forms are restricted to pterosaur material and rare representatives of two groups of herbivorous dinosaurs. Most specimens were collected during the last century from working chalk pits, long since closed. The Chalk fauna as a whole has been largely neglected since the original descriptive work, mainly by Richard Owen (1850, 1851, 1861, 1864), except for a short review of the vertebrates by Woodward (1889*a*) and a checklist of genera in Woodward and Sherborn (1890). Chalk localities and taxa were reviewed in Benton and Spencer (1995) and material listed for 50 localities.

Isolated teeth are the most commonly found specimens in the Chalk. Reptile teeth are usually diagnostic at least to ordinal and superfamilial level, and in many cases to generic and specific level. A simple key is included below as an aid to the identification of teeth; it applies only to Chalk material.

KEY TO IDENTIFICATION OF CHALK REPTILE TEETH

1. Tooth crown conical   2
   Tooth crown leaf-shaped and with deeply serrated margin, or rounded and blunt, or with a prominent groove and showing wear facets   10
2. Crown with prominent, longitudinal, ridged ornament   3
   Crown finely ornamented or smooth   4
3. Ornament of regularly spaced smooth longitudinal ridges converging towards but not reaching tip of the crown   *Platypterygius*
   Ornament of sharp-ribbed longitudinal ridges, some persisting to the tip of the crown   *Polyptychodon*
4. Crown divided into two faces by a longitudinal carina   5
   Crown undivided   9
5. Crown divided symmetrically   6
   Crown divided asymmetrically into buccal and lingual faces   8

6. Crown with a single anterior carina   *Clidastes*
    Crown completely divided into two faces and with
    smooth or finely ridged enamel   7
7. Crown short and robust   *Prognathodon*
    Crown long and buccolingually compressed   *Leiodon*
    Buccal face faceted, lingual face ridged   Plioplatecarpinae
    Buccal face flat, lingual face U-shaped   *Tylosaurus*
9. Tooth slender, with a finely pointed tip and with
    fine striate ornament   Plesiosauroidea
    Tooth very long rooted, crown slightly to strongly
    buccolingually compressed   *Ornithocheirus*
    Tooth implantation sub-pleurodont   11
10. Crown buccolingually compressed, leaf-shaped
    with a deeply serrated margin and a cingulum   *Acanthopholis*
    Crown triangular in cross-section with a prominent
    median ridge on heavily enamelled face, grooved
    marginal denticles and fine grainy enamel
    patterning   Euhadrosauria
11. Crown bluntly conical to slightly buccolingually
    compressed   *Coniasaurus gracilodens*
    Crown buccolingually inflated, constricted at the
    base, and with wear facets and an asymmetric
    longitudinal groove   *Coniasaurus crassidens*

DESCRIPTIONS

TURTLES

Remarks. Fragmentary remains of large turtles have been found at many
localities, particularly in the Cenomanian Chalk. Complete but unassoci-
ated skulls and shells of a much smaller turtle are also known. Most
marine turtles have a reduced shell compared with that of tortoises and
freshwater turtles. The carapace is lightened by reduction of the costal
plates, leaving a series of fontanelles between them and the peripheral
plates through which are exposed the distal portions of the ribs. In
specialized Cretaceous forms the carapace is effectively reduced to a
series of struts consisting only of a median series of neural plates, vestigial
costals fused to the proximal ribs, and a ring of peripheral plates. The
plastron consists of two pairs of large, heavy plates, linked anteriorly and
posteriorly by smaller paired and unpaired plates.

Two families, Cheloniidae and Protostegidae, are recorded from the English Chalk. Both are widely distributed in the Upper Cretaceous of the eastern and southern USA (Zangerl 1953, 1960). In addition, there is a possible record of a toxochelyid humerus fragment (Hirayama, pers. comm. 1999); the Toxochelyidae is the only other family of marine turtles that occurs in the Cretaceous.

Subclass TESTUDINATA
Order CRYPTODIRIDA
Family CHELONIIDAE

Genus CHELONIA

Chelonia sp.
not figured

Description. The skull is distinguishable from those of protostegids by its reduced nasals and the presence of a secondary palate. A number of characters of the carapace and plastron distinguish cheloniids from proto-stegids. In the cheloniid carapace the neural plates are not usually keeled, the interneural and intercostal sutures are offset, and the medial edges of the peripheral plates are smooth. The two largest pairs of plastral plates, the hyoplastrals and hypoplastrals, are saddle- or Y-shaped.

Remarks. The only known skull was ascribed to the protostegid *Rhino-chelys* by Collins (1970) (Hirayama, pers. comm. 1999; see below).

Occurrence. Lower Cenomanian, *M. mantelli* or *M. dixoni* Zone, to Lower Campanian, *G. quadrata* Zone; Cenomanian of Kent, Surrey and Sussex; Turonian of Kent; Lower Campanian of Hampshire; Coniacian–Campanian of Sussex.

Family PROTOSTEGIDAE

Genus PROTOSTEGA

Protostega sp.
not figured

Description. Shell fragments of large protostegids can be distinguished from those of cheloniids on several differences in the carapace and plastron. In the protostegid carapace, alternate neural plates are keeled and the interneural sutures coincide with the intercostals. The costal plates are vestigial so that the carapace has large fontanelles. The median edges of the peripherals, bordering the fontanelles, are digitate or coarsely crenel-late. In the plastron, the hyo- and hypoplastrals are comparatively very

large, subrectangular plates with deeply digitate medial and lateral margins.
Occurrence. Middle Cenomanian, *H. subglobosus* Zone; Kent and Dorset.

Genus CIMOCHELYS

Cimochelys benstedi (Mantell)
Plate 66, figure 1

Description. Several small complete and fragmentary shells of juvenile turtles have been referred to this species. The medial borders of the peripherals are not digitate, the fontanelles are small and the plastron is little reduced. The hyo- and hypoplastrals are typically subrectangular but form a solid bony plate fused to the carapace.

Remarks. No skull material is associated, although Collins (1970) has suggested that *C. benstedi* could be the 'missing' postcranial material of *Rhinochelys* (see below).

Occurrence. Cenomanian, *H. subglobosus* Zone, and Turonian; Cenomanian of Kent, Sussex and Cambridgeshire; Turonian of Kent.

Genus RHINOCHELYS

Rhinochelys pulchriceps (Owen)
Plate 66, figure 2A–B

Description. Skull with a short, blunt snout or 'beak' and no secondary palate. The triturating surface of the upper jaw has a broad lingual ridge and a deep midline groove that corresponds to a sharp median crest on the symphysis of the lower jaw. *R. pulchriceps* is distinguishable from the other species of *Rhinochelys* by, among other characters, the lack of a hooked tip to its upper jaw and the relative height of its skull (Collins 1970).

Remarks. Based on small skulls from the Cenomanian Cambridge Greensand and one skull from the Lower Chalk. Several isolated mandibular symphyses have been referred to *Rhinochelys* but they are not diagnostic at species level.

Occurrence. Cenomanian, *M. mantelli–H. subglobosus* zones; Kent, Sussex and Dorset.

ICHTHYOSAURS

Remarks. The Cretaceous ichthyosaur species, five in number, were included in the single genus *Platypterygius* von Huene by McGowan

(1972), who confirmed that *P. campylodon* stands as the valid specific name for English ichthyosaurs of the Albian and Cenomanian. *Platypterygius* ranges from the Neocomian to the Upper Campanian and is distributed almost world-wide (McGowan 1978). The more complete specimens indicate that some individuals exceeded 10 m in length.

The English material is very incomplete and consists of rostral fragments, isolated teeth, a few isolated vertebrae and femora, and numerous phalanges. The most complete rostrum indicates a skull length of about 900 mm.

Order ICHTHYOSAURIA
Family STENOPTERYGIIDAE

Genus PLATYPTERYGIUS

Platypterygius campylodon (Carter)
Plate 64, figure 1; Plate 66, figure 3

Description. The sole diagnostic character is the presence of deep, external, longitudinal grooves running laterally along the upper and lower jaws. A continuous groove or a series of pits is characteristically present in ichthyosaurs but it is exceptionally deep and well-marked in this species. Tooth characters have been shown to be very unreliable in the systematics of ichthyosaurs, but the following features serve to distinguish the teeth from those of other Chalk reptiles. The teeth, with roots enveloped in a thick layer of cement, assume a more or less bulbous form. The crowns are short, up to 30 mm in height, conical and 'fluted' with regularly spaced smooth longitudinal ridges. The ridges converge distally but do not reach the tip of the crown, which is smooth.
Occurrence. Upper Albian–Cenomanian; Kent, Surrey, Sussex, Cambridgeshire, Norfolk.

PLESIOSAURS

Remarks. Plesiosaurs are divided into two superfamilies, Plesiosauroidea for the small-headed, long-necked forms, and Pliosauroidea for the large-headed, short-necked forms. Members of both groups are widely distributed in the Upper Cretaceous of North America, Europe and New Zealand, and are also recorded from South America, Africa and East Asia (Persson, 1963). Most of the English Chalk material, consisting of isolated teeth, jaw fragments and a few vertebrae and paddle bones, belongs to the Pliosauroidea.

Subclass SAUROPTERYGIA
Order PLESIOSAURIA
Superfamily PLESIOSAUROIDEA

Remarks. Indeterminate plesiosauroid remains include a single plesiosauroid tooth recorded from the English Chalk. The crown is long, slender and tapering with very fine striate ornament running to the tip, in contrast to the sharp ridge pattern of pliosauroids. Isolated centra are known from several localities in the Cenomanian and Turonian of southern England.

Family ELASMOSAURIDAE

Genus ELASMOSAURUS

Elasmosaurus constrictus (Owen)
not figured

Description. Based on cervical centra only. The vertebrae show features characteristic of elasmosaurs, but they may not be diagnostic at genus and species level. The centra are about twice as long as high and bear sharp lateral ridges running the length of the centrum just below the neural arch attachment. In contrast, the cervical centra of pliosauroids are generally less than half the length of those of similar-sized elasmosaurs, and their length and height are approximately equal.
Occurrence. Lower or Middle Turonian; Sussex.

Superfamily PLIOSAUROIDEA
Family PLIOSAURIDAE

Genus POLYPTYCHODON

Polyptychodon interruptus Owen
Plate 64, figure 2

Description. The species is based on teeth only. The teeth have long, slightly lingually curved conical crowns, 15–60 mm in height, which are ornamented by longitudinal, closely spaced, sharp and frequently ribbed ridges. Most ridges stop on the distal third of the crowns but a few persist to the tips. There is some ontogenetic variation in the ornament pattern, the ridges being more prominent and widely spaced in older individuals. The ornament pattern also varies with respect to the position of teeth within the tooth row.
Remarks. Indeterminate pliosauroid vertebrae, paddle bones and jaw fragments have also been recorded.

Occurrence. Lower Cenomanian–Lower Santonian. *Polyptychodon* teeth have been recorded from all the following localities. In addition, records of other pliosauroid material are indicated by superscripts[1-3] keyed below. Lower Cenomanian, Kent and Cambridgeshire; Middle–Upper Cenomanian, Kent[1-3], Sussex[3], Cambridgeshire, Norfolk; Upper Cenomanian, *S. gracile* Zone, Kent; Lower–Middle Turonian, Kent and Sussex; Santonian, Kent (*M. coranguinum* Zone) and Sussex[2] (indeterminate pliosauroids: [1] jaw fragments; [2] vertebrae; [3] paddle bones).

MARINE LIZARDS

Remarks. Several marine lizards occur in the English Chalk, including fragmentary remains of mosasaurs. Most mosasaurs are very large forms completely adapted to a marine existence; they swam by lateral undulation of the body and tail, with limbs modified as steering paddles. They fed on other vertebrates, echinoderms and molluscs. Mosasaurs diversified and became very abundant during the Late Cretaceous. They inhabited sub-tropical epicontinental seas of less than about 180 m depth and of variable salinity. The geographical range of many species was wide and apparently little affected by changes in depositional environments (Russell 1967).

Mosasaurs are widely agreed to have evolved from early varanoid-like lizards (Russell 1967), but there is no consensus over their classification. In many classifications, e.g. Carroll (1988), the family is included in the superfamily Varanoidea, whereas Russell (1967) and Caldwell and Cooper (1999) accorded it separate superfamilial status.

In additional to Chalk mosasaurs, three taxa of smaller marine lizards, with sister-group relationships to mosasauroids (Caldwell 1999), are known from partial skeletons. They share features in common with mosasaurs that are associated with an aquatic mode of life and inhabited coastal nearshore environments of the Tethys and the Western Interior Seaway in North America.

Order SQUAMATA
Family CONIASAURIDAE

Remarks. Coniasaurs are small marine lizards with relatively elongate heads and a total body length probably not exceeding 0·5 m (Caldwell and Cooper 1999). Relatively little is known of the postcranial skeleton, but the body and limb proportions were probably similar to that of dolicho-saurs (q.v.), indicating a similar mode of life. Coniasaurs are known from the Cenomanian in England and France and Cenomanian–Turonian in Texas and the mid-western United States (Caldwell and Cooper 1999).

PLATE 64

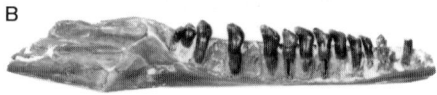

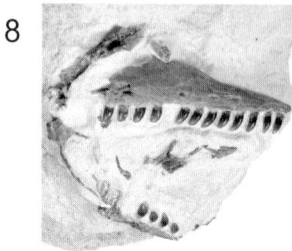

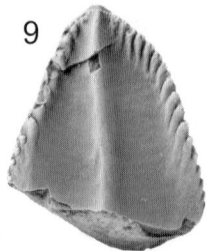

Genus CONIASAURUS

Coniasaurus crassidens Owen
not figured
Coniasaurus cf. *C. crassidens* Owen
Plate 64, figure 7A–B

Description. *Coniasaurus crassidens* was originally described by Owen on the basis of a maxilla and associated vertebrae, together with isolated mandibles and vertebrae from several localities in England. The teeth of *C. crassidens* are characteristic with a prominent longitudinal groove on the anterolateral surface and distinct wear facets that are very unusual among lizards. The teeth are subpleurodont and increase in length and thickness posteriorly to a bulbous shape, with slightly constricted bases and recurved crowns. The crowns, which are 1·5–4 mm in height, are expanded buccolingually. The type specimen contains maxillary teeth only and no lower dentition. The lack of strictly comparable teeth led Caldwell and Cooper (1999) to refer mandible fragments, all from larger individuals, to *Coniasaurus* cf. *C. crassidens* Owen. The vertebrae are procoelous with large flat zygapophyses and a zygantrum-zygosphene. Hypapophyses are present on the cervical vertebrae and, in the trunk, the ventral faces of the centra are somewhat flattened.

Occurrence. Cenomanian–Turonian; Cenomanian of Sussex; Turonian of Sussex (*C.* cf. *C. crassidens* only).

EXPLANATION OF PLATE 64

Fig. 1. *Platypterygius campylodon* (Carter). Cenomanian, Folkestone, Kent; tooth; ×1.

Fig. 2. *Polyptychodon interruptus* Owen. Santonian, *M. coranguinum* Zone, Gravesend, Kent; tooth crown in buccal view; ×1.

Fig. 3. *Leiodon anceps* Owen. Campanian, *B. mucronata* Zone, Norwich, Norfolk; tooth in A, anterior, and B, lingual views; both ×1.

Fig. 4. *Clidastes* sp. ?Upper Turonian, Dorking, Surrey; anterior fragment of right dentary; ×0·5.

Fig. 5. Plioplatecarpinae *incertae sedis*. Upper Chalk, Sussex; tooth in A, anterior, and B, posterolingual views; both ×1.

Fig. 6. *Ornithocheirus* sp. Upper Chalk, Kent; tooth; ×1.

Fig. 7. *Coniasaurus* cf. *C. crassidens* Owen. Middle Cenomanian, *H. subglobosus* Zone, Peters Pit, Wouldham, Kent; left mandible in A, lateral, and B, medial views; both ×1.

Fig. 8. *Coniasaurus gracilodens* Caldwell. Cenomanian, *H. subglobosus* Zone, Lydden Spout, Folkestone, Kent; skull showing right (upper) and left (lower) maxillae; ×1·25.

Fig. 9. Euhadrosauria *incertae sedis*. Cenomanian, *A. jukesbrownei* Zone, Hitchin, Hertfordshire; dentary tooth crown in lingual view; ×2.

Coniasaurus gracilodens Caldwell
Plate 64, figure 8

Description. Distinguished from *C. crassidens* by a shorter, more robust maxilla and difference in tooth morphology. The tooth crowns are not more than 2 mm long, recurved and slightly buccolingually compressed anteriorly. The more posterior crowns are thickened but retain a blunt-tipped conical shape and lack a longitudinal groove and wear facets.

Occurrence. Middle Cenomanian, *H. subglobosus* Zone; Kent; known from a single specimen consisting of a partial skull, shoulder girdle and four dorsal vertebrae.

Family DOLICHOSAURIDAE

Remarks. Dolichosaurs are relatively small, slender marine lizards up to about 1 m in length. They are known from the Neocomian of Yugoslavia and the Cenomanian–Turonian of England and Texas. They were adapted to anguilliform swimming by virtue of elongated trunks with at least 40 presacral vertebrae, and long flattened tails consisting of at least 70 caudals. The limbs and limb girdles are small and somewhat reduced. The generally small skulls of dolichosaurs, together with the increase in the number of their presacral vertebrae (particularly in the neck) and reduced forelimbs, exclude them from immediate ancestry of the mosasaurs.

Genus DOLICHOSAURUS

Dolichosaurus longicollis Owen
Plate 65, figures 3–4

Description. Known from several articulated postcranial skeletons, one with a few diagnostic skull elements, and isolated vertebrae. *D. longicollis* is an exceptionally long-necked form with at least 19 cervical vertebrae (Caldwell 2001). The vertebrae are similar to those of *Coniasaurus* but may be distinguished by the presence of a striate ornament pattern on the centra.

Occurrence. Cenomanian–Lower Turonian, *I. labiatus* Zone; Cenomanian of Kent, Surrey, Sussex and Cambridgeshire; Lower Turonian of Surrey.

Superfamily MOSASAUROIDEA
Family MOSASAURIDAE

Remarks. Only vertebrae, teeth and occasional jaw fragments are known from the English Chalk. Mosasaur vertebrae are very characteristic in appearance; they are procoelous and the articular faces are very smoothly finished. The shape of the articular faces varies with the position in the vertebral column, being circular or oval in cervical and dorsal vertebrae, and subtriangular tending to subhexagonal in the caudals. Identification of isolated and fragmentary vertebrae to generic level is not usually possible.

Mosasaur teeth are thecodont, unlike those of all other lizards, and become ankylosed to the jaws when mature. The tooth crowns are often generically diagnostic; they are divided into lingual and buccal surfaces by a longitudinal carina, and the surfaces may be smooth or divided into facets by vertical ridges. The tips of the teeth are usually more posterolingually inclined than the main axis of the crowns. The crowns are set in barrel-shaped bony bases. The grain of the bone in the tooth base is finer than that of the surrounding jaw and runs parallel to the longitudinal axis of the tooth base instead of the jaw.

Many specimens in old collections are labelled as '*Mosasaurus*', a term used indiscriminately to cover both determinate and indeterminate material. In fact, the genus *Mosasaurus* has not been recognized from the English Chalk, although it occurs in the Santonian–Maastrichtian of Europe (Russell, 1967).

Subfamily MOSASAURINAE

Genus CLIDASTES

Clidastes sp.
Plate 64, figure 4; Plate 65, figure 1

Description. Very few specimens have been identified. Among them is an anterior fragment of a dentary in which two tooth bases and two partial crowns are preserved. The teeth are broad-based and triangular in lateral aspect, and the crowns are slightly recurved. The crowns have a single anterior carina and are triangular in cross-section. A small projection of the dentary, anterior to the first tooth socket, is characteristic of *Clidastes*. Two articulated anterior caudal vertebrae are also identified as belonging to *Clidastes*. The caudal vertebrae have fused haemal spines that are diagnostic of mosasaurine mosasaurs. The haemal arch surface continues straight on to the lateral surface of the haemal spine with little or no curvature, and the haemal spine slopes ventromedially at an angle of about 40 degrees to the horizontal vertebral axis, a characteristic feature of *Clidastes*.

Occurrence. Santonian–Campanian; Surrey and Sussex; indeterminate mosasaurine caudal vertebrae have also been recorded from Norfolk.

Subfamily PLIOPLATECARPINAE

Genus PROGNATHODON

Prognathodon sp.
not figured

Description. Known only from teeth. The crowns are 30–40 mm in height, very robust, symmetrically bicarinate with smooth or finely ridged enamel. The tips of the crowns are blunt, barely recurved and the crown bases are subcircular to circular in cross-section.

Remarks. *Prognathodon* is known from good skull and postcranial material from Belgium, but the English Chalk has yielded only isolated tooth crowns from the Norwich area.

Occurrence. Upper Campanian, *B. mucronata* Zone; Norfolk.

Plioplatecarpinae *incertae sedis*
Plate 64, figure 5

Description. A few tooth crowns and one dentary fragment, although characteristic of most plioplatecarpines, are not identifiable to generic level. The crowns are slender, about 20 mm in height, and the bases are subcircular in cross-section. The crowns are asymmetrically bicarinate; the smaller buccal surface is faceted, the larger lingual surface is ornamented with closely spaced ridges.

Remarks. The sole Sussex specimen was originally identified as '*Plate-carpus*(?) sp.', but according to Russell (1967), the isolated tooth crowns are indistinguishable from those of *Plioplatecarpus*, which also occurs in the Campanian of Europe.

Occurrence. Santonian–Campanian; Sussex, Hampshire.

Subfamily TYLOSAURINAE

Genus TYLOSAURUS

cf. *Tylosaurus*
Plate 65, figure 5

Description. A few records of vertebrae and teeth are known from Chalk. The tooth crowns are robust and very asymmetrically bicarinate, being divided into a U-shaped lingual surface and an almost flat buccal surface that is restricted to the antero-lateral edge in anterior teeth. Fine vertical ridging is present on both surfaces and the buccal surface is lightly faceted.

The articular faces of both cervical and dorsal vertebrae are circular. All mosasaurs have a single fused peduncle on the ventral surface of the anterior cervical vertebrae that articulates with a hypapophysis; in tylosaurines the peduncle is diagnostically 'tear-drop'-shaped. The caudal vertebrae bear paired posteriorly-inclined peduncles which articulate with the chevrons, in contrast to the fused chevrons seen in mosasaurines. Caudal peduncles are also found in plioplatecarpine mosasaurs, but the vertebrae, at least in the anterior half of the tail, can generally be distinguished by the height of the articular surface; it is relatively high (i.e. height and width about the same) in tylosaurines and much lower in plioplatecarpines.

Occurrence. Upper Santonian, *M. testudinarius* Zone, to Lower Campanian, *G. quadrata* Zone; Hampshire (*G. quadrata* Zone), Kent and North Yorkshire (*M. testudinarius* and *O. pilula* zones).

Subfamily *incertae sedis*

Genus LEIODON

Leiodon anceps Owen
Plate 64, figure 3A–B

Description. This genus is based on a fragment of jaw with two teeth. A large number of isolated teeth with crown heights of between 20 and 30 mm have been referred to *L. anceps*. The crowns are slightly to strongly buccolingually compressed, symmetrically bicarinate and the enamel surfaces are smooth and shining.

Remarks. The status and relationships of *L. anceps* are uncertain. On the basis of the dentition it has been suggested that *Leiodon* may be synonymous with the tylosaurine *Hainosaurus* Dollo from Belgium, but Russell (1967) retained the genus in the Mosasaurinae.

Occurrence. Coniacian–Campanian, including Upper Campanian, *B. mucronata* Zone; Sussex, Essex, and Norfolk (*B. mucronata* Zone). In addition, indeterminate mosasaur teeth, jaw fragments and vertebrae have been found in a wide range of Chalk horizons. Unfortunately, the material is too poorly preserved to be identified beyond family level. Their distribution is as follows: Middle Cenomanian, *H. subglobosus* Zone, Kent[1]; Middle Turonian, *T. lata* Zone, Kent (jaw fragments) and Surrey[1]; Lower Campanian, *G. quadrata* Zone, Hampshire[1]; Lower Campanian, *O. pilula* Zone, Sussex[1, 2] and Wiltshire[1]; Upper Campanian, *B. mucronata* Zone, Kent[1] and Norfolk[1, 2] ([1] teeth; [2] vertebrae).

PLATE 65

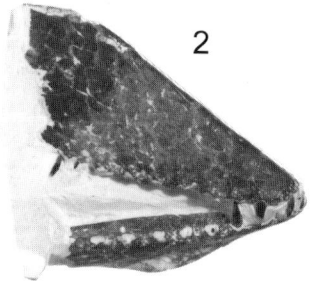

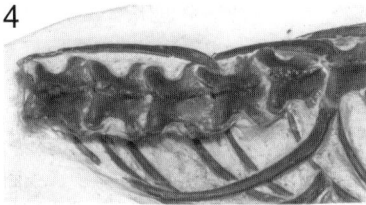

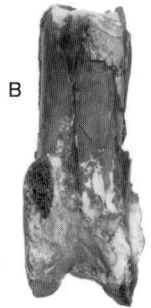

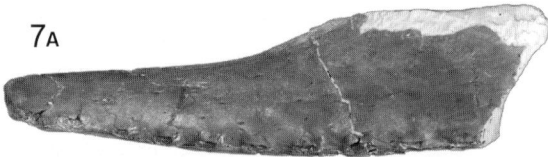

PTEROSAURS

Remarks. The English Chalk has yielded only very fragmentary pterosaur fossils, particularly jaws, together with other skull and postcranial remains. Pterosaur bones are distinctive and easily recognized by their delicate, extremely thin-walled structure, and pterosaur teeth are also easily distinguished from those of other Chalk reptiles. Isolated wing elements are the most commonly found remains in the Chalk but they cannot be identified at species level (Wellnhofer, 1978).

Order PTEROSAURIA
Suborder PTERODACTYLOIDEA
Family ORNITHOCHEIRIDAE

Genus ORNITHOCHEIRUS

Remarks. Members of the genus *Ornithocheirus* are medium-sized pterosaurs with a skull length of 0·2–0·5 m. They are generally long-snouted forms with a distinct median keel on the anterior palate. Three species, based on different snout proportions have been recorded from the Chalk.
Occurrence. Berriasian–Campanian.

Ornithocheirus giganteus (Bowerbank)
Plate 65, figure 2

Description. Based on the anterior part of a snout and attached lower jaws together with an associated scapulocoracoid and possible sternal fragment. The snout is short, deep and steeply angled in contrast to the other Chalk species. Lydekker (1888) estimated the skull length at 253 mm. The teeth

---

EXPLANATION OF PLATE 65

Fig. 1. *Clidastes* sp. Upper Chalk, Sussex; right lateral view of anterior caudal vertebrae; ×0·5.

Fig. 2. *Ornithocheirus giganteus* (Bowerbank). Cenomanian, Burham, Kent; anterior fragment of snout and lower jaws in right lateral view; ×0·67.

Figs 3–4. *Dolichosaurus longicollis* Owen. Lower Chalk, Cenomanian, Burham, Kent; articulated posterior trunk vertebrae in 3, ventral, and 4, dorsal views (anterior to the right); ×1.

Fig. 5. cf. *Tylosaurus*. Santonian, *M. testudinarius* Zone, Forness Point, east of Margate, Kent; anterior caudal vertebra in posterior view; ×0·5.

Fig. 6. Ornithocheiridae *incertae sedis*. Turonian, Burham, Kent; distal end of left metacarpal IV (wing metacarpal) in A, posterior, and B, dorsal views; ×1.

Fig. 7. *Ornithocheirus cuvieri* (Bowerbank). Cenomanian, Burham, Kent; anterior end of snout in A, left lateral, and B, ventral views; ×0·5.

are short, sharply pointed and slightly buccolingually compressed. The scapulocoracoid is typically pterodactyloid, the glenoid fossa lying equally between the two elements. A prominent acromion process on the scapula may be diagnostic of *Ornithocheirus*.

Occurrence. Cenomanian; Kent.

Ornithocheirus cuvieri (Bowerbank)
Plate 65, figure 7A–B

Description. Based on a long, slender, blunt-ended snout with a steeply rising bony crest set back from the tip. The lateral walls of the alveoli are swollen laterally and bulge prominently on the jaw margins. The alveoli extend right to the tip of the snout and are oriented slightly antero-ventrally, indicating that the teeth pointed slightly forwards. A prominent median keel is present on the palate.

Occurrence. Cenomanian; Kent.

Ornithocheirus compressirostris (Owen)
not figured

Description. Long, shallow, laterally compressed snout lacking a nasal crest, about half the depth of the anterior portion of *O. cuvieri*, and with small elliptical alveoli confluent with the lateral margin of the jaw (known from fragments only).

Occurrence. Turonian; Kent.

Indeterminate ornithocheirids
Plate 64, figure 6; Plate 65, figure 6A–B

Description. Isolated teeth, some associated with indeterminate skull fragments, are known from several localities. They are long-rooted, with conical and buccolingually compressed or very compressed crowns, oval to elliptical in cross-section. The crowns may show very fine striate orna-ment. Most of the material from the Chalk consists of fragmentary limb bones, particularly from the wing, and, although characteristically ptero-

---

EXPLANATION OF PLATE 66

Fig. 1. *Cimochelys benstedi* (Mantell). Lower or Middle Turonian, Kent; carapace in dorsal view; ×0·5.

Fig. 2. *Rhinochelys pulchriceps* (Owen). Middle or Upper Cenomanian, Folkestone, Kent. A, palate; B, skull roof; both ×1.

Fig. 3. *Platypterygius campylodon* (Carter). Cenomanian, Kent; section from mid-region of rostrum (anterior to the left); ×0·5.

PLATE 66

1

2A

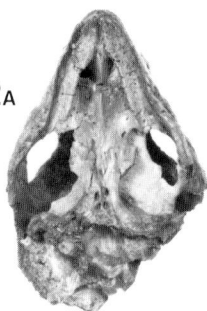

B

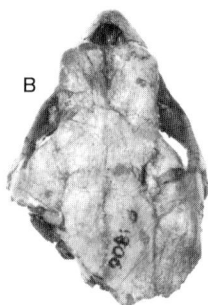

3

dactyloid, is not identifiable beyond family level. Fragmentary vertebrae
are known from the Santonian of Kent.

Occurrence. Lower Cenomanian–Santonian; Cenomanian of Kent, Surrey,
Sussex and Cambridgeshire; Turonian and Santonian of Kent.

DINOSAURS

Remarks. There are only two dinosaur records from the Chalk. Both are
herbivores, readily identifiable at family level but indeterminate generically. Remains of wholly terrestrial animals in any marine sediments are
rare and represent accidental wash-ins from nearby land areas.

DINOSAURIA
Order ORNITHISCHIA
ORNITHOPODA
Family HADROSAURIDAE

Remarks. Hadrosaurs (duck-billed dinosaurs) were the commonest and
most diverse herd-dwelling herbivores of the Late Cretaceous (Cenomanian–
Maastrichtian) in Laurasia (Weishampel and Horner 1990). The Lower
Cenomanian English occurrences are among the earliest records of the
group.

EUHADROSAURIA *incertae sedis*
Plate 64, figure 9

Description. A single unworn tooth, characteristic of crown group
hadrosaurs, is recorded from the Chalk. It is triangular in cross-section,
thickly enamelled on one face which bears a prominent median ridge (see
Key above). Originally described as *Iguanodon hilli* by Newton (1892)
and listed as a hadrosaurid *nomen dubium* by Weishampel and Horner
(1990), this tooth is from a dentary (D. B. Norman, pers. comm. 1999) and
thus the enamelled face is on the lingual side of the tooth.

Occurrence. Cenomanian, *A. jukesbrownei* Zone; Hertfordshire.

THYREOPHORA
ANKYLOSAURIA
Family NODOSAURIDAE

NODOSAURIDAE *incertae sedis*
not figured

Description. An indeterminate nodosaurid ankylosaur, *Acanthopholis
horridus* Huxley, noted as a *nomen dubium* by Pereda Suberbiola and

Barrett (1999) is represented by skull fragments, teeth, vertebrae and dermal armour plates and spikes. The teeth are particularly characteristic as described in the Key above. The keeled armour plates are flat or slightly convex medially, as figured by Pereda Suberbiola and Barrett (1999, text-fig. 2).

Occurrence. Lower Cenomanian; Kent.

Acknowledgements. I thank Mr J. Cooper (Booth Museum of Natural History, Brighton) and Dr D. Price (Sedgwick Museum, Cambridge) for access to collections in their care, and for assistance during my visits to their respective institutions: also Dr S. Molyneux for the loan of material from the British Geological Survey. I am grateful to Mr C. A. Walker, formerly of The Natural History Museum, London, and Dr D. B. Norman (Sedgwick Museum, Cambridge) for advice on the turtles and hadrosaurs respectively. Professor A. S. Gale (University of Greenwich) provided stratigraphical information on pterosaurs. I am indebted to Mr P. R. Crabb and Mr P. Hurst of The Natural History Museum Photographic Unit for photographing the specimens shown in the plates, and to Ms Pat Hart for producing the plates electronically.

REFERENCES

ABBASS, H. L. 1962. The English Cretaceous Turritellidae and Mathildidae (Gastropoda). *Bulletin of the British Museum (Natural History), Geology Series*, **7**, 173–196.

—— 1973. Some British Cretaceous gastropods belonging to the families Procerithiidae, Cerithiidae and Cerithiopsidae (Cerithiacea). *Bulletin of the British Museum (Natural History), Geology Series*, **23**, 103–175.

AGASSIZ, L. 1833–45. *Recherches sur les poissons fossiles.* Five volumes, Neuchâtel.

ARCHIAC, A. D' 1847. Rapport sur les fossiles du Tourtia. *Mémoires de la Société Géologique de France, Series 2*, **2**, 291–351.

ARKELL, W. J., KUMMEL, B. and WRIGHT, C. W. 1957. Mesozoic Ammonoidea. *In* MOORE, R. C. (ed.). *Treatise on invertebrate paleontology. Part L, Mollusca 4, Cephalopoda.* Geological Society of America, Boulder, and University of Kansas Press, Lawrence, L80–L465.

ARNOLD, W. H. 1965. A glossary of a thousand-&-one terms used in conchology. *Veliger*, **7** (Supplement), 50 pp.

BAILEY, H. W., GALE, A. S., MORTIMORE, R. N., SWIECICKI, A. and WOOD, C. J. 1983. The Coniacian–Maastrichtian stages of the United Kingdom with particular reference to southern England. *Newsletters on Stratigraphy*, **12**, 29–42.

—— —— —— —— —— 1984. Biostratigraphical criteria for the recognition of the Coniacian to Maastrichtian stage boundaries in the Chalk of north-west Europe, with particular reference to southern England. *Bulletin of the Geological Society of Denmark*, **33**, 31–39.

BENTON, M. J. and SPENCER, P. S. 1995. *Fossil reptiles of Great Britain.* Geological Conservation Review Series, Chapman & Hall, London, xii + 386 pp.

BRETON, G. 1992. Les Goniasteridae (Asteroidea, Echinodermata) jurassiques et crétacés de France: taphonomie, systématique, biostratigraphie, paléogeographie, évolution. *Bulletin Trimestriel de la Société Géologique de Normandie et Amis du Muséum du Havre*, **1992**, 590 pp.

BROMLEY, R. G. 1967. Some observations on burrows of thalassinidean Crustacea in chalk hardgrounds. *Quarterly Journal of the Geological Society of London*, **123**, 157–182.

—— 1978. Hardground diagenesis. *In* FAIRBRIDGE, R. W. and BOURGEOIS, J. (eds). *The encyclopedia of sedimentology.* Dowden, Hutchinson & Ross, Stroudsburg, PA, 397–400.

—— and GALE, A. S. 1982. The lithostratigraphy of the English Chalk Rock. *Cretaceous Research*, **3**, 273–306.

BROOKS, H. K., GLAESSNER, M. F., HAHN, G., HESSLER, R. R., HOLTHUIS, L. B., MANNING, R. B., MOORE, R. C. and ROLFE, W. D. I. 1969. Malacostraca. *In* MOORE, R. C. (ed.). *Treatise on invertebrate paleontology. Part R, Arthropoda 4.* Geological Society of America, Boulder, and University of Kansas Press, Lawrence, K295–K566.

BRYDONE, R. M. 1906. Further notes on the stratigraphy and fauna of the Trimmingham Chalk. *Geological Magazine* (5), **3**, 289–300.

—— 1909. Notes on new or imperfectly known Chalk Bryozoa (Polyzoa). *Geological Magazine* (5), **6**, 337–339, 398–400.

—— 1910. Notes on new or imperfectly known Chalk Polyzoa. *Geological Magazine* (5), **7**, 4–5, 76–77, 145–147, 258–260, 390–392, 481–483.

—— 1911. Notes on new or imperfectly known Chalk Polyzoa. *Geological Magazine* (5), **8**, 153–156.

—— 1912. Notes on new or imperfectly known Chalk Polyzoa. *Geological Magazine* (5), **9**, 7–8, 145–147, 294–296, 433–435.

—— 1913. Notes on new or imperfectly known Chalk Polyzoa. *Geological Magazine* (5), **10**, 97–99, 196–199, 248–250, 436–438.

—— 1914. Notes on new or imperfectly known Chalk Polyzoa. *Geological Magazine* (6), **1**, 97–99, 345–347, 481–483.

—— 1916. Notes on new or imperfectly known Chalk Polyzoa. *Geological Magazine* (6), **3**, 97–100, 241–243, 337–339, 433–435.

—— 1917. Notes on new or imperfectly known Chalk Polyzoa. *Geological Magazine* (6), **4**, 49–53, 145–148, 492–496.

—— 1918. Notes on new or imperfectly known Chalk Polyzoa. *Geological Magazine* (6), **5**, 1–4, 97–100.

—— 1929. *Further notes on new or imperfectly known Chalk Polyzoa. Part I.* Dulau, London, 1–38.

—— 1930. *Further notes on new or imperfectly known Chalk Polyzoa. Part II* (*Vincularia, Onychocella, Rhagasostoma, Porina, etc.*). Dulau, London, 39–60.

—— 1936. *Further notes on new or imperfectly known Chalk Polyzoa. Part III* (*Semieschara, Micropora, Cryptostoma, etc.*). Dulau, London, 61–88.

CALDWELL, M. W. 1999. Description and phylogenetic relationships of a new species of *Coniasaurus* Owen, 1850 (Squamata). *Journal of Vertebrate Paleontology*, **19**, 438–455.

—— 2001. On the aquatic squamate *Dolichosaurus longicollis* Owen, 1850 (Cenomanian, Upper Cretaceous), and the evolution of elongate necks in squamates. *Journal of Vertebrate Paleontology*, **20**, 720–735.

—— and COOPER, J. A. 1999. Redescription, palaeobiogeography and palaeoecology of *Coniasaurus crassidens* Owen, 1850 (Squamata) from the Lower Chalk (Cretaceous: Cenomanian) of SE England. *Zoological Journal of the Linnean Society*, **127**, 423–452.

CANU, F. 1900. Revision des Bryozoaires du Crétacé figurés par d'Orbigny. Deuxième Partie-Cheilostomata. *Bulletin de la Société Géologique de France* (3), **28**, 334–463.

CARROLL, R. L. 1988. *Vertebrate paleontology and evolution.* W. H. Freeman and Company, New York, xiv + 698 pp.

CARTER, D. J. and HART, M. B. 1977. Aspects of Mid-Cretaceous stratigraphical micropalaeontology. *Bulletin of the British Museum* (*Natural History*), *Geology Series*, **29**, 1–135.

CARTER, R. M. 1968. Functional studies on the Cretaceous oyster *Arctostrea*. *Palaeontology*, **20**, 454–487.

—— 1972. Adaptations of British Chalk Bivalvia. *Journal of Paleontology*, **46**, 325–340.

CHRISTENSEN, W. K. 1975. Upper Cretaceous belemnites from the Kristianstad area in Scania. *Fossils and Strata*, **7**, 1–69.

—— 1976. Palaeobiology of Late Cretaceous belemnites of Europe. *Paläontologische Zeitschrift*, **50**, 113–129.

—— 1991. Belemnites from the Coniacian to Lower Campanian chalks of Norfolk and southern England. *Palaeontology*, **34**, 695–749.

—— 1993. *Belemnocamax boweri* Crick, an unusual belemnite from the Cenomanian of northwest Germany and eastern England. *Bulletin of the Geological Society of Denmark*, **40**, 157–166.

—— 1995. *Belemnitella* from the Upper Campanian and Lower Maastrichtian chalk of Norfolk, England. *Special Papers in Palaeontology*, **51**, 1–84.

—— 1996. A review of the Upper Campanian and Lower Maastrichtian belemnite biostratigraphy of Europe. *Cretaceous Research*, **17**, 751–766.

—— 1997*a*. The Late Cretaceous belemnite family Belemnitellidae: taxonomy and evolutionary history. *Bulletin of the Geological Society of Denmark*, **44**, 59–88.

—— 1997*b*. Palaeobiogeography and migration in the Late Cretaceous belemnite family Belemnitellidae. *Acta Palaeontologica Polonica*, **42**, 457–495.

CLEEVELY, R. J. 1980. Two new British Cretaceous Epitoniidae (Gastropoda): evidence for evolution of shell morphology. *Bulletin of the British Museum (Natural History), Geology Series*, **34**, 235–249.

COLLINS, J. I. 1970. The chelonian *Rhinochelys* Seeley from the Upper Cretaceous of England and France. *Palaeontology*, **13**, 355–378.

COOPER, M. R. 1992. Pycnodonteine oysters from the Upper Cretaceous of Zululand. *Durban Museum Novitates*, **17**, 23–57.

COX, L. R. 1960. British Cretaceous Pleurotomariidae. *Bulletin of the British Museum (Natural History), Geology Series*, **4**, 385–423.

—— 1962. *British Mesozoic fossils*. British Museum (Natural History), London, 205 pp.

CRICK, G. C. 1906. Note on a rare form of *Actinocamax* (*A. grossouvrei*) from the Chalk of Yorkshire. *Naturalist*, **1906**, 155–158.

—— 1907. Note on two rare forms of *Actinocamax* from the English Upper Chalk. *Geological Magazine* (5), **4**, 389–395.

—— 1910. On *Belemnocamax boweri* n. g., n. sp., a new cephalopod from the Lower Chalk of Lincolnshire. *Proceedings of the Geologists' Association*, **21**, 360–365.

DAVID, B. and FOURAY, M. 1984. Variabilité et disjonction évolutive des caractéres dans les populations de *Micraster* (Echinoidea, Spatangoida) du Crétacé Supérieur de Picardie. *Geobios*, **17**, 447–476.

DAVIDSON, T. 1852–54. A monograph of British Cretaceous Brachiopoda, 2. *Palaeontographical Society* [*Monographs*], London, 117 pp.

—— 1874. A monograph of the fossil Brachiopoda, 4 (1), Supplement to the Recent, Tertiary and Cretaceous species. *Palaeontographical Society* [*Monographs*], London, 72 pp.

DHONDT, A. V. 1971. Systematic revision of *Entolium, Propeamussium* (Amusiidae) and *Syncyclonema* (Pectinidae, Bivalvia, Mollusca) of the European Boreal Cretaceous. *Bulletin de l'Institut Royale des Sciences Naturelles de Belgique*, **47** (32), 1–95.

—— 1972*a*. Systematic revision of the Chlamydinae (Pectinidae, Bivalvia, Mollusca) of the European Cretaceous. Part 1: Camptonectes. *Bulletin de l'Institut Royale des Sciences Naturelles de Belgique*, **48** (3), 1–60.

—— 1972*b*. Systematic revision of the Chlamydinae (Pectinidae, Bivalvia,

Mollusca) of the European Cretaceous. Part 2. *Lyropecten. Bulletin de l'Institut Royale des Sciences Naturelles de Belgique*, **48** (7), 1–81.

—— 1973*a*. Systematic revision of the Chlamydinae (Pectinidae, Bivalvia, Mollusca) of the European Cretaceous. Part 3: *Chlamys* and *Mimachlamys. Bulletin de l'Institut Royale des Sciences Naturelles de Belgique*, **49**, 1–134.

—— 1973*b*. Systematic revision of the subfamily Neitheinae (Pectinidae, Bivalvia, Mollusca) of the European Cretaceous. *Mémoires de l'Institut Royale des Sciences Naturelles de Belgique*, **176**, 1–101.

—— 1976. Systematic revision of the Chlamydinae (Pectinidae, Bivalvia, Mollusca) of the European Cretaceous. Part 4: *Merklinia. Bulletin de l'Institut Royale des Sciences Naturelles de Belgique*, **51** (7), 1–38.

—— 1983. Campanian and Maastrichtian inoceramids: a review. *Zitteliana*, **10**, 689–701.

—— 1984. The unusual Cenomanian oyster *Pycnodonte biauriculatum. Geobios, Mémoire Spécial*, **8**, 53–61.

—— 1989. Late Cretaceous *Limea* (*Pseudolimea*) species of Europe. *Bulletin de l'Institut Royale des Sciences Naturelles de Belgique, Sciences de la Terre*, **59**, 105–125.

—— 1993. Upper Cretaceous bivalves from Tercis, Landes, SW France. *Bulletin de l'Institut Royale des Sciences Naturelles de Belgique, Sciences de la Terre*, **63**, 211–259.

—— and DIENI, I. 1990. Unusual inoceramid-spondylid association from the Cretaceous Scaglia Rossa of Passo del Brocon (Trento, N. Italy) and its palaeoecological significance. *Memoire di Scienze Geologiche*, **42**, 155–187.

—— and JAGT, J. W. M. 1987. Bivalvia uit de Kalksteen van Vijlen in Hallembaye (Belgie). *Grundboor en Hamer*, **41**, 78–90.

DIBLEY, G. E. 1911. The teeth of *Ptychodus* and their distribution in the English Chalk. *Quarterly Journal of the Geological Society of London*, **67**, 263–277.

DIXON, F. 1850. *The geology and fossils of the Tertiary and Cretaceous formations of Sussex.* Published privately, London, xxi + 422 pp.

DOBROV, S. A. and PAVLOVA, M. M. 1959. Inoceramidae. *In* MOSKOVIN, M. M. *Atlas of the Upper Cretaceous fauna of Northern Causasus and Crimea.* Trudy Vsesoyuznyi Nauchno-Issledovatel'skii Institut, Prirodnykh Gazov, 130–165.

DOYLE, P. 1992. A review of the biogeography of Cretaceous belemnites. *Palaeogeography, Palaeoclimatology, Palaeoecology*, **92**, 207–216.

DUNCAN, P. M. 1869. A monograph of the British fossil corals. Part II, no. 1. Corals from the White Chalk, the Upper Greensand, and the Red Chalk of Hunstanton. *Palaeontographical Society* [*Monographs*], London, 46 pp., 15 pls.

EASTON, W. H. 1960. *Invertebrate paleontology.* Harper, New York, xii + 701 pp.

EDWARDS, H. M. and HAIME, J. 1850–54. A monograph of the British fossil corals. First part. Introduction. Corals from the Tertiary and Cretaceous formations. *Palaeontographical Society* [*Monographs*], London, i–lxxv, 1–71, pls 1–11.

ERNST, G. 1968. Die Oberkreide-Aufschlusse im Raume Braunschweig, Hannover und ihre stratigraphische Gleiderung mit Echinodermen und Belemniten. 1, Die jungere Oberkreide (Santon–Maastricht). *Beihefte zu den Berichten den Naturhistorische Gesellschaft zu Hannover*, **5**, 235–284.

—— 1970*a*. The stratigraphical value of the echinoids in the boreal Upper Cretaceous. *Newsletters on Stratigraphy*, **1**, 19–34.

—— 1970*b*. Zur Stammesgeschichte und stratigraphischen Bedeutung der Echiniden-Gattung *Micraster* in der nordwestdeutschen Oberkreide. *Mitteilungen der Geologisch-Paläontologischen Institut der Universität Hamburg*, **39**, 117–135.

—— 1971. Biometrische Untersuchungen über die Ontogenie und Phylogenie der *Offaster/Galeola* Stammesreihe (Echinoidea) aus der nordwesteuropaischen Oberkreide. *Neues Jahrbüch für Geologie und Paläontologie, Abhandlungen*, **139**, 169–225.

—— 1972. Grundfragen der Stammesgeschichte bei irregularen Echiniden der nordwesteuropaischen Oberkreide. *Geologisches Jahrbüch, Reihe A*, **4**, 63–175.

—— and SEIBERTZ, E. 1977. Concepts and methods of echinoid biostratigraphy. *In* KAUFFMAN, E. G. and HAZEL, J. E. (eds). *Concepts and methods of biostratigraphy*. Dowden, Hutchinson & Ross, Stroudsburg, PA, 541–563.

—— and SCHULZ, M. G. 1974. Stratigraphie und Fauna des Coniac und Santon im Schreibkreide-Richtprofile von Lägerdorf (Holstein). *Mitteilungen der Geologisch-Paläontologischen Institut der Universität Hamburg*, **43**, 5–60.

FOORD, A. H. 1891. *Catalogue of the fossil Cephalopoda in the British Museum (Natural History), part 2, containing the remainder of the families Lituitidae, Trochoceratidae, and Nautilidae, with a supplement*. Longman, London, 407 + 15 pp.

FOURAY, M. 1981. L'évolution des *Micraster* (Echinides; Spatangoides) dans le Turonien–Coniacien de Picardie Occidentale (Somme), interêt biostratigraphique. *Annales de Paléontologie, Invertebrés*, **67**, 81–134.

FRENEIX, S. 1986. Huîtres du Crétacé supérieur du Bassin de Challans Commequiers (Vendée). *Bulletin Trimestriel de la Société Géologique de Normandie*, **73**, 13–79, 6 pls.

FRETTER, V. and GRAHAM, A. 1962. *British prosobranch molluscs, their functional anatomy and ecology*. Ray Society, London, xvi + 755 pp.

FRITSCH, A. 1872. *Cephalopoden der böhmischen Kreideformation*. Fr. Rivá, Prague, 52 pp.

GALE, A. S. 1986. Goniasteridae (Asteroidea, Echinodermata) from the Late Cretaceous of north-west Europe. 1. Introduction. The genera *Metopaster* and *Recurvaster*. *Mesozoic Research*, **1**, 1–69.

—— 1987. Goniasteridae (Asteroidea, Echinodermata) from the Late Cretaceous of north-west Europe. 2. The genera *Calliderma, Crateraster, Nymphaster* and *Chomaster*. *Mesozoic Research*, **1**, 151–186.

—— and SMITH, A. B. 1982. The palaeobiology of the Cretaceous irregular echinoids *Infulaster* and *Hagenowia*. *Palaeontology*, **25**, 11–42.

GALSTOFF, P. S. 1964. The American oyster *Crassostrea virginica* Gmelin. *Fishery Bulletin, Fish and Wildlife Series*, **64**, 1–480.

GEINITZ, H. B. 1849–50. *Das Quadersandsteingebirge oder Kreidegebirge in Deutschland*. Dresden; 1849: 1–96; 1850: 97–292.

—— 1875. Das Elbthalgebirge in Sachsen. *Palaeontographica*, **20**, Abteilung 1, 319 pp.; Abteilung 2, 245 pp.

GIERS, R. 1964. Die Grossfauna der Mukronatenkreide (unteres Obercampan) im Östlichen Münsterland. *Fortschritte in der Geologie von Rheinland und Westfalen*, **7**, 213–294.

GOLDFUSS, A. 1833–44. *Petrefacta Germaniae*. Volume 2, 1833: 68 pp., pls 72–97; 1835: 69–140, pls 1–122; 1844: 128 pp., pls 166–200.

GRADSTEIN, F. M., AGTERBERG, F. P., OGG, J. G., HARDENBOL, J., VEEN, P. VAN,

THIERRY, J. and HUANG, Z. 1994. A Mesozoic time scale. *Journal of Geophysical Research*, **99**, 24,051–24,074.

GREGORY, J. W. 1899. *Catalogue of the fossil Bryozoa in the Department of Geology, British Museum (Natural History). The Cretaceous Bryozoa. Volume 1.* British Museum (Natural History), London, 457 pp.

—— 1907. The rotiform Bryozoa of the Isle of Wight. *Geological Magazine* (5), **4**, 442–443.

—— 1909. *Catalogue of the fossil Bryozoa in the Department of Geology, British Museum (Natural History). The Cretaceous Bryozoa. Volume 2.* British Museum (Natural History), London, 346 pp.

HÅKANSSON, E. and VOIGT, E. 1996. New free-living bryozoans from the northwest European Chalk. *Bulletin of the Geological Society of Denmark*, **42**, 187–207.

HARTMAN, W. D., WENFT, J. W. and WEDENMAYER, F. 1960. Living and fossil sponges. *Sedimenta*, **8**, 1–274.

HEINZ, R. 1932. Aus der neuen systematik der Inoceramen. *Mitteilungen aus dem Mineralogisch-Geologischen Staastinstitut in Hamburg*, **13**, 1–26.

HERM, D., KAUFFMAN, E. G. and WIEDMANN, J. 1979. The age and depositional environment of the "Gosau Group" (Coniacian–Santonian) Brandenberg/Tirol, Austria. *Mitteilungen aus dem Bayerischen Staatssamlung für Paläontologie und Historische-Geologie*, **19**, 27–92.

HERMAN, J. 1975. Les Sélaciens des terrains néocrétacés et paléocènes de Belgique et des contrées limotrophes. *Mémoires du Service Carte Géologique et Mineralogique de Belgique*, **15**, 1–450.

HINDE, G. J. 1883. *Catalogue of the fossil sponges of the British Museum.* British Museum (Natural History), London, 248 pp.

—— 1904. On the structure and affinities of the genus *Porosphaera* Steinmann. *Journal of the Royal Microscopical Society*, **1**, 1–25.

JAGER, M. 1983. Serpulidae (Polychaeta sedentaria) aus der norddeutschen Hoheren Oberkreide. *Geologisches Jahrbüch, Reihe A*, **68**, 1–222.

JAGT, J. W. M. 1999. Late Cretaceous–early Palaeogene echinoderms and the K/T boundary in the southeast Netherlands and northeast Belgium. Part 1: Introduction and stratigraphy; Part 2: Crinoidea. *Scripta Geologica*, **116**, 1–255.

JARVIS, I. 1980. Palaeobiology of Upper Cretaceous belemnites of the phosphatic Chalk of the Anglo-Paris Basin. *Palaeontology*, **23**, 889–914.

JEFFERIES, R. P. S. 1962. The palaeoecology of the *Actinocamax plenus* Subzone (lowest Turonian) in the Anglo-Paris Basin. *Palaeontology*, **4**, 609–647.

—— 1963. The stratigraphy of the *Actinocamax plenus* Subzone (Turonian) in the Anglo-Paris Basin. *Proceedings of the Geologists' Association*, **74**, 1–34.

JUKES-BROWNE, A. J. and HILL, W. 1903–04. The Cretaceous rocks of Britain. Vol. II: The Lower & Middle Chalk of England, 568 pp. Vol. III: The Upper Chalk of England, 566 pp. *Memoirs of the Geological Survey*, HMSO, London.

KAUFFMAN, E. G., HATTIN, D. E. and POWELL, J. D. 1977. Stratigraphic, paleontologic and paleoenvironmental analysis of the Upper Cretaceous rocks of Cimarron County, northwestern Oklahoma. *Memoir of the Geological Society of America*, **149**, 47–150.

KENNEDY, W. J. 1969. The correlation of the Lower Chalk of south-east England. *Proceedings of the Geologists' Association*, **80**, 459–560.

—— 1970. A correlation of the uppermost Albian and the Cenomanian of south-west England. *Proceedings of the Geologists' Association*, **81**, 613–677.

—— 1971. Cenomanian ammonites from southern England. *Special Papers in Palaeontology*, **8**, 1–133.

—— and GARRISON, R. E. 1975*a*. Morphology and genesis of nodular phosphates in the Cenomanian Glauconitic Marl of south-east England. *Lethaia*, **8**, 339–360.

—— —— 1975*b*. Morphology and genesis of nodular chalks and hard-grounds in the Upper Cretaceous of southern England. *Sedimentology*, **22**, 311–386.

KUMMEL, B. 1956. Post-Triassic nautiloid genera. *Bulletin of the Museum of Comparative Zoology*, **114**, 324–494, 28 pls.

KVATCHKO, V. I. 1997. The mode of life of *Volviflustrellaria*. *In* DOBROVOLSKY, A. A. and TAYLOR, P. D. (eds). *Bryozoa of the world: Abstracts of the Russian and International Conference, Zoological Institute, Russian Academy of Sciences, St Petersburg, 1997*. Zoological Institute, Russian Academy of Sciences, St Petersburg, p. 15.

LANG, W. D. 1914*a*. On *Herpetopora*, a new genus containing three new species of Cretaceous cheilostome Polyzoa. *Geological Magazine* (*6*), **1**, 5–8.

—— 1914*b*. Some new genera and species of Cretaceous cheilostome Polyzoa. *Geological Magazine* (*6*), **1**, 436–444.

—— 1915. On some new uniserial Cretaceous cheilostome Polyzoa. *Geological Magazine* (*6*), **2**, 496–504.

—— 1916. A revision of the 'cribrimorph' Cretaceous Polyzoa. *Annals and Magazine of Natural History* (*8*), **18**, 81–112, 381–410.

—— 1919*a*. The Kelostominae: a subfamily of Cretaceous cribrimorph Polyzoa. *Quarterly Journal of the Geological Society of London*, **74**, 204–220.

—— 1919*b*. The Pelmatoporinae, an essay on the evolution of a group of Cretaceous Polyzoa. *Philosophical Transactions of the Royal Society of London*, **209**, 191–228.

—— 1921. *Catalogue of the fossil Bryozoa (Polyzoa) in the Department of Geology, British Museum (Natural History). The Cretaceous Bryozoa (Polyzoa). Volume 3*. British Museum (Natural History), London, 269 pp.

—— 1922. *Catalogue of the fossil Bryozoa (Polyzoa) in the Department of Geology, British Museum (Natural History). The Cretaceous Bryozoa (Polyzoa). Volume 4*. British Museum (Natural History), London, 404 pp.

LARWOOD, G. P. 1962. The morphology and systematics of some Cretaceous cribrimorph Polyzoa (Pelmatoporinae). *Bulletin of the British Museum (Natural History), Geology Series*, **6**, 1–285.

—— 1973. New species of *Pyripora* d'Orbigny from the Cretaceous and the Miocene. *In* LARWOOD, G. P. (ed.). *Living and fossil Bryozoa*. Academic Press, London, 463–473.

—— 1985. Form and function of Cretaceous myagromorph Bryozoa. *In* NIELSEN, C. and LARWOOD, G. P. (eds). *Bryozoa: Ordovician to Recent*. Olsen and Olsen, Fredensborg, 169–174.

LYDEKKER, R. 1888. *Catalogue of the fossil Reptilia and Amphibia in the British Museum (Natural History), Part 1. Ornithosauria, Crocodilia, Dinosauria, Squamata, Rhynchocephalia and Proterosauria*. British Museum (Natural History), London, xxviii + 309 pp.

—— 1889*a*. *Catalogue of the Fossil Reptilia and Amphibia in the British Museum (Natural History). Part 2, Ichthyopterygia and Sauropterygia*. British Museum (Natural History), London, xxi + 307 pp.

—— 1889*b*. *Catalogue of the Fossil Reptilia and Amphibia in the British Museum (Natural History). Part 3. Chelonia.* British Museum (Natural History), London, xviii + 239 pp.

MALCHUS, N. 1990. Revision der Kreide-Austern (Bivalvia: Pteriomorpha) Ägyptens (Biostratigraphie, Systematik). *Berliner Geowissenschaftliche Abhandlungen, A,* **125,** 1–231.

—— 1995. Larval shells of Tertiary *Cubitostrea* Sacco, 1879, with a revision of larval shell characters in the subfamilies Ostreinae and Crassostreinae. *Bulletin de l'Institut Royal des Sciences Naturelles de Belgique, Sciences de la Terre,* **65,** 187–239.

MANTELL, G. A. 1822. *The fossils of the South Downs, or Illustrations of the geology of Sussex.* London, 320 pp.

McGOWAN, C. 1972. The systematics of Cretaceous ichthyosaurs with particular reference to the material from North America. *Contributions to Geology, University of Wyoming,* **11,** 9–29.

—— 1978. Further evidence for the wide geographical distribution of ichthyosaur taxa (Reptilia: Ichthyosauria). *Journal of Paleontology,* **52,** 1155–1162.

McKINNEY, F. K. 1995. One hundred million years of competitive interactions between bryozoan clades: asymmetrical but not escalating. *Biological Journal of the Linnean Society,* **56,** 465–481.

—— and JACKSON, J. B. C. 1989. *Bryozoan evolution.* Unwin Hyman, London, 238 pp.

—— and TAYLOR, P. D. 1997. Life histories of some Mesozoic encrusting cyclostome bryozoans. *Palaeontology,* **40,** 515–556.

MEDD, A. W. 1965. *Dionella* gen. nov. (Superfamily Membraniporacea) from the Upper Cretaceous of Europe. *Palaeontology,* **8,** 492–517.

—— 1966*a*. The zoarial development of some membranimorph Polyzoa. *Annals and Magazine of Natural History* (*13*), **9,** 11–22.

—— 1966*b*. *Callopora lyra* (von Hagenow), a Cretaceous membranimorph polyzoan. *Paläontologische Zeitschrift,* **40,** 108–117.

—— 1972. The R. M. Brydone Collection of Cretaceous membranimorph Bryozoa. *Geological Magazine,* **109,** 141–148.

—— 1979. *Ellisina* Norman and *Periporosella* Canu & Bassler (Superfamily Membraniporacea) from the Upper Cretaceous of Europe. *Report of the Institute of Geological Sciences,* **78/25,** 1–29.

MILLER, J. S. 1826*a*. Observations on belemnites. *Transactions of the Geological Society, London* (*2*), **2,** 45–62.

—— 1826*b*. Observations on the genus *Actinocamax. Transactions of the Geological Society, London* (*2*), **2,** 63–67.

MOORE, R. C. (ed.) 1960. *Treatise on invertebrate paleontology. Part L, Mollusca. Gastropoda.* Geological Society of America, Boulder, and University of Kansas Press, Lawrence, xiii + 351 pp.

—— 1969. *Treatise on invertebrate paleontology. Part N, Mollusca, Part 6: Bivalvia.* Geological Society of America, Boulder, and University of Kansas Press, Lawrence, 951 pp.

MORTIMORE, R. N. 1986. Stratigraphy of the Upper Cretaceous White Chalk of Sussex. *Proceedings of the Geologists' Association,* **97,** 97–139.

MURRAY, J. W. (ed.) 1985. *Atlas of invertebrate macrofossils.* Longman Group Ltd and the Palaeontological Association, London, xiii + 241 pp.

NEWMAN, W. A., ZULLO, V. A. and WITHERS, T. H. 1969. Cirripedia. *In* MOORE, R. C. (ed.). *Treatise on invertebrate paleontology. Part R, Arthropoda 4:1.* Geological Society of America, Boulder, and University of Kansas Press, Lawrence, K207–K296.

NEWTON, E. T. 1892. Note on an iguanodont tooth from the Lower Chalk ('Totternhoe Stone'), near Hitchin. *Geological Magazine* (*3*), **9**, 49–50.

OAKELY, K. D. 1937. Cretaceous sponges: some biological and geological considerations. *Proceedings of the Geologists' Association*, **98**, 330–348.

ORBIGNY, A. D' 1844–47. *Paléontologie française. Terrains Crétacés. III, Lamellibranches; IV. Gasteropodes.* Masson, Paris; 1844: 1–288; 1845: 289–448; 1846: 449–520; 1847: 521–807.

—— 1848–51. *Paléontologie française, Terrrains Crétacés. Volume 4: Brachiopodes.* Masson, Paris, 390 pp.

—— 1850. *Prodrome de Paléontologie. Volume 2.* Masson, Paris, 428 pp.

—— 1851–54. *Paléontologie française. Terrains Crétacé. Volume 5. Bryozoaires.* Masson, Paris, 1191 pp.

OWEN, E. F. 1962. The brachiopod genus *Cyclothyris. Bulletin of the British Museum* (*Natural History*), *Geology Series*, **7**, 39–63.

—— 1970. A revision of the brachiopod subfamily Kingeninae Elliott. *Bulletin of the British Museum* (*Natural History*), *Geology Series*, **19**, 27–83.

—— 1977. Evolutionary trends in some Mesozoic Terebratellacea. *Bulletins of the British Museum* (*Natural History*), *Geology Series*, **28**, 205–253.

—— 1988. Cenomanian brachiopods from the Lower Chalk of Britain and northern Europe. *Bulletin of the British Museum* (*Natural History*), *Geology Series*, **44**, 65–175.

OWEN, R. 1841. *Odontography: or a treatise on the comparative anatomy of the teeth, etc. Part 2.* Hippolyte Bailliere, London, 113–288.

—— 1850. Description of the fossil reptiles of the Chalk formations. *In* DIXON, F. *The geology and fossils of the Tertiary and Cretaceous formations of Sussex.* Longman, Brown, Green and Longman, London, 378–404.

—— 1851. A monograph on the fossil Reptilia of the Cretaceous formations, Part 1. *Palaeontographical Society* [*Monographs*], *London*, 118 pp.

—— 1861. Ibid. Supplement 3. Pterosauria (*Pterodactylus*) and Sauropterygia (*Polyptychodon*). *Palaeontographical Society* [*Monographs*], *London*, 25 pp.

—— 1864. Ibid. Supplement 4, Sauropterygia (*Plesiosaurus*). *Palaeontographical Society* [*Monographs*], *London*, 18 pp.

—— 1884. *A history of British fossil reptiles.* Vol. 4, Atlas. Cassell & Co., London, vi + 101 pls.

PALMER, C. P. 1989. Larval shells of four Jurassic bivalve molluscs. *Bulletin of the British Museum* (*Natural History*), *Geology Series*, **45**, 57–69.

PEAKE, N. B. and HANCOCK, J. M. 1961. The Upper Cretaceous of Norfolk. *Transactions of the Norfolk and Norwich Naturalists' Society*, **19**, 293–339.

PEREDA SUBERBIOLA, X. and BARRETT, P. M. 1999. A systematic review of ankylosaurian dinosaur remains from the Albian–Cenomanian of England. *In* UNWIN, D. M. (ed.). Cretaceous fossil vertebrates. *Special Papers in Palaeontology*, **60**, 177–219.

PERGENS, E. 1890. Revision des Bryozoaires du Crétacé figurés par d'Orbigny, I. Cyclostomata. *Bulletin de la Société Belge de Géologie de Paléontologie et d'Hydrologie*, **3** (for 1889), 305–400.

PERSSON, P. O. 1963. A revision of the classification of the Plesiosauria, with a synopsis of the stratigraphical and geographical distribution of the group. *Acta Universitatis Lundensis (2)*, **59**, 1–60.

PETTITT, N. E. 1949. A monograph on the Rhynchonellidae of the British Chalk. I. *Palaeontographical Society [Monographs], London*, 1–26.

—— 1954. A monograph of the Rhynchonellidae of the British Chalk. II. *Palaeontographical Society [Monographs], London*, 27–52.

PITT, L. J. and TAYLOR, P. D. 1990. Cretaceous Bryozoa from the Faringdon Sponge Gravel (Aptian) of Oxfordshire. *Bulletin of the British Museum (Natural History), Geology Series*, **46**, 61–152.

RASMUSSEN, H. W. 1950. Cretaceous Asteroidea and Ophiuroidea with special reference to the species found in Denmark. *Danmarks Geologiske Undersøgelse*, **77**, 1–134.

—— 1961. A monograph on the Cretaceous Crinoidea. *Biologiske Skrifter udgivet af Det Kongelige Danske Videnskabernes Selskab*, **12** (1), 428 pp., 60 pls.

REGENHARDT, H. 1961. Serpulidae (*Polychaeta sedentaria*) aus der Kreide Mitteleuropas, ihre Okologische, taxonomische, und stratigraphische Bewertung. *Mitteilungen aus dem Geologischen Staatinstitut in Hamburg*, **30**, 5–115.

REID, R. E. H. 1958–64. Monograph of the Upper Cretaceous Hexactinellida of Great Britain and Northern Ireland. *Palaeontographical Society [Monographs], London*, i–cliv, 1–48, pls 1–11.

—— 1958. Remarks upon the Upper Cretaceous Hexactinellida of County Antrim. *Irish Naturalists' Journal*, **12**, 236–243, 261–268.

—— 1961. Notes on the hexactinellid sponges. III, Seven Hexactinosa. *Annals and Magazine of Natural History*, **13**, 739–747.

—— 1962*a*. Notes on the hexactinellid sponges. IV, Nine Cretaceous Lychniscosa. *Annals and Magazine of Natural History*, **14**, 33–45.

—— 1962*b*. Sponges and the Chalk Rock. *Geological Magazine*, **99**, 273–278.

REUSS, A. E. 1846. *Die Versteinerungen der Böhmischen Kreideformation. Volume 2*. Schweizerbart'sche, Stuttgart, 148 pp.

RIEGRAF, W. and SCHEER, U. 1991. *Clemens August Schlüter. Cephalopoden der oberen deutschen Kreide. Reprint des in drei teilen von 1867–1876 erschienenen Werkes. Nomenklatorisch überarbeit und ergänzt von Wolfgang Riegraf und Udo Scheer*. Goldschneck-Verlag, Werner K. Wiedert, Korb., 454 pp.

ROBINSON, N. D. 1986. Lithostratigraphy of the Chalk Group of the North Downs, southeast England. *Proceedings of the Geologists' Association*, **97**, 141–170.

ROEMER, F. A. 1840. *Die Versteinerungen des norddeutschen Kreidegebirges*. Hanover, 145 pp.

ROWE, A. W. 1900–08. The zones of the White Chalk of the English Coast. *Proceedings of the Geologists' Association*, **16** (1900), 283–368; **17** (1902), 1–76; **18** (1903), 1–52; **18** (1904), 193–296; **20** (1908), 209–352.

RUSSELL, D. A. 1967. Systematics and morphology of American mosasaurs (Reptilia, Sauria). *Bulletin of the Peabody Museum of Natural History*, **23**, 1–237.

SAHNI, M. R. 1929. A monograph of the Terebratulidae of the British Chalk. *Palaeontographical Society [Monographs], London*, 62 pp.

SAVAZZI, E. 1984. Functional morphology and autoecology of *Pseudoptera* (bakevelliid bivalves, Upper Cretaceous of Portugal). *Palaeogeography, Palaeoclimatology, Palaeoecology*, **46**, 312–324.

SCHLOENBACH, U. 1867. Über die Brachiopoden der norddeutschen Cenoman-Bildungen. *In* BENECKE, E. W. (ed.). *Geognostica Paläontologia Beitrage.* Munich, 403–713.

SCHLÜTER, C. 1871–76. Die Cephalopoden der oberen deutschen Kreide. *Palaeontographica*, **21**, 1–24, pls 1–8 (1871); **21**, 25–120, pls 9–35 (1872); **24**, 121–264, pls 36–55 (1876).

SCHULZ, M. G. 1985. Die Evolution der Echiniden-Gattung *Galerites* im Campan und Maastricht Norddeutschlands. *Geologisches Jahrbüch, Reihe A*, **80**, 3–93.

—— and WEITSCHAT, W. 1975. Phylogenie und Stratigraphie der Asteroideen der nordwestdeutschen Schreibkreide. Teil I: *Metopaster/Recurvaster* und *Calliderma/Chomataster* Gruppe. *Mitteilungen aus dem Geologisch-Paläontologisches Institut der Universität Hamburg*, **44**, 249–284.

SEITZ, O. 1961. Die Inoceramen des Santon von Nordwestdeutschland. 1. (Die Untergattungen *Platyceramus, Cladoceramus*, und *Cordiceramus*). *Beihefte zum Geologischen Jahrbüch*, **46**, 1–186.

—— 1965. Die Inoceramen des Santons und Unter-Campan von Nordwestdeutschland. II. Biometrics, Dimorphismus und Stratigraphie der Untergattung *Sphenoceramus* J. Böhm. *Beihefte zum Geologischen Jahrbüch*, **69**, 1–194.

SHARPE, D. 1853–57. Description of the fossil remains of Mollusca found in the Chalk of England. Cephalopoda. *Palaeontographical Society* [*Monographs*], *London*, 68 pp, 27 pls: 1–26, pls 1–10 (1853); 27–36, pls 11–16 (1855); 37–68, pls 17–27 (1857).

SIVERSON, M. 1999. A new large lamniform shark from the uppermost Gearle Siltstone (Cenomanian, Late Cretaceous) of Western Australia. *Transactions of the Royal Society of Edinburgh: Earth Sciences*, **90**, 49–66.

SLADEN, W. P. and SPENCER, W. K. 1891–1908. A monograph of the British fossil Echinodermata from the Cretaceous formations. Vol. II, The Asteroidea and Ophiuroidea. *Palaeontographical Society* [*Monographs*], *London*, 138 pp.

SMISER, J. S. 1935. A revision of the echinoid genus *Echinocorys* in the Senonian of Belgium. *Mémoires du Musée Royal d'Histoire Naturelle de Belgique*, **67**, 1–52.

SMITH, A. B. and WRIGHT, C. W. 1989. British Cretaceous echinoids. Part 1, Introduction and Cidaroida. *Palaeontographical Society* [*Monographs*], *London*, 1–101, 32 pls.

—— —— 1990. British Cretaceous echinoids. Part 2, Echinothurioida, Diadematoida and Stirodonta (1, Calycina). *Palaeontographical Society* [*Monographs*], *London*, 101–198, pls 33–72.

—— —— 1994. British Cretaceous echinoids. Part 3, Stirodonta 2 (Arbacioida and Phymosomatoida 1). *Palaeontographical Society* [*Monographs*], *London*, 199–267, pls 73—92.

—— —— 1996. British Cretaceous echinoids. Part 4, Stirodonta 3 (Phymosomatoida 2) and Camarodonta. *Palaeontographical Society* [*Monographs*], *London*, 268–341, pls 93–114.

—— —— 1999. British Cretaceous echinoids. Part 5, Holectypoida and Echinoneoida. *Palaeontographical Society* [*Monographs*], *London*, 342–390, pls 115–129.

—— —— 2000. British Cretaceous echinoids. Part 6, Cassiduloida. *Palaeontographical Society* [*Monographs*], *London*, 391–439, pls 115–129.

SMITH, A. M. 1995. Palaeoenvironmental interpretation using bryozoans: a review. *Geological Society, London, Special Publication*, **83**, 231–243.

SOBETSKI, V. A. 1977. Bivalve mollusks from the Late Cretaceous platform sea. *Akademie Nauk SSSR, Trudy Paleontologisheskogo Institut*, **159**, 3–155. [In Russian].

SORNAY, J. 1966–73. For list of papers published during this period, see DHONDT, A. V. 1983.

SOWERBY, J. 1815–18. *The mineral conchology of Great Britain.* Volume 2. London, 251 pp.

SOWERBY, J. DE C. 1826–29. *The mineral conchology of Great Britain.* Volume 6. London, 236 pp.

—— 1836. *In* FITTON, W. H. Observations on some of the strata between the Chalk and the Oxford Oolite in the south-east of England. *Transactions of the Geological Society, London* (*2*) **4**, 333–342, 353–361.

STANLEY, S. M. 1970. Relation of shell form to life habits of the Bivalvia (Mollusca). *Memoir of the Geological Society of America*, **125**, 1–296.

STENZEL, H. B. 1971. *Treatise on invertebrate paleontology, Part N, Mollusca 6, Volume 3: Oysters.* Geological Society of America, Boulder, and the University of Kansas Press, Lawrence, N953–N1224.

STOKES, R. B. 1975. Royaumes et provinces fauniques du Crétacé établis sur la base d'une étude systématique du genre *Micraster. Mémoires du Muséum National d'Histoire Naturelle Serie C, Science de la Terre*, **31**, 1–94.

SZÁSZ, L. 1985. Coniacian *Inoceramus* from the Babadag Basin (North Dobrogea). *Memoriile, Institutuli de Geologie si Géophysique, Bucarest*, **32**, 137–184.

TAYLOR, P. D. 1988. Colony growth pattern and astogenetic gradients in the Cretaceous cheilostome bryozoan *Herpetopora. Palaeontology*, **31**, 519–549.

—— 1994. Systematics of the melicerititid cyclostome bryozoans; introduction and the genera *Elea, Semielea* and *Reptomultelea. Bulletin of The Natural History Museum, London* (*Geology Series*), **50**, 1–103.

—— 1999. Bryozoa. *In* SAVAZZI, E. (ed.). *Functional morphology of the invertebrate skeleton.* Wiley, Chichester, 623–706.

—— and SEQUIROS, L. 1982. Toarcian bryozoans from Belchite in north-east Spain. *Bulletin of the British Museum* (*Natural History*), *Geology Series*, **36**, 117–129.

TEICHERT, C., KUMMEL, B., SWEET, W. C., STENZEL, H. B., FURNISH, W. M., GLENISTER, B. F., ERBEN, H. K., MOORE, R. C. and NODINE ZELLER, D. E. 1964. *Treatise on invertebrate palaeontology. Part K, Mollusca 3. Cephalopoda-General Features-Endoceratoidea-Actinoceratoidea-Nautiloidea-Bactritoidea.* Geological Society of America, Boulder, and University of Kansas Press, Lawrence, 519 pp.

THOMAS, H. D. 1939. On *Trochiliopora humei* Gregory and *T. gasteri* sp. nov. *Proceedings of the Geologists' Association*, **50**, 527–529.

—— and LARWOOD, G. P. 1960. The Cretaceous species of *Pyripora* d'Orbigny and *Rhammatopora* Lang. *Palaeontology*, **3**, 370–386.

TOULMIN SMITH, J. 1847–48. The Ventriculidae of the Chalk. *Annals and Magazine of Natural History*, **20**, 73–97, 170–191, 203–220, 279–295, 352–372.

TRÖGER, K.-A. 1967. Zur Paläontologie, Biostratigraphie und faziellen Ausbildung der unteren Oberkreide (Cenoman bis Turon). 1. Paläontologie und Bio-stratigraphie der Inoceramen des Cenomans bis Turons Mitteleuropas. *Abhandlungen der Staatlichen Museums für Mineralogie und Geologie zu Dresden*, **12**, 13–207.

—— 1989. Problems of Upper Cretaceous inoceramid biostratigraphy and paleobiostratigraphy in Europe and western Asia. *In* WIEDMANN, J. (ed.). *Cretaceous of the Western Tethys*. E. Schweitzerbart'sche, Stuttgart, 911–930.

—— and CHRISTENSEN, W. K. 1991. Upper Cretaceous (Cenomanian–Santonian) inoceramid bivalve fauna from the island of Bornholm, Denmark. *Danmarks Geologiske Undersøgelse, A*, **28**, 1–47.

VOIGT, E. 1958. Untersuchungen an Oktocorallen ause der oberen Kreide. *Mitteilungen aus dem Geologischen Staatinstitut in Hamburg*, **27**, 5–49.

—— 1959. Die Ökologische Bedeutung der Hartgründe ('Hardgrounds') in der oberen Kreide. *Paläontologisches Zeitschrift*, **33**, 129–147.

—— 1979. Vorkomen, Geschichte und Stand der Erforschung der Bryozoen des Kreidesystems in Deutschland und benachbarten Gebieten. *Aspekte der Kreide Europas. IUGS Series A*, **6**, 171–210.

—— 1982. Répartition et utilisation stratigraphique des Bryozoaires du Crétacé Moyen (Aptien–Coniacien). *Cretaceous Research*, **2** (for 1981), 439–462.

—— 1983. Zur Biogeographie der europaïschen Oberkreide Bryozoenfauna. *Zitteliana*, **10**, 317–347.

—— 1991. Mono- or polyphyletic evolution of cheilostomatous bryozoan divisions? *Bulletin de la Société des Sciences Naturelles de l'Ouest de la France, Mémoire, Hors Série*, **1**, 505–522.

WALASZCZYK, I. and COBBAN, W. A. 2000. Inoceramid faunas and biostratigraphy of the Upper Turonian–Lower Coniacian of the Western Interior of the United States. *Special Papers in Palaeontology*, **64**, 1–118.

WALLER, T. R. 1981. Functional morphology and development of veliger larvae of the European oyster *Ostrea edulis* Linn. *Smithsonian Contributions to Zoology*, **328**, 1–70.

WEISHAMPEL, D. B. and HORNER, J. R. 1990. Hadrosauridae. *In* WEISHAMPEL, D. B., DODSON, P. and OSMÓLSKA, H. (eds). *The Dinosauria*. University of California Press, Berkeley, 534–561.

WELLNHOFER, P. 1978. Teil 19. Pterosauria. *In* WELLNHOFER, P. (ed.). *Handbüch der Paläoherpetologie*. Gustav Fischer Verlag, Stuttgart, x + 82 pp.

WENZ, W. 1969. *Handbüch der Paläozoologie*. 6, *Gastropoda*. 7 volumes, Borntraeger, Berlin.

WHITTLESEA, P. S. 1983. Observations on the stratigraphical range and morphology of the Cretaceous cribrimorph bryozoan: *Ubaghsia crassa* (Lang). *Bulletin of the Geological Society of Norfolk*, **33**, 27–31.

—— 1985. Notes on Chalk fossils. *Bulletin of the Geological Society of Norfolk*, **35**, 39–46.

—— 1991. The Maastrichtian in Norfolk. *Bulletin of the Geological Society of Norfolk*, **40**, 33–51.

—— 1996a. Temporary Chalk exposures in east Norfolk (Upper Campanian, zone of *Belemnitella mucronata sensu lato*) 1989–1990. *Bulletin of the Geological Society of Norfolk*, **43** (for 1993), 3–24.

—— 1996b. The palaeoecology of two chalks in the Upper Campanian of Norfolk, England. *Bulletin of the Geological Society of Norfolk*, **43** (for 1993), 25–44.

—— 1996c. Effects of sedimentary regimes on relative abundance of bryozoan colony forms: an example from the Late Campanian allochthonous chalks of Norfolk, England. *In* GORDON, D. P., SMITH, A. M. and GRANT-MACKIE, J. A. (eds). *Bryozoans in space and time*. NIWA, Wellington, 377–382.

WIEDMANN, J. 1960. Zur systematik Jungmesozoischer Nautiliden unter besonderer Berücksichtigung der iberischen Nautilinae d'Orb. *Palaeontographica, A*, **115**, 144–206, pls 17–27.

WILLIAMS, A. *et al.* 2000. *Treatise on invertebrate paleontology. Part H. Brachiopoda revised.* Geological Society of America, Boulder, and University of Kansas Press, Lawrence, 919 pp.

WITHERS, T. H. 1935. *Catalogue of fossil Cirripedia in the Department of Geology. II, Cretaceous.* British Museum (Natural History), London, 534 pp.

WOODS, H. 1896. The Mollusca of the Chalk Rock. Pt. 1. *Quarterly Journal of the Geological Society, London*, **52**, 68–98.

—— 1897. The Mollusca of the Chalk Rock. Pt. 2. *Quarterly Journal of the Geological Society, London*, **53**, 377–403.

—— 1899–1913. A monograph of the Cretaceous Lamellibranchia of England. *Palaeontographical Society* [*Monographs*], *London*, volume 1, xliii + 1–232; volume 2, 1–473.

—— 1922–29. A monograph of the fossil macrurous Crustacea of England. *Palaeontographical Society* [*Monographs*], *London*, 122 pp.

WOODWARD, A. S. 1889. A synopsis of the vertebrate fossils of the English Chalk. *Proceedings of the Geologists' Association*, **10**, 273–338.

—— 1902–12. The fossil fishes of the English Chalk. *Palaeontographical Society* [*Monographs*], *London*, 257 pp.

—— 1904. The jaws of *Ptychodus* from the Chalk. *Quarterly Journal of the Geological Society, London*, **60**, 133–135.

—— and SHERBORN, C. D. 1890. *A catalogue of British fossil Vertebrata.* Dulau & Co., London, xxxv + 396 pp.

WORSSAM, B. C. and TAYLOR, J. H. 1969. Geology of the country around Cambridge. *Memoir of the Geological Survey of Great Britain*, Sheet 188, 159 pp.

WRIGHT, C. W. 1979. The ammonites of the English Chalk Rock (Upper Turonian). *Bulletin of the British Museum (Natural History), Geology Series,* **31**, 281–332.

—— 1996. Cretaceous Ammonoidea (with contributions by J. H. CALLOMAN and M. K. HOWARTH) *Treatise on invertebrate paleontology. Part L, Mollusca 4, revised.* Geological Society of America, Boulder, and University of Kansas Press, Lawrence, 362 pp.

—— and COLLINS, J. S. H. 1972. British Cretaceous crabs. *Palaeontographical Society* [*Monographs*], *London*, 114 pp.

—— and KENNEDY, W. J. 1981. The Ammonoidea of the Plenus Marls and the Middle Chalk. *Palaeontographical Society* [*Monographs*], *London*, 148 pp.

—— —— 1984–96. The Ammonoidea of the Lower Chalk. Parts 1–5. *Palaeontographical Society* [*Monographs*], *London*, 403 pp.

—— and WRIGHT, E. V. 1940. Note on Cretaceous Asteroidea. *Quarterly Journal of the Geological Society, London*, **96**, 231–248.

—— —— 1951. A survey of the fossil Cephalopoda of the Chalk of Great Britain. *Palaeontographical Society* [*Monographs*], *London*, 1–40.

WRIGHT, T. 1864–82. A monograph on the British fossil Echinodermata from the Cretaceous Formations. 1, The Echinoidea. *Palaeontographical Society* [*Monographs*], *London*, 371 pp.

YAMAGUCHI, K. 1994. Shell structure and behaviour related to cementation in oysters. *Marine Biology*, **118**, 89–100.

SYSTEMATIC INDEX

References to plates are in bold (**43.2**=Pl. 43, fig. 2); text-figure references are in *italics*.

scriptum, Pachinion 39, **2.1–2**
scutata, Echinocorys 287, **59.1–2**,
     13.1A–D
Scyliorhinidae 311
Scyliorhinus antiquus 311, *14.2A*
     dubius 311, *14.2B–C*
Sedentaria 47
Sellithyridinae 83
semiglobosa, Gibbithyris 87, **12.2**
semiplana, ?*Hyotissa* 148, **24.11–13**
semiplicata, Cretodus 308, *14.2U–V*
semisulcata, Limatula 141
septemplicatus, Microchlamys 126
'*Septifer' lineatus* 104, **17.4**
serpentinus, Onchotrochus 43, **5.4**
Serpulidae 47
Serpulinae 48
serratus, Spondylus 134, **22.7–8**
sexcostata, Neithea 128, **21.7–9**
simplex, Hamites 207, **41.5**
     Porochonia 31, **1.3**
Siphonia koenigi 41, **4.1–2**
socialis, Uintacrinus 262, **50.1**
Solenoidea 155
?Solenidae 155
sorigneti, Tylocidaris 271, **53.2–3**
sornayi, Thomelites 195, **35.6**
Spatangoida 291
Sphenoceramus cardissoides 114, 115
     pachtii 114, **19.1**
     patootensis 114, **19.8**
     pinniformis 115, **19.3–4**
     steenstrupi 115, **19.9**
Spinax major 303
spinosus, Spondylus 129, **22.1**
Spirorbinae 52
Spondylidae 129
Spondylus aequalis 129
     brightonensis 129
     duplicatus 129
     dutempleanus 132
     fimbriatus 131, **22.3**
     gaedorupus 129
     latus 132, **22.2, 4–5**
     serratus 134, **22.7–8**
     spinosus 129, **22.1**
     striatus 131, **22.6**
Squalicorax falcatus 303, *14.2CC*
     kaupi 304, **61.6**, *14.2DD*

Squalus 296
Squamata 331
squamosa, Capillithyris 90, **13.5**
squat lobsters 241
Squatina cranei 342, *14.2K–L*
Squatinidae 302
Squatiniformes 302
Stauranderasteridae 263
Stauronema carteri 31, **3.3–4**
steenstrupi, Sphenoceramus 115, **19.9**
steinlai, Calliomphalus 164, **27.8**, *8.1D*
Stegochoncha sp. 102, **17.1**
stellata, Guettardiscyphia 34, **1.2**
Stenopterygiidae 329
Stereocidarinae 269
(*Stereocidaris*) *carteri, Temnocidaris*
     269, **53.11**
     intermedia 269, **53.5**
     sceptrifera 270, **53.6–7**
Sternotaxis gregoryi 284, **57.10–11**
placenta 285, **58.8–9**
plana 285, **58.5–7**
trecensis 285, **57.8–9**
stewarti, Pholadomya (*Pholadomya*)
     158, **26.10**
Stichomicropora marginula 72, **9.8**
Stichophyma tumidum 37, **2.3**
Stomatoporidae 58
strahani, Pharetrospongia 35, **3.1–2**
Stramentidae 250
Stramentum pulchellum 250, **47.11**
strehlensis, Arca 101
striatula, Terebratulina 92, **11.4**
striatus, Apateodus 319, **63.3**
     Periaulax 164, *8.1E–F*
     Spondylus 131, **22.6**
Stromboidea 172
subacuta, Chlamys? 119, **21.6**, *7.4C*
'*subconicula', Echinocorys* 13.1E–F
subcylindricum, Ophiomusium 268,
     52.10
'*subglobosus', Echinocorys* 13.1O–P
     Holaster 284, **59.3–4**
sublaevigatum, Eutrephoceras 226,
     10.1A
suborbiculatum, Amphidonte 145
subovalis, Lima 137
Subprionocyclus branneri 205, **37.8–9**
     hitchinensis 205